HELPING YOUR CHILD WITH MATH

Sol Weiss

Professor Emeritus, Mathematical Sciences
West Chester University
West Chester, Pennsylvania

Published by Prentice Hall Press
New York, New York 10023

Photo on page 35 courtesy of the Moore School, University of Pennsylvania.

Photos on pages 36, 37, and 39 courtesy of the IBM Corporation.

First paperback edition, 1987

Published by Prentice Hall Press
A Division of Simon & Schuster, Inc.
Gulf + Western Building
One Gulf + Western Plaza
New York, New York 10023

Manufactured in the United States of America

1 2 3 4 5 6 7 8 9 10

Library of Congress Cataloging-in-Publication Data
Weiss, Sol.
Helping your child with math.

1. Mathematics—Study and teaching (Elementary)
I. Title.
QA135.5.W383 1986 372.7 85-11151
ISBN 0-13-386343-3 (cloth)
ISBN 0-13-386301-8 (paper)

Contents

For

Preface

Among the more common anxieties of our time is the one produced by a fear of mathematics. How many times have you heard,

> "I'm terrified of mathematics."
>
> "I freeze up the moment numbers are mentioned."
>
> "When I look at numbers, my mind immediately goes blank. A curtain comes down over my eyes.'

A well-known newspaper reporter, who thought she had a terminal case of math anxiety, confessed: "For as long as I could remember, I suffered from panic, embarrassment, resentment, and self-loathing when faced with the simplest problem in mathematics. . . . And then I would pull down the little window shade in my mind until the problem went away."

I once asked 15 guests at a party how they could prove by logical reasoning that at least two of us had our birthdays in the same month. One guest later said she had no problem with the question until she realized she was doing math.

"Mathophobia," the intense and fearful emotional reaction to any contact with mathematics, often goes back to the earliest unhappy experiences with math. The child who is unsuccessful and miserable with mathematics—struck by headaches, sweaty palms, diarrhea, knots in the stomach, and general mental disorganization every time he steps into a math class—becomes the adult "mathophobe." As an adult, he may find his dread of mathematics forcing him to reroute his career so as to avoid encounters with that intimidating subject.

Unhappily, too many children are taught math by elementary-school teachers who are themselves mathophobic. Their own fear of the subject is transmitted, like some communicable disease, to generations of children. In their pathetic "teaching" of mathematics, such teachers cannot transmit the wonder in mathematics. They can only kill it.

Many children—and their parents—are convinced they do not do well in math because they are "naturally" poor at it. They say they have a "mental block" or lack a "mathematical mind." But a "mathematical mind" is not needed to learn elementary-school mathematics. Unless you're brain-damaged, the intelligence you show in many things you do routinely every day is enough to carry you through this level of the subject.

When, in 1933, Shinichi Suzuki, a gentle Japanese school teacher, observed the natural and unforced way four-year-old children learned to sort out the complexities of the Japanese language before they learned to read, he suspected they could learn to play the violin the same way. Insisting on strong parental support and a home life filled with music, Suzuki was soon graduating musicians who had grown up with violins and Teddy bears side by side. As much can be done with learning mathematics.

Growing up in a mathematically hospitable environment—in a home filled with plenty of math books, games, and puzzles—a child will not be stumped by elementary-school mathematics. When dinner-table conversation encourages mathematical thinking with fun questions (like, "What's the biggest number?"), a child will develop at least acceptable competence in mathematics. Home environment and active parental involvement are extremely important in shaping a child's interest and performance in mathematics.

With low SAT scores now a national cliché, and with widespread concern over the difficulties children are having learning math, it is well to remember that these difficulties often begin in the earliest elementary grades, when the child fails to acquire basic concepts and skills needed for more advanced work. When instead of moving forward, the child feels he's descending into an abyss, and when instead of seeing the light, he sees only another tunnel at the end of the tunnel, failure leads to frustration, then to fear, and, ultimately, to deep hostility toward mathematics.

The problem with mathophobes is essentially psychological. Fear, anxiety, and a lack of self-confidence are the culprits, not stupidity. That the problem is mostly psychological has been borne out in my experience with thousands of school children, college students, teachers, parents, and even with a sampling of hard-core criminals in a large state correctional facility.

A major premise of this book is that with good teaching and learner cooperation, anyone who is not brain-damaged can learn at least the basics of mathematics. Success will then feed on success, and understanding will lead to enjoyment.

Helping Your Child with Math can serve as a parent sourcebook. In it you will find basic elementary-school mathematics explained in a way that you, the parent, can understand. You will become comfortable with the subject and gain the confidence to help your child with it.

Part I tells you, among other things, why children need to study mathematics, and how to get them interested in it. It suggests ways to help low achievers as well as mathematically talented children to live up to their potential.

Part II deals with the *content* of elementary-school mathematics and specific ways of helping your child learn and enjoy it. The basic mathematics of each topic is explained, and a detailed plan for teaching it is presented. The "Highlights" section at the end of each chapter spells out briefly and simply the main ideas your child should be left with.

You will also find in this part of the book a wealth of teaching strategies, activities, games, and visual aids that have proved successful with thousands of children and hundreds of teachers.

Scattered throughout the book are interesting historical notes and practical applications of the ideas in the book to everyday affairs. Sections titled **For the Curious** offer interested readers an optional higher-level development of some topics. Sections titled **For the Daring** pose interesting problems and puzzles, with solutions at the back of the book.

A major purpose of the book is to help you help your child with mathematics, right from the beginning, thus saving him or her from the dismal cycle of frustration, fear, and hostility by building understanding and confidence.

An ounce of prevention is worth a pound of cure in education as in health and industry. A Japanese industrialist, asked to define the difference in quality control in Japan and the United States, said the Japanese try to control quality by prevention—by making things work the first time around—while Americans try to do it by detection after the product has been manufactured. To undo the damage is more costly and frustrating than to do it right the first time.

A recent study reports that greater parental involvement in schooling gives Japanese and Taiwanese children an edge over their American counterparts, with the differences appearing as early as the first grade.

The study contends that "Our national problem lies not only in American schools but in American homes. The average American family does not take the responsibility to provide the informal introduction to reading and mathematics that you find in other countries."

The purpose of this book is to help you, the parent, accomplish this goal. Another purpose is to provide you and other adults with a chance to take another look at an awesome, nonlethal weapon in humankind's arsenal.

Part I

PUTTING YOUR CHILD AT EASE WITH MATH

Chapter 1

Why Study Mathematics?

A fair question for a child to ask is why he or she needs to study mathematics. It's an equally fair question for *anyone* required to study mathematics to ask—except, perhaps, for the few who have a natural love for the subject.

Let us, at the outset, distinguish between *needing* mathematics and *enjoying* it. It's obviously possible to need to do things we don't especially enjoy, as it is to enjoy doing things we don't need. The happiest combination—enjoying what we need to do—does not happen often when children study mathematics. So why study it? Why do they need it?

They need it to gain literacy in a technological world and as a gateway to future careers. It can also enrich their lives.

Everyone sees the need for computational skills, but to understand what you hear, see, and read in the 1980s requires more mathematics than that. Pick up your newspaper, turn on your radio, watch television, and observe how often graphs, ratios, statistics, probability, computers, mathematical models, and a host of other mathematical references are included in their reports.

Roger Bacon, the thirteenth-century English philosopher and scientist, pointed to the need for mathematics in his day because, he said, "Mathematics is the gate and key of the sciences . . . . Neglect of mathematics works injury to all knowledge, since he who is ignorant of it cannot know the other sciences or things of the world." Seven hundred years later, Bacon's judgment is still right on target.

Every occupation and profession requires some knowledge of mathematics. Though the amount varies with each occupation, certainly more than the basic computational skills are usually needed, especially by those who wish to enter the fast-changing high-technology industries.

A recent Harvard and Massachusetts Institute of Technology study of the long-range employment outlook for the United States concluded that schools will have to shift from narrow vocational training toward producing workers with greater versatility in mathematics and language. Without such a shift, the high-technology industries that provide millions of jobs will continue their rapid evolution but there will be no workers competent to deal with it.

Other studies have shown that inadequate mathematical training has been a serious obstacle to equal opportunity for women who wanted to become chemists, physicists, architects, and doctors. For instance, today's discoveries in chemistry are increasingly based on mathematics. In a striking affirmation of this fact, the 1985 Nobel Prize in chemistry was awarded to two mathematicians.

In recent years, colleges have found it necessary to increase the mathematics entrance requirements for such "nonmathematical" careers as biology, food science, nursing, and psychology.

The Mathematical Association of America recommends that no less than four years of high-school mathematics be taken for careers in the environmental sciences, business management, economics, engineering, biochemistry, biology, computer science, dentistry, medical technology, optometry, pre-medicine, pharmacy, psychology, and any of the mathematical and physical sciences. No less than three years of high-school mathematics is recommended even for the study of law, linguistics, nursing, physical therapy, geography, political science, and sociology. Higher mathematics requirements for high-school graduation are being implemented across the country.

Since no child knows precisely what he will be doing later in life—and is likely to change occupations several times in his lifetime—investing in the study of mathematics is like investing in an insurance policy that will pay off when he is confronted with unforeseeable employment requirements of the future.

"Bread-and-butter" needs are not the only reasons for studying mathematics. There are other less practical yet equally powerful reasons. Call them aesthetic and intellectual. Relax with mathematics, and it can become interesting, even exciting. Give it time, and it will provide you with a fresh prism through which to view the world.

With it, you can analyze and solve problems—some intriguing, some perplexing. At times the solutions are elegant; once in a while, astonishing. Mathematics can enrich our aesthetic life no less than music and art.

Is it not remarkable to contemplate that the distance around the ring on your finger, around the moon, or around *any* circle—no matter how large or small—will *always* be a little more than three times the distance across its center? That the number π (*pi*) helps measure a circle, forecast a presidential election, and rate a television show? Or that by the sheer power of mathematics we can, with only pencil and paper, compute the distance to the sun and predict the precise moment a space ship will return to earth?

Not everyone will see beauty in these reflections, just as not everyone sees beauty in music, dancing, horses, or ships. But those who cultivate such appreciations can enrich their lives.

Not all these reasons for studying mathematics apply to everyone. But among them, everyone will find *some* that connect with their own lives. As for children, no argument about long-range need will lead them to study mathematics and enjoy it. Good teaching can.

Chapter 2
Motivation

Motivation, a vital force in human behavior, plays a key role in how well you do in your job as well as in your bridge lessons, and it plays a crucial role in how well your child learns mathematics. The presence or absence of motivation can spell the difference between excitement and boredom, between really trying and giving up, between success and failure. Highly motivated children with low mathematical ability have been known to outperform unmotivated children with greater ability. The question: How do you ignite the spark?

Obviously you, the parent, are but one among other influences shaping your child's attitude and performance in mathematics. But your influence can be powerful and decisive.

Before you can encourage a positive attitude toward anything, you yourself must set the example. If in the relationship with your child you transmit the message that mathematics can be interesting, useful, and fun, you will have taken an important step in the right direction. In a good parent-child relationship, the desire to please the parent is a powerful motivator.

The manner in which you work with your child strongly affects the result. What you say and how you say it is particularly important. One investigator studied children's reactions to verbal praise, physical contact, and facial expressions indicating approval. He found that when such encouragement was frequent the level of mathematical performance by the children increased; when the encouragement was stopped, performance level was reduced; and when encouragement was reintroduced, the performance level again rose. So—criticize less; praise more.

Though a good parent-child relationship can do wonders for motivation, perhaps the best motivator for a child (as for an adult) is *success*. By providing your child with many opportunities for success, you are capitalizing on the intrinsic motivation of success. These opportunities will be there if you adjust your expectations to a realistic performance level that is challenging without being frustrating.

Socrates, one of the great teachers of all time, motivated his students by plucking insights out of their minds through incisive questioning. In mathematics, there are two types of questions you can ask: drill, fact, or memory questions like "What?" "Which?" "Where?" usually involving only recall or recognition of information; and thought-provoking questions like "Why?" or "What happens if . . . ?" requiring higher-level thinking. While fact questions are a necessary part of learning, thought questions are the ones that release the adrenaline, stimulate the child's thinking, and ignite the imagination.

In an age of passive spectatorship, we must persist in keeping the child actively involved in the learning process. Active involvement means that instead of giving your child long explanations, you develop concepts *with* him through questions and, where applicable, through manipulation of physical materials. The child's active involvement is a sure-fire way of arousing interest.

Whatever explanations you do give must be comprehensible to your child, and this can happen only if (1) you, the parent, understand the concept you're trying to explain, (2) your language is geared to the child's level, and (3) the child has the prerequisite understandings and skills for that concept—without which even a highly motivated child will be frustrated.

To make a new concept clear and meaningful, define the new words, explain why you're teaching the concept, link it to what the child already knows, use illustrations from his personal experiences, and use lots of physical materials in your explanation.

To stimulate interest in a new topic, try to "turn on" your child by first posing an interesting problem or puzzle, relating a story, using an appealing visual aid, or playing a game—*directly related to the topic you wish to teach.* In Part II of this book you will find specific suggestions for doing this.

Imitate glamorous entertainers: provide fun, mystery, and drama. How? By posing problems with humorous situations or solutions, presenting "magic" problems and puzzles, and creating dramatic settings for a concept (like an imaginary conversation between the number 2 and the number 3). Where do you find such problems? Many are presented in Part II of this book.

Games—directly related to the topic—are an excellent way to stimulate a child and to get him actively involved in the learning process. They are particularly useful in providing a fresh and exciting setting for practice in needed skills. But games can be overdone. They should not be played too often, nor for too long in any one session. It's wise to end a game at a point where the child hungers for more. Many games are suggested in Part II in the context of specific subject matter.

Merely being inspirational and exhorting your child to try harder is not enough. You are pushing buttons that are not connected to anything. But igniting and sustaining a child's interest by keeping him actively involved in the development of ideas and providing him with opportunities to be successful are powerful ways to light the fuse.

Chapter 3

Instructional Aids

A few years ago, a well-known mathematics educator watched his young daughter do her homework. She completed a page of subtraction problems this way:

(1)	(2)	(3)
33	81	134
−17	−28	−86
24	67	152

Obviously, she subtracted the smaller number from the larger one in each column.

Looking at her first answer, he turned to his daughter in frustration and asked her to take out a box of toothpicks and count out 33.

"Now give me 17," he said. When she did, she found there were 16 toothpicks left.

"Before you got 24, and now you get 16. Which answer is right?" he asked.

"They are both right!" came the confident reply. Pointing to her first answer, she explained, "That's the way you do it with pencil and paper." Then pointing to the other answer, she added, "And that's the way you do it with toothpicks."

The purpose of instructional aids is to increase the chances that your child sees the connection between the mathematics she is learning and the world of toothpicks. Concrete objects communicate mathematics in a way that words and symbols do not. With them, concepts come to life. As one child, studying how to find areas of geometric figures with a pegboard, exclaimed excitedly, "I love these pegboard activities. They're more fun than math."

The importance of physical materials in learning mathematics, especially in the early years, will be repeated over and over again in this book because young children do not think in abstractions; they think in terms of concrete objects. The suggested teaching strategies in Part II are mostly based on such materials. They permit the child to observe, explore, and discover, and provide an intuitive basis for grasping abstract concepts.

Good physical materials appeal to as many senses as possible, are simple to use, can be used to clarify a variety of concepts, and can be used over and over again. Nothing more complicated than a set of poker chips, for instance, can be used—as you will see—to explain a variety of mathematical concepts.

Most of these materials are available commercially; some are simple household items or can easily be constructed by the parent and, even, by the child. Where and how to use them will be suggested in Part II of the book.

On page 8 are lists of materials used with elementary-school children.

Simple household items

Buttons
Drinking straws
Counters (disks or poker chips)
Containers
Tongue depressors
Checkers
Lids of containers or cans
Toothpicks
Bottle tops
Ice cream sticks

Items available commercially

Magnetic board (with cutouts)
Place value devices
Hundred chart
Number lines
Cuisenaire rods
Geoboard
Pegboard
A compass (for drawing circles)
Dienes materials
Tangrams
Attribute blocks
Spinners
Dice
Play money
Flash cards for basic arithmetic facts
Fraction chart and strips (see page 146)

Other materials you will need for this book

A set of triangles, squares, rectangles, parallelograms, trapezoids, and circles
A set containing a rectangular prism, cube, pyramid, cone, cylinder, and sphere
A pocket calculator
A meter stick; a metric ruler
A transparent grid ruled up in square units
Sheets of 10 × 10 squared graph paper
Sheets of "bird arrays" (described on page 72)
A set of oak-tag strips measuring 1 to 9 units (described on page 83).

PEGBOARD

A pegboard is a piece of masonite or wood in which 100 holes are bored—10 rows, 10 holes per row. Pegboards with the holes already drilled are widely available commercially.

Golf tees or pegs are also needed to fill the pegboard holes. You will need pegs in two different colors, 25 of each color. The different colors add clarity to explanations.

The pegboard can be used to teach counting and the basic arithmetic operations, number properties, and number patterns.

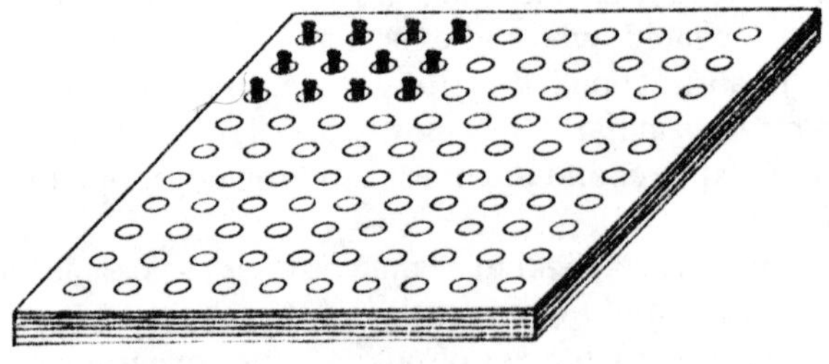

Pegboard

GEOBOARD

A geoboard is a board with a square arrangement of nails, spaced about 2 inches apart. Shown below is a board with a 6 × 6 square array of nails; that is, 6 rows of nails, each row containing 6 nails.

Rubber bands of different colors are stretched across the nails to explore various geometric shapes and the concepts of perimeter, area, and symmetry. The geoboard is also used to teach the basic arithmetic operations, fractions, and number patterns.

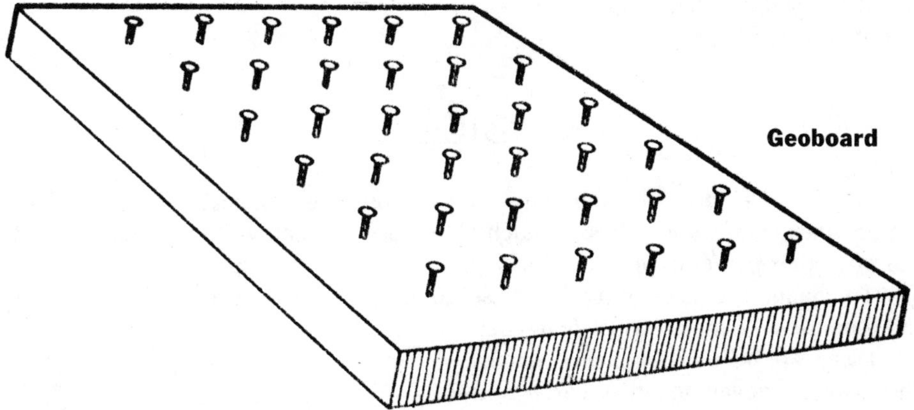

Geoboard

HUNDRED BOARD

A hundred board can be made out of a piece of plywood, with 10 rows of 10 hooks each for hanging small tags. The numerals 1 to 100 are written on the tags.

The board is used for learning counting and the basic arithmetic facts; searching for number patterns; developing the meaning of fractions, decimals, and percents.

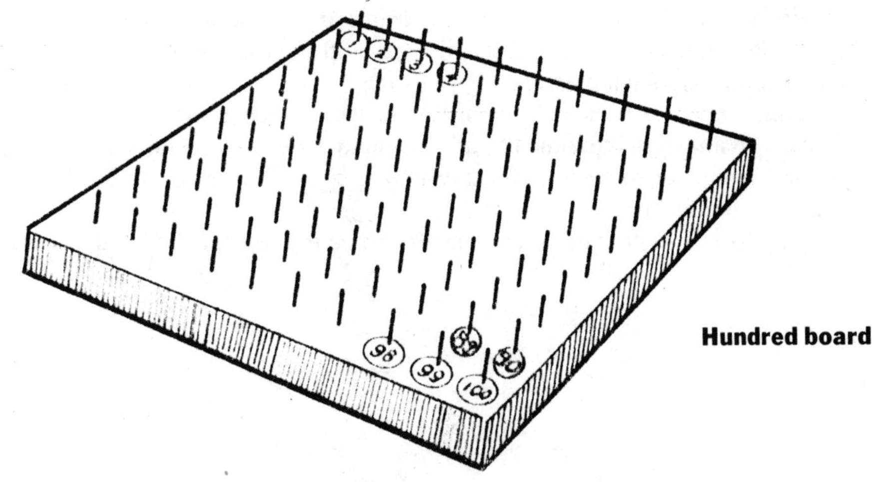

Hundred board

SPINNER

Spinners are used in many mathematical games and other activities throughout Part II. In some instances you'll need two or three spinners.

You can make a spinner with a 5-inch × 5-inch piece of wood, a Popsicle stick (cut to the shape of a pointer), a washer, and a screw.

Drill a small hole in the Popsicle stick. Screw the stick to the center of the board with the washer so that it can spin freely.

The face of the spinner can contain the numerals 0 through 9, or any other numerals needed for a particular game. You can prepare changeable tagboard faces for the spinner that can be mounted on the board.

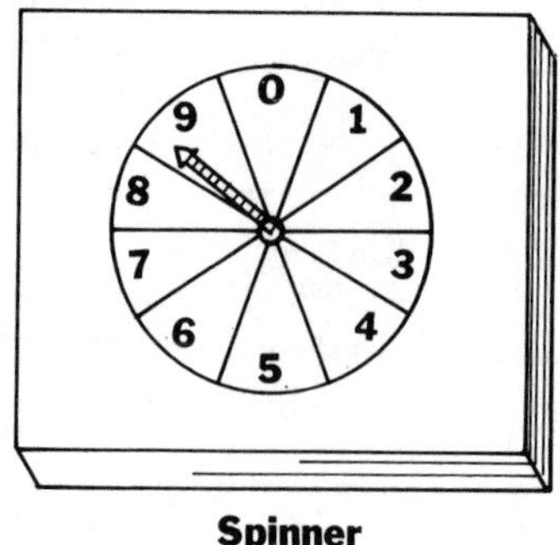

Spinner

DICE

Dice will also be used frequently. You can buy them or make your own. Dice can be made from wood cubes; on each blank face you can write the numerals 1 to 6 or any other numerals you choose.

To change the faces of the dice, use gummed labels, or start with blank faces and then use a "write-on wipe-clean" pen.

Dice can be used in many games. They can aid development of early number concepts and number combinations, such as recognizing all the two-number combinations that add up to a given number.

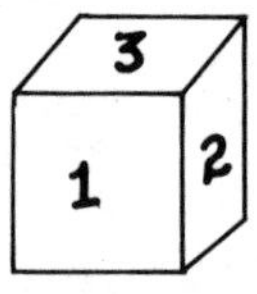

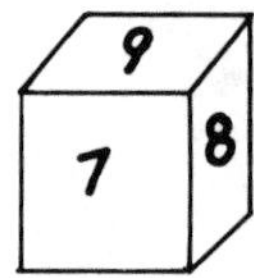

Dice

TANGRAMS

A tangram is a kind of jigsaw puzzle, consisting of several pieces arranged to produce a desired geometric shape. Tangrams are available commercially; they can also be made by parent or child. Children (and adults) find them fun as well as a challenge to spatial imagination.

Tangrams date back to ancient China, where, according to legend, a Chinese scholar had a favorite ceramic tile he cherished. One day, the tile fell from his hands and broke into seven pieces, and the scholar spent the rest of his life trying to put them together again.

Fig. 3-1 shows how the seven pieces can be put together to form a square.

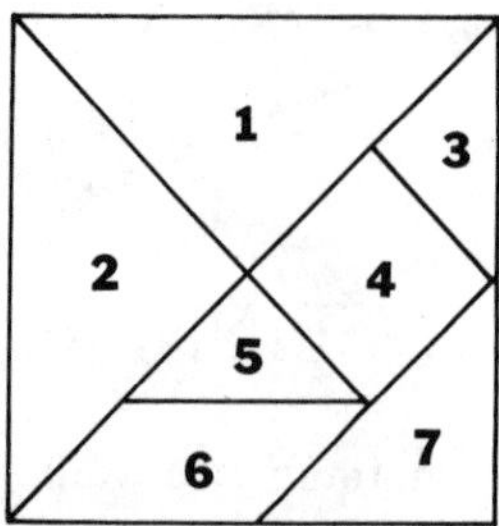

Figure 3-1

To use tangrams with your child, start with figures that are broken up into no more than two or three pieces. For instance, the two pieces in Fig. 3-2 can produce a square (Fig. 3-3), a triangle (Fig. 3-4), a trapezoid (Fig. 3-5), and a parallelogram (Fig. 3-6).

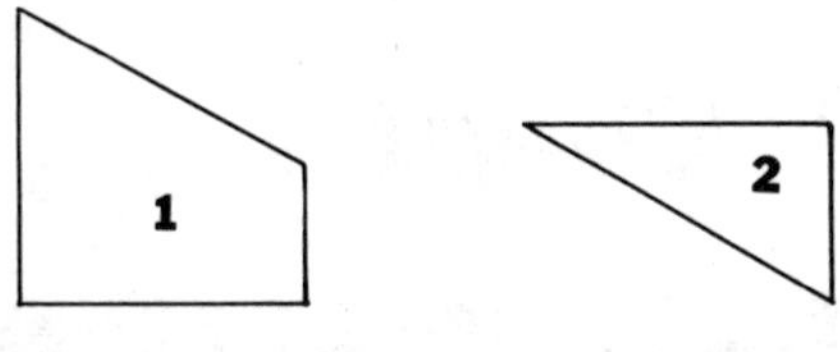

Figure 3-2

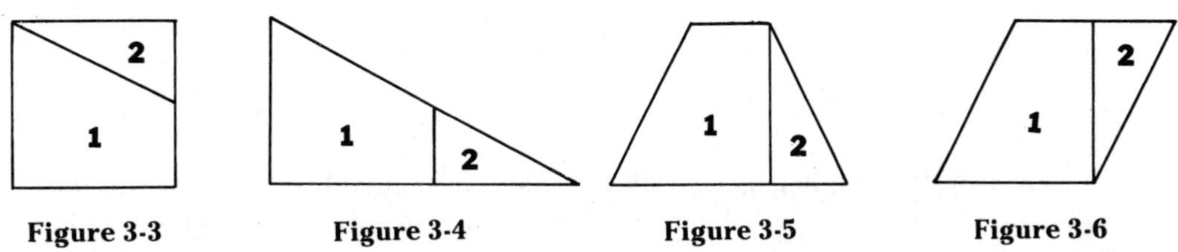

Figure 3-3 **Figure 3-4** **Figure 3-5** **Figure 3-6**

Chapter 4

Making Drill More Interesting

For two decades, beginning with the mid-1950s, one of the dirtiest five-letter words in education was *Drill.* It conjured up a taskmaster with a mission to take the joy out of learning.

Maybe used-car salesmen knew something educators didn't when they changed their pitch from "used" Cadillacs to "previously owned" ones. Maybe *drill*, too, should have been scrapped for a more alluring word.

Apart from the word itself, there is nothing wrong with drill. It's essential for accomplishing important objectives. But it must be made to sound like an adventure, and evoke anticipation, not an "Aw shucks, that again!" groan.

Mastery of a skill such as computation requires practice. Drill is a way to provide this practice. But turning it into an endless string of unmotivated exercises destroys it as an aid to learning. Children are more likely to accept drill if they are made aware of what it can do for them. Show them how drill can develop accuracy and efficiency and strengthen retention. This is a good way to nourish a child's self-confidence.

Below are suggestions a parent can follow to make drill more interesting to a child. Some are commonsense procedures; others are specific illustrations of what can be done. With a little forethought, the parent can add many variations and extensions to these suggestions:

- Make drill exercises brief to avoid fatigue and boredom, and give at spaced intervals to produce better retention.

- Avoid the kind of drill where the responses become mechanical. You can ease this problem by:

1. Providing drill only *after* the child understands the skill being taught. Otherwise, the drill becomes a procedure for following memorized directions.

2. Including in the drill more than just one type of exercise. For instance, drill in division may also include a multiplication question or a verbal problem. Varying the material forces the child to stop and think about each problem.

- Let the child know as quickly as possible whether his answers are correct.
- Review skills and provide practice in these skills right before they are needed. For instance, practice multiplication immediately before teaching division.
- Give the child several problems with "solutions" provided and ask him to mark each answer as correct or incorrect. For each incorrect answer, the child is to find the correct answer and identify the error that produced the wrong answer.
- Drill with games and puzzles—but with moderation.
- Let the child use visual aids and manipulative materials freely in drill work.
- Disguise drill in fresh and interesting settings. For instance, ask your child to pretend he is a "secret agent" called upon to code and decode secret messages. Let's create a simple code:

  Each letter of the alphabet, a comma, a period, and a space will be represented by the numbers shown in the table below:

A	0	H	7	O	14	V	21
B	1	I	8	P	15	W	22
C	2	J	9	Q	16	X	23
D	3	K	10	R	17	Y	24
E	4	L	11	S	18	Z	25
F	5	M	12	T	19	,	26
G	6	N	13	U	20	.	27
						space	28

  The "key" to the code is to *add* 1 to the number value of each letter in the message. We shall now write in *code* the message "SPY CAUGHT":

Word Message:	S	P	Y		C	A	U	G	H	T
Number Value: (from table)	18	15	24	28	2	0	20	6	7	19
Add 1:	19	16	25	29	3	1	21	7	8	20

  Therefore, the message S P Y C A U G H T, written in code, is: 19 16 25 29 3 1 21 7 8 20.

  To *decode* this message, reverse the procedure: *Subtract* 1 from each number; then translate from the table:

Coded Message:	19	16	25	29	3	1	21	7	8	20
Subtract 1:	18	15	24	28	2	0	20	6	7	19
From table:	S	P	Y	*space*	C	A	U	G	H	T

  The code just used is a simple one. It involves adding and subtracting 1 to the numbers in the table. Other codes might involve adding and subtracting other numbers, or multiplying and dividing by various numbers. In this way, the "secret agent" setting opens up countless opportunities for drill with the basic operations, and even with more advanced work.

- Use *palindromes* to provide practice in addition as well as fascination with the properties of numbers. This exercise is discussed in Chapter 11, page 93.

- Make drill in subtraction more interesting by a procedure similar to that used with palindromes, but this time practice *subtracting* rather than adding. For instance, start with the number 83. Reverse the digits and subtract the smaller number from the larger. Repeat the process until the answer is 0:

(1)
83 given number
38 digits reversed
45 difference

(2)
45 difference (from #1)
54 digits reversed
9 difference

(3)
9 difference (from #2)
9 digits reversed
0 difference

Let's now try the number 875. Reverse the digits and subtract the smaller number from the larger. Repeat the process until the answer is 0:

(1)
875 given number
578 digits reversed
297 difference

(2)
297 difference (from #1)
792 digits reversed
495 difference

(3)
495 difference (from #2)
594 digits reversed
99 difference

(4)
99 difference (from #3)
99 digits reversed
0 difference

Ask the child whether the difference will *always* come out to 0, or whether one must choose special numbers. Let her experiment with as many numbers as possible to find out. With numbers that have more than three digits a calculator can be used. (Yes: the difference, when the procedure is carried far enough, will always come out to 0.)

- Analyze the mathematics behind a number trick to serve as a stimulus for drill. For instance, ask the child to explain why the following number trick works:

	EXPLANATION Let n be the number selected.
(1) Take a number.	(1) n
(2) Subtract 2.	(2) $n - 2$
(3) Multiply by 3.	(3) $3n - 6$
(4) Add 6.	(4) $3n$
(5) Add the original number.	(5) $4n$
(6) Divide by 4.	(6) n

Your answer will always be the original number.

- Use number "tricks," such as guessing a birthday, as drill exercises.

To guess someone's birthday, the "someone" must keep in mind two numbers: the number of the *month* in which he was born, and the number of the *day of the month* on which he was born. The months are numbered 1 to 12, beginning with January. Then give these directions:

1. Multiply the number of the month in which you were born by 5.
2. Add 7.

3. Multiply by 4.
4. Add 12.
5. Multiply by 5.
6. Add the number of the day on which you were born.

To "guess" the birthday, mentally subtract 200 from the result in step (6). In the answer you get, the last two digits give you the day and the remaining digits give you the month of the birthday.

For example, suppose the birthday is April 19. This is month #4, day #19. Now let's work through the six steps in the directions and see what happens.

(1) $4 \times 5 = 20$	(4) $108 + 12 = 120$
(2) $20 + 7 = 27$	(5) $120 \times 5 = 600$
(3) $27 \times 4 = 108$	(6) $600 + 19 = 619$

When you subtract 200 from 619, you get 419. So you know the child was born on the nineteenth day (last two digits) of the fourth month (the remaining digit)—that is, April 19.

Explanation: The trick is based on the fact that the directions you give to the person are a disguised way of adding 200, the day number, and 100 times the month number. When you subtract 200, the day number and 100 times the month number are left. Because the month number was multiplied by 100, it appears as the number of hundreds. So you find it to the left of the figures that represent the day number.

- Drill with games. Below is a game you can adapt to almost any topic in mathematics. Many other games, applicable to specific topics, are suggested in Part II.

BINGO

MATERIALS NEEDED

1. A set of *question cards* containing questions on any topic, including the basic operations with whole numbers, fractions, percent, geometry, or any combination of these topics.
2. A set of *answer cards* (Fig. 4-1) resembling the usual bingo cards, containing the answers to questions on the question cards. Each card has different answers, as do regular bingo cards.
3. *Counters* (or buttons) to cover the spaces on the answer cards.

GAME ACTION

Played like bingo. A question is selected at random from the question cards and read to the players. The question is preceded by one of the call letters—B, I, N, G, O—indicating the column in which the answer is to be located.

Correct answers found on the answer cards are covered. Winner is the first to have a vertical, horizontal, or diagonal line completely covered. Check all winning cards for errors.

B	I	N	G	O
		free		

Figure 4-1

- Try "mental arithmetic." Here, the child performs calculations without paper and pencil or calculator. One investigator estimated that about 75 percent of all nonoccupational arithmetic in everyday life is performed mentally. If this is true, then the need to train children in mental arithmetic from an early age is obvious.

  A few minutes of oral practice as often as possible can go a long way in strengthening this skill in your child and can, at the same time, provide a welcome diversion from the more tedious forms of practice with computation. The exercises should include basic computations as well as verbal problems, and should become progressively more advanced.

  Depending on the child's mathematical level, the following are examples of suitable exercises for mental arithmetic. (Answers are in brackets following each question.)

1. *Practice with the basic arithmetic operations.*
   For example: What is the sum of 8 and 5 ? [13]
   How many threes are there in 24 ? [8]

   Name a pair of factors whose product is 24. [3, 8]
   Name as many other pairs as you can. [(6, 4), (12, 2), (24, 1)]

   Count backwards by fours, starting with 32. [32, 28, 24, 20, 16, 12, 8, 4, 0]

   How much is 23 rounded off to the nearest 10 ? [20]

   Round off 1 and seven-eighths to the nearest whole number. [2]

   What's the lowest common denominator of 2 and 3 ? [6] Of 6 and 8 ? [24]

   What's the greatest common divisor of 12 and 30 ? [6]

   How much must I add to 65 to get 90 ? [25]

2. *Estimating answers.* For instance: About how much is 59, 32, and 18 ? [$60 + 30 + 20 = 110$] About how much is 2 1/4 × 3 7/8 ? [$2 \times 4 = 8$]

3. *Verbal problems.* For example: Cough drops come 6 in a package, and 12 packages in a box. How many cough drops are there in a box? [72]

4. *Puzzles*. For example: I'm thinking of two numbers whose sum is 7 and whose product is 12. What numbers am I thinking of ? [3 and 4]

Drill need not be tedious, boring, or any of the other terrible things often associated with it. Don't call it "drill," if that helps. Design activities in a motivational setting, and dress them up in alluring attire, and you will find your child can have his drill and enjoy it, too.

Chapter 5
Problem Solving

A burglar alarm goes off in a home. A neighbor sees two men run from the house and then take off in a large white car. Police are notified and, a mile away, stop two men in a large white car for investigation. At the same time, other officers are dispatched to investigate the alarm.

They find out the following:

1. The home was forcibly entered through a broken basement window.
2. On the way to the basement window, the burglar walked through mud.
3. The burglar slid down a whitewashed basement wall and was attempting to jimmy open the door leading to the kitchen when the alarm went off.

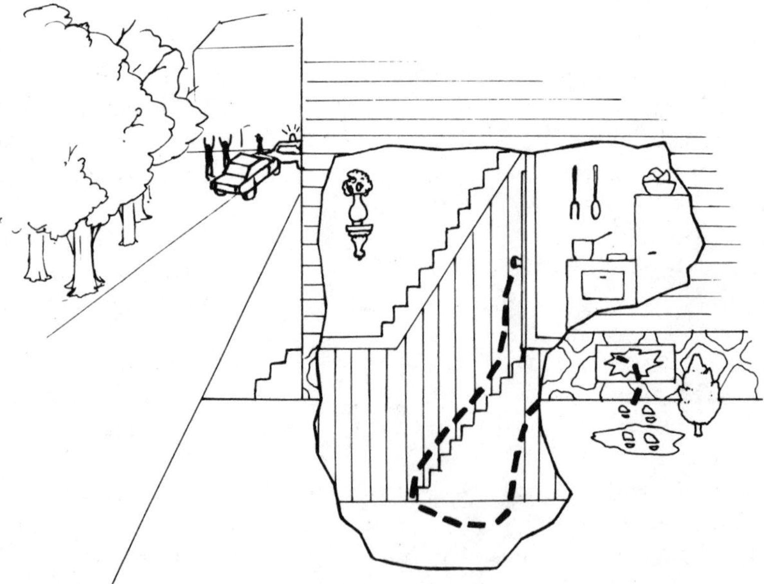

How did the police solve this crime?

According to the detective in charge, he essentially worked out the solution backward. He reasoned that if the two men in the large white car were the burglars, then

1. Their shoes should be encrusted in mud.

2. Their clothing might show bits of a white substance coming from the freshly whitewashed walls they slid down.

3. Their clothing might contain bits of glass from the window they broke to gain entry into the basement.

When the two men were examined, the officers observed that both had a white substance on their clothing and fresh mud on their shoes. The tip of a large screwdriver found on the front floor of their car had small pieces of fresh white paint chips on it. The jimmy marks on the basement door matched the 1-inch tip on the screwdriver. Later, laboratory tests showed that the glass from the broken basement window matched the glass found on their clothing.

When confronted with this evidence, the men pleaded guilty.

* * *

A couple return home at 10 P.M. from an early movie, only to surprise a burglar who, upon seeing them, dashes out of the house. The neighborhood is dotted with expensive single homes separated by shrubbery, driveways, and yards.

Police respond to a call from the homeowner. Then a detective arrives. No suspicious cars are seen in the area, the only car there being that of the homeowner.

To solve this crime, the detective decided to proceed on a hunch. The absence of any suspicious cars on the scene indicated to him that the burglar had to have been dropped off by an accomplice, or pickup man. So the detective acted on the hunch that the burglar was (1) scared and running through backyards unfamiliar to him, and (2) looking for his pickup man.

The detective decided to drive through the neighborhood slowly, with lights off, pretending to be the pickup man. In a matter of minutes, he heard someone call out "Dan!" from the bushes along the roadway. The detective responded with a noncommittal "Yeh." The man then whistled, and the detective responded with his own whistle, enticing the burglar to within 10 feet of the detective's car—when, suddenly and unexpectedly, the police radio in his unmarked car was activated. The burglar took to his heels, but was caught 50 yards away.

These are two examples of strategies used by police to solve crimes. While these incidents actually occurred, there is a whole world of crime and detective *fiction*—a gripping form of problem solving—that relies on these and a range of similar strategies to catch the one who dunit. The detective always analyzes the facts and, by deductive reasoning, draws the inescapable conclusions so dear to the heart of Sherlock Holmes. He sometimes works on a hunch, always looks for patterns of behavior, and often goes at the problem backward: "If Jones is the killer, what would have to be true?"

The process of elimination is a standby tactic, known to every good detective, in keeping with Sherlock Holmes's dictum: "When you have eliminated the impossible, whatever remains—however improbable—must be the truth." When the crime is especially intractable, the more imaginative sleuth goes at it with less routine and more creative methods. Sometimes the cleverest thing he does is to look for the obvious, sitting right under his nose.

Problem solving in mathematics employs exactly the same strategies.

What Is Problem Solving?

Problem solving is looking for an answer to a question for which you don't have a readily available strategy. It usually requires applying what you already know to new and unfamiliar situations. If a solution is immediately apparent, then for *you* this is not solving a problem. What is a problem for one person may be a routine exercise for another.

For elementary-school children there are routine and more advanced problems. An example of a routine problem is, "If you travel 140 miles by bus in $3\frac{1}{4}$ hours, what is your average speed?" More advanced problems involve more complex situations requiring more original and complex strategies for their solution. An example might be, "How many different triangles do you see in Fig. 5-1?" [8] Both kinds of problems are important for children.

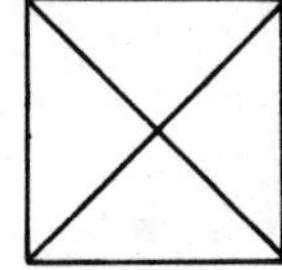

Figure 5-1

Because of its central importance in mathematics, problem solving has been the subject of much research and study over the years. One study found that the primary factors related to success in problem solving are computational skill, reading with the ability to absorb details, and an understanding of fundamental mathematical concepts. Other researchers say the basic ingredient for success is the ability to recognize the "action" in the problem situation. The good problem solver knows "what's going on" in the problem.

It is generally accepted that IQ is significantly related to problem-solving ability, that sex differences are not, and that socioeconomic status alone does not appear to be significant.

Strategies for Problem Solving

Like athletes, some problem solvers are "naturals," while others have to work hard at it. But everyone, from young children to adults, can improve his skill enormously by learning a few basic problem-solving strategies and then using them to solve many varied problems. An important companion strategy is learning to recognize and avoid *ineffective* problem-solving methods.

Overall strategy for solving problems involves:

- Understanding the problem
- Devising a plan for solving it
- Carrying out the plan
- Checking the solution; seeing whether it makes sense

We will soon illustrate how to use this strategy to solve specific problems. But first, let's expand a bit on what each of these four steps entails.

Understanding the problem. To test understanding,

Try to restate the problem in your own words.

Identify what you're asked to find.

Identify the pertinent facts you're given, including the relations between them.

Make sure you're not making unwarranted assumptions about the problem.

Determine what information, if any, is missing or not needed, and whether there are any contradictions in the given facts.

Devising a plan for solving it. This is the guts of problem solving.

A few questions:

Have you seen this problem, or a similar one, before? If yes, how was it solved?

In mulling over the problem are you taking all the relevant information into consideration?

A couple of routine moves:

Give the unknown—the answer you're looking for—a name, such as the letter x or some other symbol. Use the name just as though you know what it stands for. Using a letter makes it easier to express relationships in the problem.

Where possible, draw a picture, an illustration, or a diagram as closely to scale as you can. (See following Problem 1.)

A few strategies:

- Try a similar but simpler problem. If the problem involves five people, reduce it to a similar problem with only two people, and then try to solve the simpler one. See whether the method for solving the simpler problem works with the more complex one. (See following Problem 14.)
- Look for a pattern. Can you draw any conclusions from this pattern? (See following Problems 2, 3, 6.)
- Organize your information by making a chart, table, or drawing a picture. This makes relationships, including patterns, easier to detect. (See following Problems 5, 6, 7, 8, 9, 10, 12.)
- Work the problem backward. Ask: To find the answer, what must be true? What would it take to get the unknown? (See following Problem 4.)
- Use a trial-and-error approach. Estimate the answer, make a reasonable guess, then check to see whether it works. (See following Problems 5, 12.)
- Break away from a "safe" method; try an off-beat, more imaginative one. (See the Gauss problem on page 88; also following Problem 13.)
- Step away from the problem to see whether there is an obvious solution sitting "right under your nose." (See following Problem 15.)

Carrying out the plan. To do this, you must understand the mathematical ideas, rules, and processes involved; have the needed computational skills; and avoid errors in reasoning.

Checking the solution; seeing whether it makes sense. Ask:

Does the answer satisfy the conditions of the problem?

Is there another way to find the same solution?

Is there more than one solution?

Looking back at the problem and your solution leads to increased power and knowledge of problem solving. Seeing the problem as a whole and how the steps in your solution fit together makes it easier to repeat the technique with other problems.

Estimating Answers

In this chapter and in others we make reference to *estimation*, the skill of making a reasonable guess. We said it was needed in problem solving to make the child aware of an unreasonable answer. We could have added that, although much time is given to learning to calculate exact answers, adults use *estimated* answers as much as 80 percent of the time.

So there is a good payoff in helping your child develop estimating skills. You can do this informally in many day-to-day situations as well as through more formal instruction. When children become comfortable with the process, they enjoy estimating.

Skill in estimating depends on the ability to deal with 10s and multiples of 10. The idea is to round off numbers to convenient multiples of 10, 100, 1000, etc., so that computation with the resulting numbers can be done mentally.

Another kind of estimating depends on a reference point. The idea is to determine whether the answer is *over* or *under* the reference point. For instance: Estimate the number of gallons of gas you can buy for $5.

You can share with your child some of your estimating techniques in innumerable day-to-day situations. In restaurants—estimate the tip; purchases—estimate the change from $10; travel—estimate travel time; cooking—estimate four times a given recipe. Is a box big enough to hold a given amount? How long will your sneakers last?...

Strategies in Action

The following problems and their solutions illustrate the problem-solving strategies just described.

PROBLEM 1 A frog has fallen to the bottom of a 5-foot-deep well and wants to get out. Each day she climbs up 2 feet but falls back 1 foot. How many days will it take her to get out of the well?

Solution: To visualize the problem better, draw a picture (Fig. 5-2).

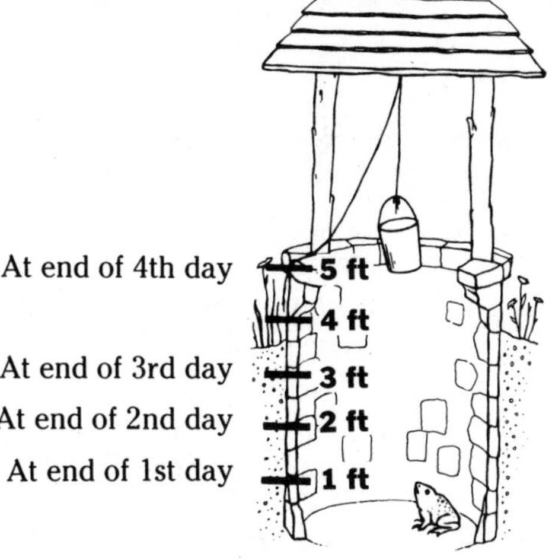

Figure 5-2

On the first day, the frog reaches the 2-ft mark but falls back to the 1-ft mark. On the second day, she reaches the 3-ft mark but falls back to the 2-ft mark. On the third day, she reaches the 4-ft mark but falls back to the 3-ft mark. On the fourth day, the frog reaches the top and jumps out of the well.

The catch in this problem is the temptation to say that on the fourth day the frog reaches the 5-ft mark and falls back to the 4-ft mark. But looking at the picture makes it clear that the leap on the fourth day gets her out of the well.

PROBLEM 2 Fill in the three missing terms:

Z1, Y2, X3, W4, ____, ____, ____.

Solution: The letters of the alphabet are listed in reverse order, while the numbers are listed in ascending order. This pattern suggests that the next three terms are:

V5, U6, T7

PROBLEM 3 Find a pattern in the *numbers* suggested by the sequence of dot arrangements in Fig. 5-3.

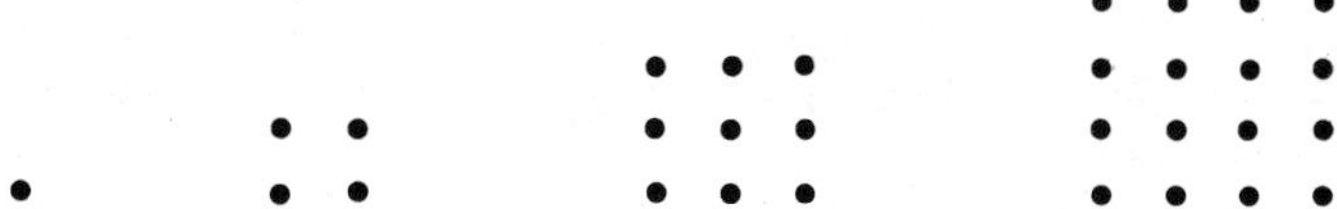

Figure 5-3

Solution: Notice that each dot arrangement forms a square containing the following numbers of dots: 1, 4, 9, 16. These are the square numbers 1^2, 2^2, 3^2, 4^2.

If this pattern continues, then the fifth number would be 5^2, or 25; the sixth number would be 6^2, or 36; the tenth number would be 10^2, or 100; and the nth number would be n^2.

PROBLEM 4 This is a game involving two players. The first player selects any whole number from 1 to 10. The second player then adds any whole number from 1 to 10 to the number selected by the first player. The play then alternates with each player adding any number from 1 to 10 to the sum left by the opponent. The game continues until one player reaches the sum of 100; this player is the winner.

The problem: If the players are A and B, how can A make sure he wins every game? That is, can you find a winning strategy for this game?

Solution: The problem can be solved by working backward.

For A to win, he must force B to obtain some total between 90 and 99 on the next-to-last play. (Then to win, A can add the appropriate number from 1 to 10 and reach 100.) So A must have a total of 89 on the third-to-last play.

For A to obtain 89, he must previously force B to reach some total between 79 and 88. Therefore, A must have 78.

By the same reasoning we can show that the winning sequence of numbers is 89, 78, 67, 56, 45, 34, 23, 12, 1. If A reaches any of these numbers, he can win. It follows that whoever starts with 1 can force a win.

PROBLEM 5 Eric was 5 years old when his father was 33. How old was Eric when his father was five times his age?

Solution: One way to organize the information for this problem is to make a table. The top row shows Eric's age and the bottom, his father's (Fig. 5-4).

Eric	**5**	**6**	**7**
Father	**33**	**34**	**35**

Figure 5-4

Starting with 5 and 33, the ages given in the problem, fill in the two ages for each successive year until reaching the ages where the father's is five times Eric's—7 and 35.

PROBLEM 6 If you fold a strip of paper in half 10 times (Fig. 5-5), into how many parts is it folded?

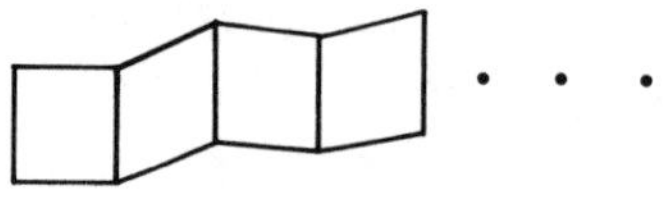

Figure 5-5

Solution: Make a table (Fig. 5-6) showing the number of folds and the resulting number of parts, beginning with one fold and two parts. Continue with two folds and four parts, three folds and eight parts, and so on.

Folds	**1**	**2**	**3**	**4**	**5**	**. . .**
Parts	**2**	**4**	**8**	**16**	**32**	**. . .**

Figure 5-6

Notice a pattern emerging. If u don't see it yet, carry the table to a few more folds. The pattern is that the number of parts doubles each time another fold is made. That is, 1 fold produces 2 parts; 2 folds produce 2^2, or 4 parts; 3 folds produce 2^3, or 8 parts; 5 folds produce 2^5, or 32 parts; and 10 folds will produce 2^{10}, or 1024 parts. More generally, n folds will produce 2^n parts.

Reminder:
$$2^2 = 2 \times 2 = 4$$
$$2^3 = 2 \times 2 \times 2 = 8$$
$$2^5 = 2 \times 2 \times 2 \times 2 \times 2 = 32$$
$$2^{10} = 2 \times 2 \times 2 \times 2 \times 2 \times 2 \times 2 \times 2 \times 2 \times 2 = 1024$$

PROBLEM 7 Jane is taller than Sue but shorter than Anne. Which of the three is the tallest?

Solution: Draw a picture showing the conditions of the problem (Fig. 5-7). The picture shows Anne to be the tallest.

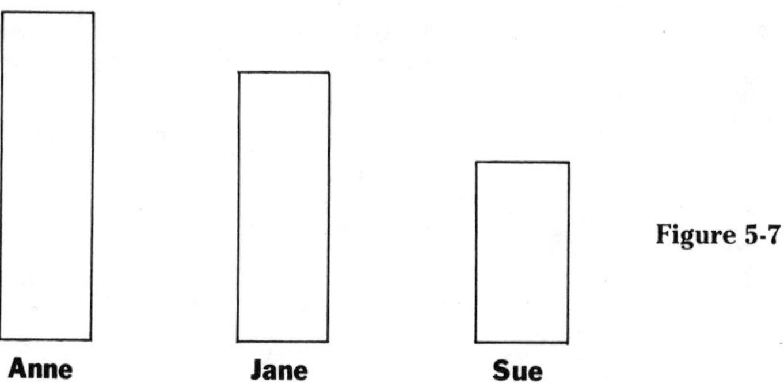

Figure 5-7

PROBLEM 8 Joe is shorter than Tony but taller than Bob. Bob is shorter than Joe but taller than Mike. Who is the tallest and who is next to the tallest?

Solution: Don't allow yourself to get confused by four names. Just proceed as you did in Problem 7, drawing a picture reflecting the conditions of the problem (Fig. 5-8). The picture shows Tony to be the tallest and Joe next to the tallest.

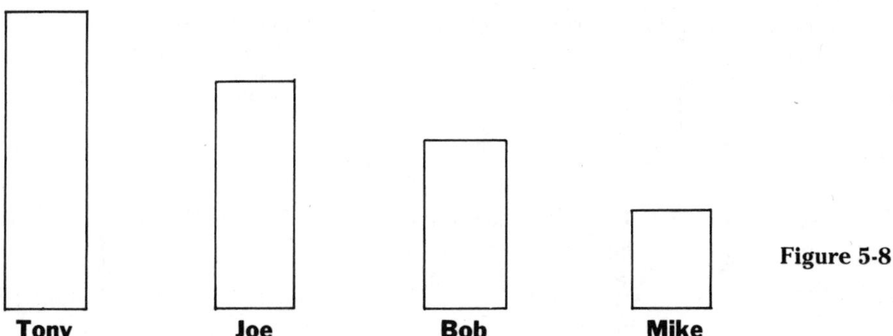

Figure 5-8

PROBLEM 9 Peach Street is parallel to Cherry Lane. Pine Street is perpendicular to Fifth Avenue, which is parallel to Cherry Lane. Is Pine Street parallel or perpendicular to Peach Street?

Solution: Draw a picture (Fig. 5-9).

Pine Street
Peach Street
Cherry Lane
Fifth Avenue

Figure 5-9

The picture shows Pine Street to be *perpendicular* to Peach Street.

PROBLEM 10 On a certain day I went to the barber, played golf, went shopping, and visited a museum. The barber is closed Sundays and Mondays; I can play golf only Sundays, Tuesdays, Thursdays, and Saturdays; stores are closed Sundays and I don't like to shop on Saturdays; and the museum is closed Thursdays. On what day of the week did I do all these things?

Solution: Don't be intimidated! Just organize the information given in the problem in a chart (Fig. 5-10). Mark with an *X* the days on which an activity *cannot* take place.

	Sun.	Mon.	Tues.	Wed.	Thur.	Fri.	Sat.
Barber	x	x					
Golf		x		x		x	
Shopping	x						x
Museum					x		

Figure 5-10

Tuesday appears to be the only free day on which I could have done all four things.

PROBLEM 11 What are the missing numbers in this series?

1, 3, 7, 13, 21, __, __, __.

Solution: Find the differences between the numbers and see whether a pattern emerges:

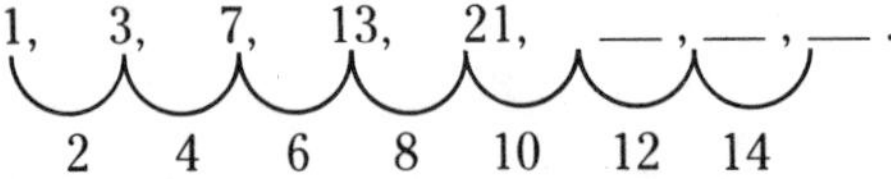

The differences appear to be increasing by two's: 2, 4, 6, 8, 10, 12, 14, suggesting that the next three numbers in the series are:

31, 43, 57

PROBLEM 12 A bridge toll is 45 cents; it can be paid only in quarters, dimes, and nickels. In how many ways can you pay the toll?

Solution: Arrange the information in a table; then write out all the combinations that work (Fig. 5-11).

	(1)	(2)	(3)	(4)	(5)	(6)	(7)	(8)
Quarters	1	1	1					
Dimes	2	1		4	3	2	1	
Nickels		2	4	1	3	5	7	9

Figure 5-11

The table shows eight ways in which the toll can be paid.

PROBLEM 13 There are 60 vehicles in a garage. Some are cars, others are motorcycles. Together they have 200 wheels. Cars, of course, have four wheels; motorcycles, two. How many cars are in the garage?

Standard Solution: The standard way of solving this problem algebraically is:

Let x = the number of cars
Let $60 - x$ = the number of motorcycles

Since each car has 4 wheels, x cars have $4x$ wheels.
Since each motorcycle has 2 wheels, $(60 - x)$ motorcycles have $2(60 - x)$ wheels.
Since the 60 vehicles together have 200 wheels, we write

$$\begin{aligned} 4x + 2(60 - x) &= 200 \\ 4x + 120 - 2x &= 200 \\ 4x - 2x &= 200 - 120 \\ 2x &= 80 \\ x &= 40 \text{ cars} \end{aligned}$$

Offbeat Solution: No, you don't need algebra to solve this problem—*if* you can break away from the "safe" solution and, instead, allow your imagination to take off:

Imagine the front wheels of each car jacked up. How many wheels would there be on the ground? Since there are 60 vehicles, each with two wheels on the ground, there would be 120 wheels on the ground.

How many wheels are in the air? If there are 200 wheels altogether, and 120 are on the ground, then there are 80 in the air.

If 80 wheels are in the air, there must be 40 cars.

PROBLEM 14 How many *squares* are there in the 4×4 array shown in Fig. 5-12?

Figure 5-12

Solution: Reduce the problem to a simpler one by starting with a 1 × 1 array, then go to a 2 × 2 array, then to a 3 × 3 array (Fig. 5-13). Then try to detect a pattern.

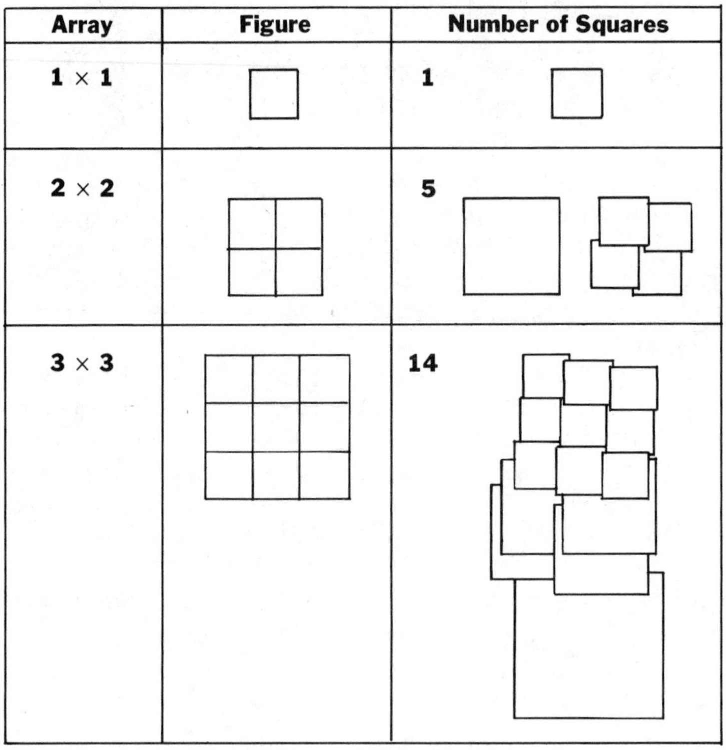

Array	Figure	Number of Squares
1 × 1		**1**
2 × 2		**5**
3 × 3		**14**

Figure 5-13

The numbers 1, 5, 14, are the sums of squares:

$$\begin{aligned}
1 \times 1 \text{ array: } 1 &= 1 &&= 1^2 \\
2 \times 2 \text{ array: } 5 &= 1 + 4 &&= 1^2 + 2^2 \\
3 \times 3 \text{ array: } 14 &= 1 + 4 + 9 &&= 1^2 + 2^2 + 3^2
\end{aligned}$$

If this pattern holds, you would expect a 4 × 4 array to contain

$$\begin{aligned}
& 1^2 + 2^2 + 3^2 + 4^2 \quad \text{squares} \\
= & \ 1 + 4 + 9 + 16 \\
= & \ 30 \text{ squares}
\end{aligned}$$

PROBLEM 15 You're the engineer on a train traveling 60 miles an hour between New York and Washington. The train takes on 85 passengers in New York on its way to Newark, where it drops 7 passengers and picks up 13. When it reaches Philadelphia, it drops 29 and picks up 63 passengers. In Baltimore, it drops 17 and picks up 8. When the train reaches Washington, all passengers get off. What's the name of the engineer?

Solution: If you step aside from all the train stops and numbers of passengers getting on and off and read the first few words in the problem, you will find, "You're the engineer."

1. **John, Tom, and Rich differ in height. Their last names are Aiken, Carey, and Baker, but not necessarily in that order. John is taller than Rich, but shorter than Tom. Aiken is the tallest of the three, and Baker is the shortest. What are John's and Rich's last names?**

2. **At a family dinner, mother wanted to serve food that everyone liked. But she had to contend with several eating idiosyncrasies: Liz didn't like pizza or steak. Arin liked pizza, chicken, fish, pasta, and steak. Benj didn't like chicken or pasta, while Jerry didn't like pizza or veal. Judy, however, liked everything. What did mother serve?**

3. **How many squares are there in an 8 $\times$ 8 checkerboard (Fig. 5-14)?**

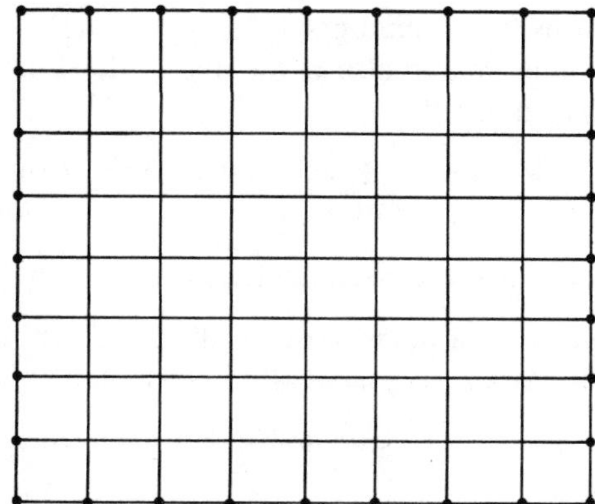

Figure 5-14

(The answers appear on page 256.)

How Can You Help Your Child with Problem Solving?

You can help, in a general way, by stressing the importance of problem solving and the satisfaction that can be derived from it. Help will be most effective when the child senses that you regard problem solving as an important skill, and that you yourself actively engage in it. Take advantage of your child's curiosity about numbers, shapes, puzzles, and brainteasers by sharing problem-solving experiences with her.

Beyond this, the parent who wishes to teach a child how to solve problems can do a number of more specific things:

- Help your child develop specific strategies for attacking problems, like those just illustrated, by doing many different types of problems. Select problems and strategies compatible with the child's level of development.
- Provide lots of experience estimating answers.
- Provide problems at varying levels of difficulty.
- Encourage your child to use concrete materials like counters—where feasible—to help visualize a problem and verify its solution.
- Ask your child to make up original problems conforming to conditions you set down.
- Use problems without numbers. For instance, ask, "If you know the product of two numbers and also know one of those numbers, how can you find the other number?"
- Train your child to note the absence of information needed to solve a problem, or the presence of unnecessary information.
- Have your child test the reasonableness of an answer by *estimating* it.
- Suggest checking an answer by reading the problem with the answer filled in. Does the answer meet all the conditions of the problem?
- Encourage your child to find alternate ways to solve a problem.
- Where helpful, ask several questions about a problem:
    - Are there any words you don't understand?
    - Can you tell the problem in your own words?
    - What's the important information in the problem?
    - What does the problem ask you to find?

Then guide your child, if necessary, to a strategy for solving it.

- Since *thinking* rather than following directions is the key to problem solving, it's essential that you analyze your child's thinking as she tries to solve a problem. Help her understand mistakes, and commend her for what is commendable.

Chapter 6
Calculators

The proliferation of inexpensive, hand-held electronic calculators has predictably led to a great debate over the virtues and dangers of using them with elementary-school children. Are we going to substitute button pushing for real learning? Are we terminally myopic if we ask that children understand the mathematics behind the answers given by a calculator? Will we take full advantage of a new technology that places incredible computational power at our fingertips?

Using calculators with very young children will "rot the mind," some fear, and produce children who neither know the basic arithmetic facts nor are able to perform the simplest computations on their own. They will become so dependent on the machine that their minds will no longer function.

The problem with the "rot-the-mind" argument is that it flies in the face of hundreds of research studies showing that children who use calculators achieve at least as well as those who don't use them. These studies also show that calculators often motivate even the most reluctant learners.

On the other extreme are those who want to do away completely with "paper-and-pencil" arithmetic and let the calculator do all the computing. There is no good reason, they say, to teach children how to add, subtract, multiply, and divide when a calculator can find the answer instantaneously.

The problem with this argument is its assumption that the only reason for teaching arithmetic is to get answers to computations. But there are other reasons, related to the wider question of why *any* mathematics should be taught. (See Chapter 1.) When arithmetic is taught properly, it provides insights into the world of mathematics. It exposes the child to mathematical reasoning and to procedures for solving problems. It develops number systems and operations, and shows the connections between them. Calculators do not do this. They weren't designed to. They provide answers, not understanding.

The ability to display numbers and perform calculations on a calculator may be incorrectly interpreted as "understanding." But understanding numbers and operations involves a slow, delicate process of concept development that cannot occur by merely grinding out answers on a calculator. Only after understanding occurs is it time to let the calculator take over.

The question is not whether calculators should not be used at all, or whether they should completely replace traditional teaching. The question is when, where, and how to use them most profitably.

A Calculator, Your Child, and You

How can you, the parent, use a calculator to help your child learn mathematics? The first step is to become familiar with the book of instructions that comes with your calculator. It's best to work through the examples in the booklet, calculator in hand. Since different calculators react differently to the same series of "commands," it is suggested that you buy one with a *constant* arithmetic feature—that is, one that continues to perform computations when the = key is pressed, as illustrated in #1 below.

Depending on your child's mathematical level, you can use a calculator in the following ways:

1. To count by ones, have child press: 1, +, =, =, =, . . .
   By counting by 1's to 100 or 1000 on a calculator, and seeing the display show the number at each step, the child acquires an appreciation for the size of 100 or 1000.

2. To count by twos, have the child press: 2, +, =, =, =, . . .
   To count by fives, have the child press: 5, +, =, =, =, . . .
   To count by tens, have the child press: 10, +, =, =, =, . . .

   The child can count backwards in a similar way, but subtracting instead of adding.
   To count backwards by twos, starting with 18, the child presses

   18, −, 2, =, =, =, . . .

   To count backwards by fives, starting with 70, the child presses

   70, −, 5, =, =, =, . . .

3. Provide drill with the basic facts: display a fact like 5×7, and ask the child for the answer before verifying it on the calculator.

4. Have the child make up his own addition (or subtraction, multiplication, division) example, work out the answer mentally or on paper, and then check it on the calculator.

5. Reinforce place value concepts with activities like this: Enter 563 on the calculator, then ask:
   (a) What number can you subtract from 563 to replace the 6 with a 0? Try it on the calculator.
   (b) How can you replace the 5 with a 2?
   The child will see that in (a), 60 must be subtracted rather than 6; in (b), 300 must be subtracted rather than 3.

6. For enrichment, ask the child to find a pattern from these entries:

$$\begin{aligned} 1 \times 9 + 2 &= 11 \\ 12 \times 9 + 3 &= 111 \\ 123 \times 9 + 4 &= 1{,}111 \\ 1234 \times 9 + 5 &= 11{,}111 \end{aligned}$$

etc.

Let the child give the answer to 123456 × 9 + 7 and then verify it on the calculator. What about the ninth number in the series? *Any* number in the series?

7. Motivate the investigation of negative numbers.
   (a) Ask the child to press: 5, −, 1, =, =, =, =, =, =.
   The last display is −1. Ask what the child thinks this number means.
   (b) Repeat this exercise with different numbers, like counting backwards by 2's, starting with 10.
   (c) Ask the child to enter on the calculator 3 − 5. When he gets the answer −2, ask what he thinks this answer means.

8. Present more challenging problems. Ask the child to determine the number that each question mark stands for below. (Answers appear on page 262.)

   (a)
   $$\begin{array}{r} 42 \\ \times\ \underline{??} \\ 966 \end{array}$$

   (b)
   $$\begin{array}{r} 1? \\ 27\overline{)4?2} \end{array}$$

9. Play "wrong key, wrong answer."
   (a) I meant to enter 69 − 15 on the calculator, but pressed one wrong key and got a wrong answer of 50. What wrong key did I press?
   (b) I thought I entered 18 × 24 on the calculator, but pressed one wrong key and got an incorrect answer of 252. What wrong key did I press?

10. Try word problems.
   (a) Find two numbers that when multiplied together equal 288. Find as many more such pairs of numbers as you can.
   (b) Enter 25 on the calculator. Find as many ways as you can of making the 25 change to 100 by addition, subtraction, multiplication, or division.
   (c) If a person's heart beats 82 times a minute on the average, how many times does his heart beat in a year? How many times will his heart beat altogether from birth to age 35?

11. Play games.

   (a) ***Blackjack.*** Display 0 on the calculator. Players take turns adding 1, 2, or 3 to the display. The player who gets the display to show 21 wins.
   Variation 1: Player who gets the calculator to show 21 *loses*.
   Variation 2: Display 21 on the calculator. Players take turns *subtracting* 1, 2, or 3 from the display. Player who gets the calculator to show 0 wins.
   Variation 3: Player in Variation 2 who gets calculator to show 0 *loses*.
   NOTE: Encourage the child to figure out a winning strategy for each game.

   (b) ***Fifty.*** Set calculator to 0 display. Players take turns adding any number from 0 to 9 to the display. Player who obtains 50 on the display wins.
   NOTE: Use variations similar to those suggested for Blackjack.

(c) ***Fade.*** A three-digit number is displayed on the calculator. The object of the game is to make one digit in the number disappear, with a single operation, without changing any of the other digits. For instance, for the number 125, a player may be required to make the "2" disappear. The single entry −20 accomplishes this by producing 105. The players can decide whether obtaining a display of 15 is also acceptable. (This result can be obtained with the entry, −110.)

The game starts with one player entering a three-digit number on the calculator. The child then directs a second player to make any of the digits (1) disappear, or (2) be changed to a different digit, without changing any of the other digits in the original number.

The play alternates from player to player, with a point scored for each correct entry. The player with the highest score at the end of a set amount of playing time is the winner.

Chapter 7
Computers

The irresistible invasion of the computer into every aspect of our lives has been truly phenomenal. It has revolutionized business and industry, education, and the arts. It has worked its way into medicine, engineering, transportation, and law enforcement. The enduring love affair Americans have had with cars and television is now being transformed into a giddy passion for the personal computer. In 1978, about 5000 desk-top computers were sold in the United States; in 1982, 5 million. By 1990, the estimate is 80 million. (If microprocessors in weapon systems and home appliances are included, there will be a *billion* computers throughout the world by that date.)

Thanks to the transistor and the silicon chip, computers began their dizzying development in the 1950s. In the past five years, they have become a hundred times faster, a thousand times smaller, and less expensive to operate than those of only a few years earlier. ENIAC, the computer developed in 1946 at the University of Pennsylvania in response to war needs, weighed 30 tons, contained 18,000 vacuum tubes (which failed at the rate of 1 every 7 minutes) and miles of wiring, filled a 30-by-50-foot room, cost $500,000 (in 1950 dollars), and could perform 5000 additions or subtractions per second (Fig. 7-1). A personal computer today (1984) costs about $2000, is run by a silicon chip the size of a child's fingernail, and, at the touch of a button, can make more than a million calculations per second. In 1981, a laboratory squeezed the equivalents of 750,000 vacuum tubes onto a chip! One computer scientist estimated that if the automobile business had developed like the computer business, a Rolls Royce would now cost $2.75 and run 3 million miles on a gallon of gas.

Figure 7-1 General View of the ENIAC

How Does a Computer Work?

A computer is a machine that makes calculations with lightning speed, remembers enormous amounts of information, makes decisions on the basis of this information, and can communicate with the external world, first to receive the data of the problem and later to yield the answers.

All its functions are built up from a few very simple activities, the most important being a capability to (1) add two numbers, (2) subtract one number from another, and (3) compare two numbers to see if they are the same. The speed with which a computer can perform millions of such simple activities and its capacity to store, sort through, and rapidly retrieve immense amounts of information are the reasons why it has such a fantastic range of capabilities.

Though no human can compete with a computer's prodigious memory or the speed with which it manipulates symbols, it should also be said that no present computer can compete with a human in recognizing patterns. A "pattern" is a group of symbols seen as a whole. An example of pattern recognition is when you see a forest in a large grouping of trees. Computers are great at counting the trees, but people are great at seeing the forest. A baby will recognize its mother instantly—something no computer can yet do.

All computers, whatever their size, consist of the same fundamental parts:

1. An *input* device, which takes information from the outside and converts it into computer language—using the *binary* numeration system (see Chapter 10) which the computer understands. In the computer shown in Fig. 7-2, the *keyboard* serves as the input device. It resembles an ordinary typewriter keyboard except that it has more keys which, when depressed, cause a series of electronic pulses to be sent to the computer.

Figure 7-2 An IBM Personal Computer

As each line is typed on the keyboard, the computer translates it, character by character, into sequences of 0s and 1s—which, in turn, are translated into sequences of electrical signals. These signals are then transformed into bursts of electrons that strike bits of phosphor coating the screen, lighting up a pattern of dots forming letters and numerals, and thus reproducing each line of text on the screen.

2. A *memory*, which can store information from outside as well as the instructions given to the computer. The information is stored as patterns of electrical charges in memory "cells," or memory locations, that can be charged by an electrical impulse. A computer's memory contains many thousands of such cells. Some of these patterns represent numbers; others represent letters and other symbols.

Computer Memories

A computer has two different kinds of memory: (1) a permanent memory programmed into the computer at the time of its manufacture, called ROM; (2) the working memory in which programs can be kept, called RAM.

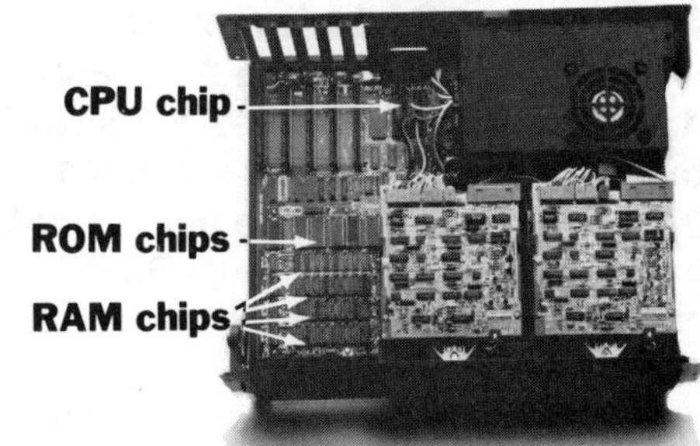

Figure 7-3 Inside an IBM Personal Computer

The information stored in ROM (Read-Only Memory) gives the computer, when turned on, its instructions from the keyboard, sending numbers, letters, and other symbols to the screen, and accepting programs from a disk. Users cannot change what's in this memory; they can "only read" it. ROM is like an instruction book, where users can look things up but can in no way change its contents. This memory is not destroyed when the power is switched off.

RAM (Random-Access Memory), also called "read/write memory," is an area in the computer where outside information is loaded, and can be read, changed, or edited, then run on the computer. Unless this information is first *saved,* or stored, it is lost when the computer's power is turned off. Each element of information in this memory has its own location from which it is easily retrieved. It's the main memory of most computers. RAM is like a blackboard on which information can be written, read, changed, or erased.

The memory of most computers is organized into "8-bit bytes." A *bit* is a binary digit—0 or 1. It's the smallest unit of information a computer uses. When a cell has an electrical charge it is considered to be holding a 1; a cell with no charge is a 0. Eight bits is called a *byte,* which has become the standard unit of information used in computers. An example of a bit is 0, or 1; an example of a byte is 0 0 1 1 0 0 1 1.

(continued)

While a bit provides only two number patterns (0 and 1), a byte can provide 2^8 or 256 different number patterns. Two bytes, when used together, can provide $2^8 \times 2^8 = 2^{16} =$ 65,536 different number patterns—which is the number of memory locations the ordinary microcomputer can handle. The processor in the computer can connect itself to any one of these 65,536 locations, and "read" or "write" the patterns in all of them in 1/60 of a second.

Computers have been programmed to treat each number pattern as a code representing a number, letter, or instruction. Though it can't deal directly with the letter *T*, for instance, a computer can deal with a byte that represents the letter *T*. By internationally agreed convention, known as the American Standard Code for Information Interchange (ASCII), the code for *T* is 0 1 0 1 0 1 0 0. The code for the letter *B* is 0 1 0 0 0 0 1 0. The code for a question mark *?* is 0 0 1 1 1 1 1 1. Every letter, number (0 to 9), and commonly used punctuation mark has its own unique byte in ASCII.

3. A *central processing unit* (CPU), the "brain" of the computer, which processes the information fed into it in accordance with the program of instructions. The processor is made up of two parts: the *control unit*, which coordinates the operations of the entire computing system; and the *arithmetic/logic unit*, which does the calculating. The CPU keeps the operations in time sequence; it also transmits data and instructions to and from the input and output (keyboard and screen) and the RAM.

The Incredible Chip

The central processing unit, the heart of the computer, is a chip that weighs about an ounce and is the size of a cornflake. It's a tiny sliver of silicon a quarter of an inch square, covered with very fine metal connections that join together thousands of transistors. A chip has no moving parts (Fig. 7-4). It is hermetically sealed in a plastic case, usually has many metal legs, and resembles a mechanical centipede.

(continued)

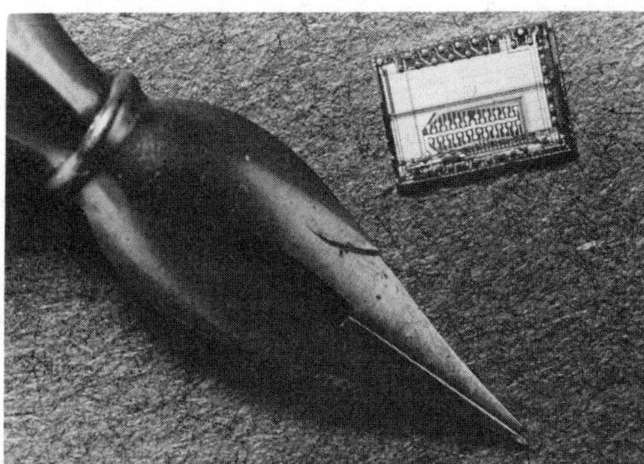

Figure 7-4 IBM 72K-bit SAMOS Chip

A chip can process thousands of bursts of information in a fraction of a second. An electrical current speeds through the chip's electronic circuits and is translated into video displays.

The computer's internal memory is likewise stored in chips, similar in appearance to the CPU chip. Each memory chip generally holds 16,384 bits, so the 65,536 bytes of a complete memory requires 32 such chips.

The technology for cramming an immense number of components into a single, tiny package—known as "integrated circuits"—started in the space programs of the 1960s, when it was necessary to develop ways to compress electronic components into small, lightweight packages.

4. An *output* device that gets information out of the computer. In Fig. 7-2, the output devices are the monitor (screen), sometimes called the cathode ray tube (CRT), and the printer. The screen displays information—a computer program or the results of that program. The printer transfers information stored in the computer onto paper.

Most computers come with a device for storing information, called a *disk drive.* The device "reads" information from a disk and copies it into the computer's memory for later use by the computer, or "writes" information from the computer's memory onto a disk so that it can be stored.

A disk, much like a record made out of recording tape, is a revolving $5\frac{1}{4}$-inch plate upon which information is stored. It can hold the equivalent of hundreds of pages of text, stored as hundreds of thousands of charged/not charged magnetic areas.

When the computer is turned on, the drive starts spinning the disk at a rate of hundreds of revolutions per minute. As the disk spins, a record-playing head moves across its surface and either creates magnetic patterns on the disk's surface as it "writes" data on it, or "reads" the patterns already there and feeds them into the computer's memory.

WHAT IS A COMPUTER?

If you open up a computer and look inside you will see a number of silicon chips fixed into a complex web of wires on a printed circuit board (Fig. 7-3). One of the bigger chips is the central processor, others are the ROM and RAM chips, and still others connect with the input and output devices which form part of the computer's complete system.

Technically speaking, the computer comprises only the central processor and the boards and chips connected to it. But when people speak of a computer, they usually mean to include the other components like disk drives, a monitor, and a keyboard. You need all these components to do anything useful with a computer.

Today, computers fall roughly into three categories:

1. *Mainframe* computers—the largest machines, costing hundreds of thousands of dollars, that have to be kept in specially constructed buildings and require a large staff of operators, programmers, and analysts. Such machines are used by large companies, the Internal Revenue Service, and the Weather Bureau.
2. *Minicomputers*—small machines that do not need a special building or the services of more than a couple of specialists. Though their power is less than that of the mainframes, there is sufficient power for a wide range of commercial and scientific applications.
3. *Microcomputers*—much smaller and cheaper machines which include the "personal computer." The system shown in Fig. 7-2 is a microcomputer.

Despite the computer's fantastic capabilities, we still can't think of it as having an intelligence of its own. It can only execute, with great precision and speed, the instructions given it by a human. A computer has no more mind of its own than your dishwasher. It still can't see as we see, hear as we hear, or speak as we speak. There is intelligence in what the computer *does*, but that intelligence is provided by the person who wrote its instructions. However, although computers cannot yet be said to possess that elusive quality called "human intelligence," computer scientists are working on it.

Artificial Intelligence

An intriguing direction taken by computer scientists in their quest for a machine that can emulate human intelligence involves the term *Artificial Intelligence* (AI). Loosely defined, "intelligence" might be thought of as the ability to adapt to new situations, to reason, to perceive relationships between facts, to derive meanings, and to learn—that is, improve performance on the basis of past experience.

In their effort to build a machine that exhibits such "intelligence," computer scientists are pursuing several specific goals. First, they want the machine to be able to *solve problems*. Though computers are already solving problems, a separate program is needed for each type of problem. You will see a program on page 44 that enables a computer to convert from degrees Fahrenheit to Celsius. But this program can be used only for converting Fahrenheit to Celsius, and for nothing more. The aim of Artificial Intelligence is to write programs that can solve many types of problems. Such a general-purpose program might be able to convert degrees Fahrenheit to Celsius, find areas of circles, play chess, and solve problems in astronomy. There already exists such a general-purpose program called "Adept" that can help military field commanders make tactical decisions in battle.

Artificial Intelligence wants a machine that *understands languages* such as English, French, and Spanish. Though it is already possible to translate about 80 percent of a story written in, say, Spanish to English, the remaining 20 percent has proven very difficult for computers to do.

Another goal of Artificial Intelligence is for a computer to be able to *detect a pattern* from a problem situation and then use this pattern to solve the problem. In Chapter 5 you saw examples of such patterns.

A *game-playing* machine is another goal of AI. In the last 20 or 30 years machines have been built to play games according to the rules and win. Today's computers can play perfect tic-tac-toe, world-class chess and checkers. But the ultimate goal is to enable a computer to accept the definition of *any* game of strategy and then, with practice, to play the game with a skill comparable to or greater than that of human beings.

What Does a Computer Program Look Like?

Despite its dazzling power and versatility, the computer cannot yet think for itself. It can only faithfully execute instructions given it by a human being. But because of its extreme speed, the computer must be given *in advance* a complete set of instructions as to what you want it to do. Such a set of instructions is called a *program*.

The program must be written in a language the computer can understand. The rules of grammar required by such a language are very detailed and must be followed exactly. Typing the program on the computer keyboard must be accurate; even a misplaced "space" could cause the program not to run. One such "bug" in an AT&T program knocked out all the long-distance telephone service to Greece in 1979. It was months before AT&T's programmers pinned down the problem.

In the early days, computers were programmed in "machine language"—the binary code using 0s and 1s. Then came "high-level" languages like BASIC, FORTRAN, ALGOL, and COBOL. These resemble everyday language, but must be translated into the binary code for the computer to understand them. BASIC is the language most often used to introduce programming; it is the language we'll be using in the rest of this chapter.

Since the computer can deal with only one instruction at a time, it needs to know in what order to deal with the instructions in a program. This is accomplished by *numbering every line in the program*. Line numbers in BASIC are usually written in increments of 10 (or 100). Numbering lines this way permits us to insert extra lines, between lines already there, at a later date.

To enter and then execute a program on the computer,

1. Type on the keyboard the line numbers and statements in the program.
2. At the end of each line, press the RETURN key and then type the next line.
3. To ask the computer to execute the program, type the word RUN and press the RETURN key. The program should now run on your screen.

EXAMPLES OF PROGRAMS

You already know that the computer has a memory where it can store information. We will instruct the computer to use part of its memory to store certain numbers in specific memory locations named by letters such as A and B, and certain names of people in specific memory locations named by a letter and a dollar sign next to it, such as M$ (read "M string"). If we want the computer to print out a particular number or name, we mention the name of its memory location and the computer will print out what's stored there.

Whenever you put a number into a particular location, you destroy any number that was there before. But numbers can be *retrieved* from the memory and used without destroying them much like the tape in a tape recorder. When you record, you erase what was on the tape before, but you can play back what's on there as often as you like without erasing it.

Let's start with a very simple program.

Program 1

```
10  PRINT "MEMBERS OF MY FAMILY"
20  INPUT M$
30  PRINT "DEAR "M$", WATCH ME HAVE FUN WITH THE COMPUTER."
40  GOTO 20
```

If you type this four-line program on your computer keyboard and then type RUN and press the RETURN key, the computer will begin executing the program, beginning with line 10.

Line 10 instructs the computer to print on the screen the material between the quotation marks. The following will appear on the screen:

MEMBERS OF MY FAMILY

In line 20, the command INPUT tells the computer to type a question mark (?) and then wait for *you* to type in the name of a member of your family. If you type in ALISSA, the computer stores ALISSA in the memory location we called M$.

Line 30 tells the computer to print whatever is between the quotation marks, and the name stored in memory location M$ which, in this instance, is ALISSA. The following will appear on the screen:

DEAR ALISSA, WATCH ME HAVE FUN WITH THE COMPUTER.

Line 40 tells the computer to ask you for another name of a member of your family by printing another*?*. If you type in JOSH, the computer will respond with

DEAR JOSH, WATCH ME HAVE FUN WITH THE COMPUTER.

If in response to the next *?* you type in LINDSAY, the computer will respond with

DEAR LINDSAY, WATCH ME HAVE FUN WITH THE COMPUTER.

The computer will keep asking for more names until you stop the program by pressing the BREAK key.

Program 2 Let's now look at a program that tells the computer to count from 1 to 60.

	EXPLANATION
10 FOR A = 1 TO 60	10 Causes memory location A to take on the values from 1 to 60.

20 PRINT A	20 Tells the computer to print the number stored in location A.
30 NEXT A	30 Tells the computer to go to the next value of A.
40 END	40 Tells the computer to terminate the program.

When you run this program, containing FOR and NEXT statements which cause the computer to go through a loop 60 times, the computer will print the numbers from 1 to 60.

Program 3 Suppose you want the computer to count from 3 to 60 in *threes*. The computer will do this if line 10 is slightly modified.

10 FOR A = 3 TO 60 STEP 3	10 Causes A to take on the values from 3 to 60,
20 PRINT A	*with A increasing by threes.*
30 NEXT A	[*NOTE*: "STEP 6" in line 10 would cause A
40 END	to increase by sixes, starting with 3.]

When you run this program, the computer will print the numbers 3, 6, 9, 12, 15, 18, 21, 24, 27, 30, 33, 36, 39, 42, 45, 48, 51, 54, 57, 60.

Program 4 This program asks the computer to list the numbers from 1 to 100, with each number followed by its square. For instance, 1 1, 2 4, 3 9, and so on.

> *NOTE*: The symbol for multiplication in BASIC is * (not ×). So "3 squared" (meaning "3 times 3") is written 3*3 (not 3 × 3), and "N squared" is written N*N.

	EXPLANATION
10 FOR N = 1 TO 100	10 Causes N to take on the values from 1 to 100.
20 PRINT N, N*N	20 Causes the first number and its square to be printed.
30 NEXT N	30 Tells the computer to go to the next value of N.
40 END	40 Tells the computer to terminate the program.

When you run this program, the computer will print in two columns (because of the comma in line 20):

```
  1   1
  2   4
  3   9
  4   16
  5   25
  .   .
  .   .
  .   .
100   1000
```

NOTE: To get such a listing for the numbers 1 to 1,000,000, all you do is change line 10 to:

10 FOR N = 1 TO 1000000

Interactive Programs

An important feature of a computer is that it allows the user to interact with the running program. The computer can be programmed to ask you for data, give you the result of the required computation, and then ask for more data. The statement in BASIC that allows user-program interaction is the INPUT statement. When the computer reaches an INPUT statement it stops running the program, types a question mark on the screen, and waits for the user to supply the necessary data. Then the computation proceeds from where it was interrupted. Program 1 on page 42 and Program 5 (which follows) are examples of interactive programs.

Program 5 A program that asks the computer to convert from degrees Fahrenheit to Celsius.

> *NOTE*: To understand this program, you need to know that
>
> 1. The formula for converting from Fahrenheit to Celsius is
>
> $$C = \frac{5}{9}(F - 32), \text{ where } F = \text{Fahrenheit and } C = \text{Celsius}$$
>
> 2. In BASIC, we indicate division by / and multiplication by *. Therefore, written in BASIC, the formula becomes
>
> C = 5/9*(F−32)

Program	EXPLANATION
10 PRINT "FAHRENHEIT"	10 Tells the computer to print on the screen the word between the quotation marks: FAHRENHEIT.
20 INPUT F	20 Tells the computer to type a question mark (?), wait for you to type in a value for F, and store that value in memory location F.
30 C = 5/9*(F−32)	30 Tells the computer to perform the indicated calculations and store the result in location C.
40 PRINT "CELSIUS:" C	40 Tells the computer to print the word between the quotation marks, CELSIUS: and the value for C.
50 PRINT	50 Tells the computer to skip a line (nothing is printed).
60 GOTO 10	60 Tells the computer to execute statements starting with line 10 instead of the next statement (line 70).
70 END	70 Tells the computer to terminate the program.

If you supplied the following values for F:

32, 212, 98.6, and −40,

the computer would print the following:

```
FAHRENHEIT   ? 32
CELSIUS:       0

FAHRENHEIT   ? 212
CELSIUS:       101

FAHRENHEIT   ? 98.6
CELSIUS:       37

FAHRENHEIT   ? -40
CELSIUS:       -40
```

Computers and Your Child

In an explosion of excitement exceeding that in science and mathematics at the time of Sputnik more than 25 years ago, schools across the country are plunging into computers. Students, dazzled by the power at their fingertips on the computer keyboard, talk with excitement of color graphics, word processing, data-base processing, and a host of other computer applications. Armies of school children now occupy their spare time playing computer games and writing programs, while the tiny fingers of four-year-olds gently press computer keys to learn the alphabet.

With headlines like "Computer Skill Is Newest Rung in Ladder to the Top," and "Ignorance of Computers Will Limit a Child's Future Options," schools and parents fear their children will be left behind in the dark ages if they are not exposed to the marvels and mysteries of the microcomputer.

Computers are used most frequently with children for "computer-assisted instruction"—using the computer to practice a range of basic skills—with the computer functioning as a kind of mechanized teacher's aide. What is sometimes overlooked is that once the novelty wears off, drill with a computer can become just as boring as drill without a computer, unless attention is paid to the ideas discussed in Chapter 4.

A more fundamental danger is that the computer—like earlier "breakthroughs" in education—will be hailed as the panacea for all our learning problems. It is easy to forget that in education the computer is only a tool, though of immense potential. The eager beavers need to be reminded periodically that there are no technological shortcuts to a good education; that good teachers, with proper support, remain at the center of it all.

Whether computers become just another fad and go the way of earlier "breakthroughs" like teaching machines and programmed learning, or whether they live up to their potential, will depend on how they are used with children and on the quality of the children's teachers.

In an earlier chapter we deplored the mathematical inadequacy of many elementary-school teachers. With the advent of computers, it becomes even more important that elementary-school children be taught mathematics and computers by trained specialists whose sole responsibility is to teach these subjects.

Even in his earliest experiences with computers, the child should not be confined merely to executing canned programs and responding to computer commands. In the primary grades these commercially prepared programs provide drill with routine skills (like the basic number facts) and structured drill that includes hints when the child gives an incorrect answer. They also illustrate concepts through "motion" (like "taking away" in subtraction) and display simulations (like "tossing a coin" many times in a few seconds when studying probability).

The child must also gain experience *writing* programs, however simple. Writing a program forces the child to think through clearly and precisely what he wants the computer to do. Running it lets him see the result of his written instructions. This experience gives the child a sense of power because he realizes he himself is in control.

A glimpse of the potential of computers in education comes from Seymour Papert, professor of mathematics and education at M.I.T., who, in his book *Mindstorms*, presents an exciting vision of how computers may affect the way children think and learn. To Papert, the phrase "computer as pencil" evokes the way in which children will be using computers in the future.

"Pencils are used for scribbling as well as writing, doodling as well as drawing, for illicit notes as well as for official assignments," writes Professor Papert. Computers, he believes, can be used by children as casually and as personally for even a wider diversity of purposes.

"In many schools today," he writes, "the phrase *computer-assisted instruction* means making the computer teach the child. One might say the computer is being used to program the child. In my vision, *the child programs the computer* and, in doing so, both acquires a sense of mastery over a piece of the most modern and powerful technology and establishes an intimate contact with some of the deepest ideas from science and mathematics."

Through his research at M.I.T.'s Artificial Intelligence Laboratory, Professor Papert led hundreds of children—some talented, others emotionally or intellectually disabled—to become quite sophisticated programmers. "There is nothing very surprising about the fact that this should happen," he says.

"Programming a computer means nothing more or less than communicating to it in a language that it and the human user can both understand. And learning languages is one of the things children do best. Every normal child learns to talk. Why should a child not learn to 'talk' to a computer?"

Part II

HOW AND WHAT TO TEACH YOUR CHILD

Chapter 8
Sets

PARENTS, NOTE: **This chapter provides background material for later chapters.**

Which contains more points: a 6-inch line segment, or a line segment stretching from the tip of your nose to the moon?

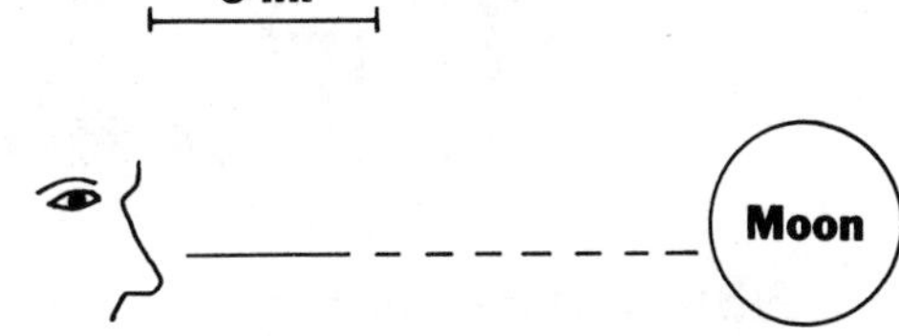

If you think the segment between your nose and the moon contains more points, you have a surprise coming.

Mathematicians grappled with this puzzle for many years until Georg Cantor (1845–1918) came up with the answer: both contain exactly the same number of points! He used a simple idea—**sets**—to arrive at this conclusion. (The explanation of Cantor's proof is beyond the scope of this book.)

In the rest of the book we won't deal extensively with sets, but will occasionally refer to them. So we'll now say a few things about them.

In mathematics, a **set** is any collection of objects. These do not have to be physical objects; they can be *abstract* ideas like numbers, the top 10 song hits of the year, the 1988 Presidential candidates, or a collection of hieroglyphics. The objects in a set are called its *members*, or its *elements*.

The set concept is useful because it enables us to think of a collection of objects as a single entity. A set, as a single entity, has characteristics not possessed by its individual members. For example, the set of people who make up the United States Senate, as a whole, has the authority to pass laws, a power not possessed by any single senator.

We name sets by capital letters like A, C, R, and list their members within braces. For example,

$$A = \{\text{Carl, Andy, Shaun, Lisa}\}$$

is read: "A is the set whose members are Carl, Andy, Shaun, and Lisa."

$$F = \{0,4\}$$

is read: "F is the set whose members are 0 and 4."

Though it is natural to assume that a "collection" means two or more, this is not the case with sets. A set may contain just one member, or even no members.

When a set has a limited number of elements, it's called a *finite* set. Sets A and F above are examples of finite sets.

A set with an unlimited number of members is called an *infinite* set. One with no members is called the *empty* set.

An example of an infinite set is the set of *counting numbers* (sometimes called the set of *natural* numbers):

$$N = \{1, 2, 3, 4, \ldots\}$$

The three dots within the braces indicate that these numbers continue indefinitely.

The set of even numbers,

$$E = \{2, 4, 6, 8, \ldots\},$$

is another example of an infinite set.

What's a Number?

A number is an abstract idea. You can't see it, touch it, or smell it. The blotch of ink you see when you look at 6 is not the number six, but our *symbol* for it. We can do as the Romans did and represent the number six with their symbol *VI*—or, like the Mayans, use the symbol $\dot{-}$. These blotches of ink representing numbers are called *numerals*. You must also have seen numerals made of paper, wood, plastic, or metal.

We first become acquainted with the concept of number by counting objects. Later we learn to think of number as a purely abstract idea without any reference to counting or objects. It's commonly believed that mathematics began when the perception of "six apples" was freed from apples and became the *number* 6.

***NOTE*: Because the distinction between a *number* and a *numeral* can be confusing, especially to children, we'll generally use "number," and confine our use of "numeral" to situations clearly involving physical symbols.**

One-To-One Matching

When you set the dinner table and lay out the soup spoons, you end up with a spoon for each person and a person for each spoon. The number of spoons and the number of people will obviously be the same. In mathematics, we describe what you've done by saying that you established a *one-to-one matching* between the set of spoons and the set of people.

If several couples go to the theater, each couple consisting of a man and woman, then there is a one-to-one matching between the set of men and the set of women; that is, for every woman there is a man, and for every man there is a woman. You don't have to *count* the men and women to know that there are as many women as men in your theater party.

An important outcome of this discussion is to realize that whenever there exists a one-to-one matching between the elements of two sets, you know without counting that both sets must contain exactly the same number of elements.

This simple concept of a one-to-one matching (more formally called a *one-to-one correspondence*) is behind what we do when we count objects. When we count—John, Sue, Frank, Evelyn, and Anne—what we are really doing is setting up a one-to-one matching between this set of people and the set of counting numbers:

John	Sue	Frank	Evelyn	Anne
$\updownarrow$	$\updownarrow$	$\updownarrow$	$\updownarrow$	$\updownarrow$
1	2	3	4	5

The last number in the matching, 5, tells us the number of people.

You Won't Believe This!

The concept of a one-to-one matching can be used to answer a tantalizing question: Which set has more numbers: the counting numbers, or the even numbers?

$$\mathbf{N} = \{1, 2, 3, 4, 5, \ldots\}$$

or

$$\mathbf{E} = \{2, 4, 6, 8, \ldots\}$$

The counting numbers, of course—many say—since they contain the even numbers *and* the odd numbers.

(continued)

But this conclusion turns out to be wrong. Both sets contain exactly the *same number* of numbers!

And how can we prove this incredible answer? By showing that there exists a one-to-one matching between the two sets. And we did say that if a one-to-one matching exists between the members of two sets then both sets must contain the same number of members.

Now, if you're starting to freeze up and thinking of pulling down the little window shade in your mind, STOP! Just relax, and let's move on together.

If we can show that for every *natural* number there exists an *even* number, and for every *even* number there exists a *natural* number—then we have it made. For then, there *must* be as many numbers in one set as in the other.

Let's pair the counting numbers and the even numbers in the following way:

$$N = \{1, 2, 3, 4, 5, \ldots\}$$
$$\updownarrow \; \updownarrow \; \updownarrow \; \updownarrow \; \updownarrow$$
$$E = \{2, 4, 6, 8, 10, \ldots\}$$

If you can now imagine these pairings going on indefinitely, you will see that to every counting number there corresponds an even number *twice* as large.

For instance, to the counting numbers 1, 6, 15, and 70, there correspond the even numbers 2, 12, 30, and 140:

N	1	6	15	70
E	2	12	30	140

And to every even number there corresponds a natural number *half* as large.

For instance, to the even numbers 6, 10, 26, and 150, there correspond the natural numbers 3, 5, 13, and 75:

E	6	10	26	150
N	3	5	13	75

Since there exists a one-to-one matching between the numbers of the two sets, we must conclude that there are as

(continued)

many even numbers as counting numbers. We should add, however, that though the two sets contain the *same number* of numbers, they do not contain the *same* numbers. The set of natural numbers contains the odd and even numbers, while the set of even numbers contains only the even numbers.

***NOTE*: If this problem reminds you of the one about the points between your nose and the moon, then you get an A for insight. Both solutions involve the same principle—one-to-one matching.**

Chapter 9
Counting

INTRODUCTION

A man, anxious to shoot a crow, hit upon a scheme for deceiving the suspicious bird. He sent two men to the watch house, one of whom left while the other remained. But the crow counted the men and kept her distance. The next day, three went to the watch house, and again the crow perceived that only two left. The same thing happened with four men. Finally, to confuse the crow, five men were sent, and this time the crow, thinking that all men had left the watch house, lost no time in returning. The scientist reporting the incident concluded that crows can count up to four.

An interesting situation arises with wasps, the carnivorous insects that supply their nests with caterpillars they kill with their formidable sting. Different families of wasps allot different numbers of caterpillars to their nests. One species of wasp considers one large caterpillar enough for its young. Another supplies 5 victims; other species supply 10, 15, and even up to 24. The number seems to be constant in each species.

How does the insect know when her task is completed? Not by the nest being filled, since if some caterpillars are removed she does not replace them. Obviously, there is some counting instinct at work in the wasp.

But there is no mystery about how humans have counted over the ages. They have used sticks, pebbles, scratches on sticks (Robinson Crusoe fashion), little heaps of grain, knots on a string, counting beads, the abacus, electronic calculators, and computers. Nevertheless, behind all these devices is the universal finger method to which every child turns instinctively.

Without counting, it is unlikely that man would have conceived of numbers. But even without counting it's still possible to determine which of two sets of objects is the greater, which the lesser, or whether they contain the same number of objects.

To show that we have the same number of fingers on both hands, we simply match finger with finger on each hand. (**One-to-one matching**, remember?) And if many people are gathered in a large room with chairs, we don't have to count the people and the chairs to determine whether there are enough chairs to go around.

We only need to ask everyone to be seated. If no chairs are left vacant and no one is left standing, then there are as many chairs as people. (Again, one-to-one-matching!)

But what counting does, which matching does not, is to tell us *how many* chairs and *how many* people there are.

It would be well for you to have on hand the following **materials** for use with this chapter:

A magnetic board, a set of magnetized shapes or disks, and a set of numerals

Coins, checkers, buttons

A pegboard (10 rows of holes, 10 holes in each row), and pegs in two colors

A set of 10 cards marked with 1 through 10 shapes:

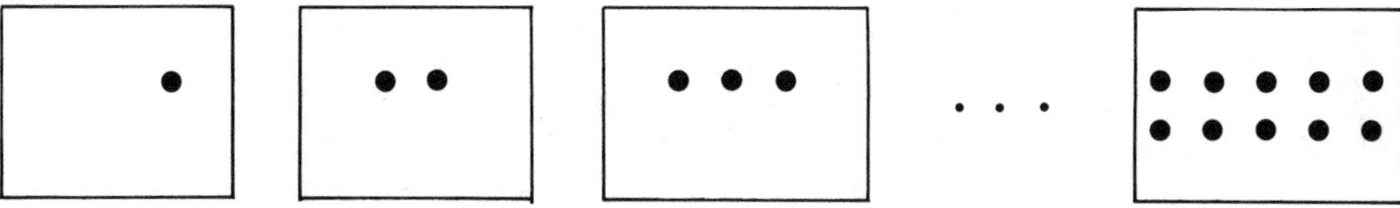

A hand-held, inexpensive electronic calculator.

THE CHAPTER IN A NUTSHELL

In this chapter you will see how to

- Get your child ready for counting
- Explain the meaning of counting
- Use a variety of activities and games to help your child learn to count
- Provide interesting problems involving counting

You will also be given

- A list of understandings and skills your child should be left with at the completion of the chapter
- A set of questions, at the end of the chapter, to test and reinforce the child's comprehension of what he learned. (The answers appear at the end of the book.)
- Several challenging problems under **For the Daring**

THE MEANING OF COUNTING

Learning to match two sets of objects is learning to count. To count the objects in a set, we match them, one-to-one, with the members of another set that is always at hand—the set of counting numbers {1, 2, 3, 4, 5, . . .} :

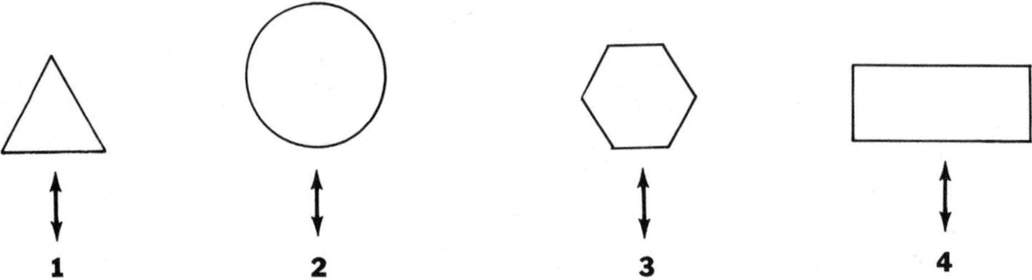

Then a wonderful thing happens. The *last* number needed to complete the matching tells us *how many* objects there are in the first set.

This is all there is to counting as practiced by primitive peoples, by us, and by the greatest mathematicians.

READINESS FOR COUNTING

> *NOTE:* The **Readiness** section for each topic lists the underlying understandings and skills your child needs in order to understand the new work. These will have been explained earlier. Before starting a new topic, test your child informally to make sure he or she has these prerequisites. If not, go back to where they were first presented.

A child can learn to count by rote, that is, reel off numbers in sequence—one, two, three, and so on—without understanding the meaning of counting. To acquire this understanding, a child must first learn to *match* objects in one set with those in another set. The activity that follows is an effective way to get across the concept of matching.

1. Lay out on the table three red and three black checkers:

2. Ask: Is there a red checker for every black one?
   Is there a black checker for every red one?

What can you say about the number of black checkers compared with the number of red ones?

Repeat this activity with different numbers and different objects.

To be ready for counting, your child should be able to

- Set up a one-to-one matching between two sets of objects
- Tell when a one-to-one matching between two given sets is possible
- Tell whether one set of objects has as many, more than, or fewer than another set
- Produce a set with as many objects as another set you have presented
- Explain what must be done to make one set have as many objects as another set

TEACHING COUNTING

Getting Started

Initial counting activities should involve small numbers of objects. Most children will already recognize a set of one or two objects.

Place two disks on the magnetic board and ask your child to tell the number of disks in the set. When it's established that it is a *set of two*, ask the child to find other sets of two in the room.

Introduce *three* by building onto a set of two. Below the set of two disks lay out a set of three disks, and then the rest of the sets through *ten*. (If you don't have a magnetic board, use the tabletop.)

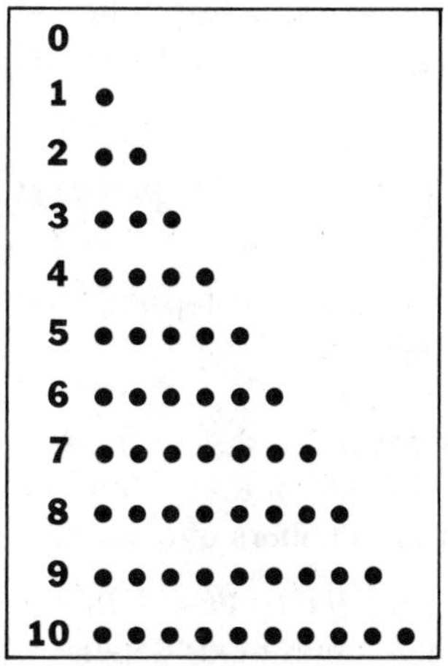

With the introduction of each number, introduce the *numerals* 1, 2, 3, 4, 5, 6, 7, 8, 9, 10.

Remove all disks from the board and introduce the idea of *zero* (and the numeral 0) as the number that tells us there are *no* disks on the board. That is, 0 represents the number of objects in the *empty* set.

0
1 •
2 • •
3 • • •
4 • • • •
5 • • • • •
6 • • • • • •
7 • • • • • • •
8 • • • • • • • •
9 • • • • • • • • •
10 • • • • • • • • • •

After the child gains an intuitive understanding of the number *one*, he can think of *two* as "one more than" one; *three* as "one more than" two; and so on. Let your child realize that each number has a successor.

Number rhymes, songs, and finger-play activities can all be used to familiarize children with number names and their sequences. Well-known examples of such rhymes are "Ten Little Indians," "One, Two, Buckle My Shoe," "Three Blind Mice," and jump-rope rhymes.

Two Uses of Numbers

Children must memorize the names and order of the numbers 1 through 10. After they can count to 10, introduce the concept of "first," "second," "third," and so on.

The child will now have seen two uses of numbers:

1. As telling *how many* objects there are in a set. When a number answers the question "How many?" it's called a *cardinal* number. In "five children," "five" is a cardinal number.
2. As telling the *order* or *position* of an object in a set. The numbers "first," "second," "third" tell us this. Such numbers are called *ordinal* numbers. In "the fourth book," "fourth" is an ordinal number.

Using the Number Line

The *number line* is a most useful device for helping a child learn the sequence of numbers and how to count. The numbers on the line are *equally spaced*, and the arrow on the right end indicates that the line—and the numbers—continue indefinitely in that direction.

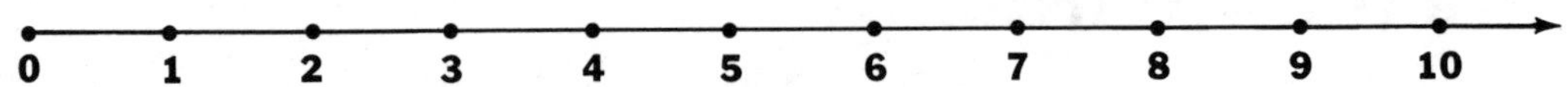

Ask your child to place his finger on the 0 point and count the *spaces* between the numbers as he moves the finger from left to right. As he reaches point 1, he counts "1"; as he reaches point 2, he counts "2"; and so on.

Later, he can use the number line to count in twos, threes, and fives, and to learn the meaning of addition, subtraction, multiplication, and division.

Activities for Teaching Counting

After these introductory steps, follow up the teaching of counting with activities like those described below.

1. Place several disks on the table. Ask the child to count out loud as he moves them, one at a time, to the other side of the table. Repeat with different numbers of objects.
2. Count out loud: "One, two, three, four." Then ask the child to say what number comes next. Repeat with other numbers.
3. Ask what number comes after three; before six. Repeat with other numbers.
4. Ask the child to hand you three checkers; seven buttons; four disks. Repeat with other numbers of objects.
5. Lay out a row of checkers, a row of buttons, and a row of disks. Ask the child to point to the *second* checker from the left; the *first* button from the left; the *third* disk; etc.

6. In #5, point to various objects and ask the child to tell in which position each object is located.

7. Ask the child to lay out sets of 1 through 10 objects (in random order). Let him show a set of 0 objects on the table.

8. Ask the child to draw pictures of sets containing 0 through 10 objects.

9. Ask the child to insert *three black pegs* in one row of a pegboard (starting at the left edge of the board and skipping no holes). Then ask for *five red pegs* in the row directly below. Ask him to count the pegs out loud as he inserts them into the holes. Then ask:

   Which row has more pegs?
   How do you know?
   Which has fewer pegs? How can you tell?

   Repeat with different combinations of pegs.

10. Place a pile of coins on the table.
   (a) Ask the child to count the coins in the pile.
   (b) Ask him to count out three coins and put them in a box.
   Repeat with different numbers of coins.

> *COMMENT:* When the child is asked to put three coins in a box, he may count "one, two, three" and then put only the *third* coin in the box instead of all three coins. This shows that he is associating a number with each coin rather than with the entire set of coins. The child is confusing the ordinal number *third* with the cardinal number *three.*
>
> A way to deal with this error is to begin counting with a set of objects that are not visible to the child. As he removes the objects from the box, one by one, the child names the number of the set counted. So when he says "one," he sees one object; when he says "two," he sees two objects; and so on.

11. Ask the child to count *backward* from 3; from 7; from 10. Repeat with other numbers. (Counting backward lays the foundation for subtraction.)

12. Ask the child to start counting from numbers other than 1 (say, from 3 or from 8).

13. Ask the child to count by twos; by threes. (This lays the foundation for multiplication.)

14. Lay out, in random order, a set of 10 cards containing 1 through 10 shapes:

□□□   □□□□□ □□□□□   . . .   □□

(a) Ask the child how many shapes there are on each card.
(b) Ask him to point to the card with
1. the greatest number of shapes
2. the smallest number of shapes
(c) Ask him to point to the second card; to the fifth.
(d) Ask the child to arrange the cards in order, from least shapes to most.
(e) Select a card at random. Ask the child to name the number *before* and the number *after* the one represented on the card.

15. Ask the child to arrange on a magnetic board rows of shapes to represent the numbers 1 through 10.
(a) Then ask him to place the appropriate numeral to the left of each set of shapes.
(b) Clear the board. Then place the numerals back on the board, one at a time, in random order. Ask the child to place next to each numeral the appropriate number of shapes.
16. Use a hand-held electronic calculator to practice counting. Ask the child to set the display on the calculator to "1." As he counts out loud let him check with the changing display.

NOTE: To count on the calculator, enter "1" and press the + button. Then, each time you press the = button, the display will advance by one.

17. Ask the child to count *backward*, beginning with 10, and keep checking with the calculator display as he calls out each number.

NOTE: To count backward on the calculator, set it to "10"; press the − button, then the 1 button. Then, each time you press the = button, the display will reduce by one.

18. Display various numbers on the calculator. Ask the child to read them.
19. Call out various numbers. Have the child display them on the calculator.
20. For practice writing numerals, have the child make a paper-plate clock.

LOOKING AHEAD:

The child will soon see that when he adds, subtracts, multiplies, and divides, he is really only counting.

HIGHLIGHTS OF COUNTING

NOTE: The **Highlights** section for each topic lists the understandings and skills your child should be left with at the completion of the chapter. To help determine the extent to which your child has succeeded, test him or her with the **Questions** that follow the **Highlights**.

Upon completion of this chapter, your child should have acquired the following understandings:

- *Counting* means matching a set of objects, one-to-one, with the set of counting numbers.
- Numbers tell us *how many* objects there are in a set (cardinal numbers). They also tell us in which *position* an object is located (ordinal numbers).
- *Zero* (0) is a number that says there are **no** objects in a set.
- Vocabulary:

set	more than
one-to-one matching	fewer than
as many as	

Also, your child should be able to

- Count at least up to 10 objects
- Write the numerals 0 to 10
- Lay out the correct number of objects corresponding to a given number
- Tell the number of objects there are in a given set
- Arrange *in order* a set of two or more given numbers
- Identify the ordinal position of an object in a set
- Count backward (from 10 or less)
- Tell whether one set of objects has as many, more than, or fewer than another set

QUESTIONS ON THE CHAPTER

NOTE: If necessary, read these questions to your child.

1. What do you do when you count? Show what you mean by counting these stars: * * * *
2. Tell me two ways you can use numbers. Give some examples.
3. What does the number zero (0) tell you about a set of objects?
4. Take some checkers and some disks and set up a one-to-one matching between them. Which set is larger? How can you make the number of objects in each set the same?
5. In some faraway land, there is a *boozie* for every *oozie*, and there is an *oozie* for every *boozie*. What can you tell me about the number of oozies compared with the number of boozies? How do you know?

6. Use checkers to show examples of one set having *more* checkers than another. What can you say about the second set?

7. Here is a set of buttons. Count them out loud.

8. Here is a pile of checkers, a pile of disks, and a pile of buttons. Write the number of objects in each set.

9. Here is a set of disks. Select eight disks.

10. Here are four cards with different numbers of circles on them. Arrange them in order, from the smallest to the largest.

11. *Lay out 10 disks in a row.*
    (a) In this row of disks, point to the third disk; to the second; to the fifth; to the ninth; to the tenth.
    Point to various disks, one at a time.
    (b) What is the position of this disk? This one? This one? . . .

12. What's a **one-to-one matching**?

1. A restaurant offers a choice of three appetizers, four entrees, and two desserts. How many different meals can be made with these choices?

2. In how many ways can five people be arranged in a line at a supermarket checkout counter?

3. A way to get rich is to select the winning three-digit number in a lottery. If each digit can be 0 to 9, how many different three-digit numbers can you form?

COMMENT: All three questions are *disguised* counting problems. **The answers appear on page 257.**

Chapter 10
Place Value

INTRODUCTION

If you haven't marveled at anything lately and you're aching to do so, try thinking about how we represent numbers. With only 10 basic symbols (0, 1, 2, 3, 4, 5, 6, 7, 8, 9), called *digits*, we can represent such vastly different quantities as the cost of a hamburger, the size of Agatha's shoes, the distance of Vega from the earth, and the weight of a speck of dust.

The system that does all this is the Hindu-Arabic numeration system, first developed by the Hindus about 2000 years ago and brought to Europe by the Arabs. It is also called the Decimal System because it uses *ten* basic symbols.

Now someone unaccustomed to the decimal system might find something puzzling, yet marvelous, about it. Wouldn't you expect ten symbols to represent just ten numbers? Wouldn't more numbers require more symbols? The essential question, then, is: What ingenious mechanism is at work in the decimal system to enable us to represent *infinitely many* numbers with just ten symbols?

So clever is the idea that, with it, we can devise other numeration systems needing even fewer symbols. The computer, for instance, uses only *two* basic symbols, 0 and 1!

We pay little attention to, and much less marvel at, this great invention of the human mind, because it has been absorbed so completely into our culture and works so efficiently. Yet, without it, none of the technological marvels surrounding us would have been possible.

Many different systems to represent numbers were developed by earlier civilizations, but those peoples never learned the secret of how to make a few symbols represent many numbers. Some of these systems used knots in a rope, piles of pebbles, or notches on a stick. Can you imagine using any of these methods to represent the population of the United States? Using the decimal system after these earlier systems is like handling rockets after sparklers.

In this chapter, we'll let you in on the secret.

A Hot Celebration

Cutting notches on sticks evolved into a widespread method of keeping records of debts and payments, and the notched pieces of wood came to be known as *tally sticks*.

There is a story concerning the Bank of London's use of tally sticks (as late as 1790) to record its transactions. Horizontal tally marks were made across the stick, which was then split vertically. Half the stick was given to the investor and half was kept by the bank.

When the bank decided to modernize its accounting system, the occasion was celebrated by setting the tally sticks on fire. But the fire got out of control and the bank itself burned down.

Why Ten Digits?

***Digit*, from the Latin word *digitus*, means "finger." It is believed the decimal system is based on 10 digits because humans, having 10 fingers, grew accustomed early in their history to counting by tens. It's entirely possible that if we had been born with four fingers on a hand, giving us eight fingers altogether, we might have used eight basic digits in our numeration system.**

The decimal system, introduced into Europe about 1300 A.D., greatly simplified the arithmetic based on the Roman system. Because it can be used to represent numbers so much more easily and to perform computations with far greater efficiency, the decimal system is clearly superior to all earlier systems.

The **materials** needed for this chapter include:

Disks

Sheets of "bird arrays" (see page 72)

A hundred chart

Spinners

THE CHAPTER IN A NUTSHELL

In this chapter you will see how to

- Get your child ready for understanding "place value"
- Explain the meaning of place value
- Use a variety of activities and games to reinforce your child's understanding of place value
- Provide interesting problems involving place value

You will also be given

- A list of understandings and skills your child should be left with at the completion of the chapter
- A set of questions to test and reinforce your child's comprehension of place value
- Insights into the decimal system and the binary system (used with computers) under **For the Curious**
- "A Weighty Problem" under **For the Daring**

THE MEANING OF PLACE VALUE

Here are some disks:

How can you determine how many there are?

One way is to count them. Another is to separate them into *groups of ten*:

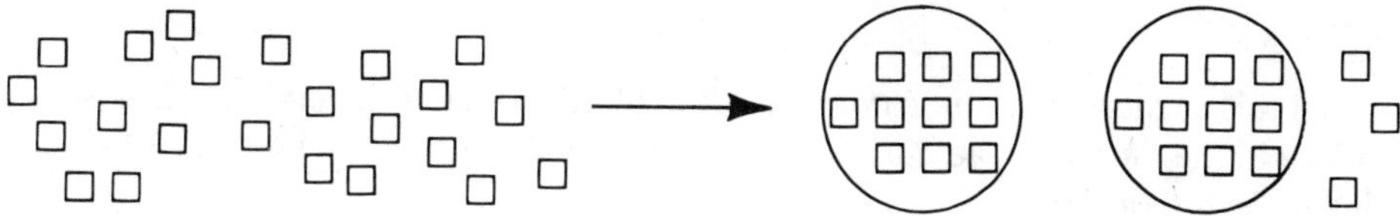

The result is 2 groups of *ten* with 3 left over, or 3 *ones*. We show this result as

tens	ones
2	3

, or

10	1
2	3

, or just 23.

Likewise, 56 means

tens	ones
5	6

, or

10	1
5	6

; that is,

56 contains 5 tens and 6 ones.

If the idea is to group the objects in sets of ten, what do we do when we get *ten* groups of ten? Of course, we assemble *them* into a single group of a *hundred:*

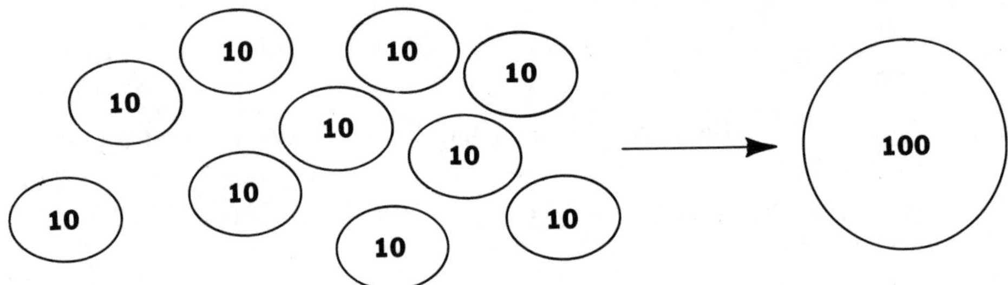

What's our next question? You guessed it: What do we do when we get *ten* hundreds? We, of course, assemble them into a single group of a *thousand:*

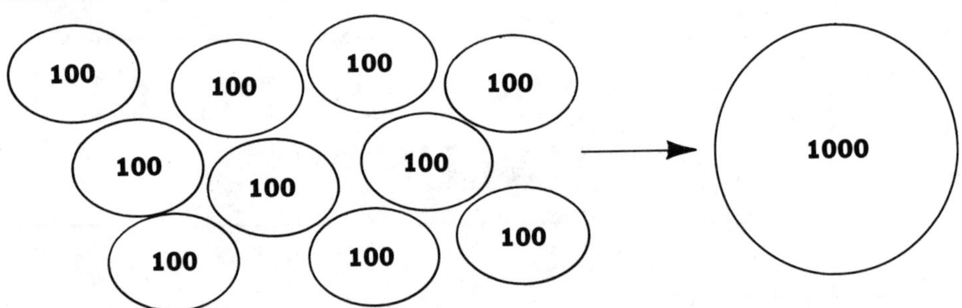

So, 347 means

100	10	1
3	4	7

; that is, 347 contains 3 groups of a hundred, 4 groups of ten, and 7 ones.

Likewise, 205 means

100	10	1
2	0	5

; that is, 205 contains 2 hundreds, *no* tens, and 5 ones.

All this suggests that the *value* of each digit in a number depends on its position or *place* in the number. The *3* in 375 has a value of 3 *hundreds*, while *3* in 139 has a value of 3 *tens*, or thirty:

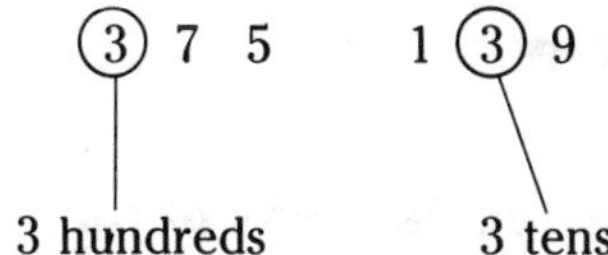

The system of making the value of a digit in a number depend upon its place in the number is called **place value.** The smallest place value, 1, is in the extreme right position. The next value is 10; the next value is 100; the next is 1000; and so on:

$$\cdots 1000 \quad 100 \quad 10 \quad 1$$

In such a system, every digit can have many different values if it occupies different positions in a number. In the example below, *2* assumes a different value in each situation:

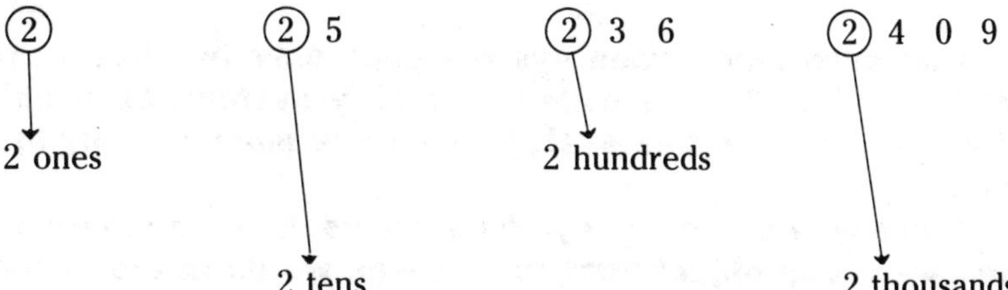

Place value is the secret behind the decimal system. When tally marks were used to represent *two* objects, the symbol // represented *only* two objects. But when the symbol 2 is used in a place value system, it can represent many different numbers of objects, as we have just seen.

We should stress the importance of zero in a place value system. Because the early Babylonians did not have a symbol for zero, they sometimes left empty spaces between digits, resulting in much confusion. In 3 6, we might not know whether the *3* represents 3 *tens* or 3 *hundreds.* By writing 306, the meaning becomes clear.

The Origin of Zero

Where and when the idea of zero originated is not certain, but the Babylonians, about 2000 years ago, introduced the symbol ≲ to denote the absence of a digit. The Hindus, about 800 years later, are known to have used a symbol for zero.

The word *zero* probably comes from the Latin *zephirum*, a translation of a Hindu word meaning "void" or "empty."

(continued)

The Binary System

A *numeration system* is a way of representing numbers. It consists of a set of basic symbols and rules for combining these symbols to represent various numbers.

The decimal system is a numeration system using 10 basic symbols (0, 1, 2, 3, 4, 5, 6, 7, 8, 9) and relying upon place value for an efficient way to represent various numbers.

We can create other numeration systems based on different numbers of basic symbols. Since an essential component of any place value numeration system is zero, we always use 0 as one of the basic symbols.

One such numeration system uses only two basic symbols, 0 and 1. This is called he *binary system*, *bi* meaning "two" in Latin. In this system, we write numbers only in 0s and 1s.

Since only two basic symbols are used in the binary system, we group objects not in tens—as we do in the decimal system—but in *twos*: 2 ones, or 2; 2 twos, or 4; 2 fours, or 8; 2 eights, or 16; and so on.

The place values in the binary system (expressed in decimal system language) are, therefore,

. . .	16	8	4	2	1

The value of each place in a number is always two times that of the next place to the right.

Numbers written in the binary system are of special interest because of their application to computers. An electric circuit has only *two* conditions: ON or OFF. In a panel of light bulbs, each light can only be either ON or OFF. And in the binary system, only *two* symbols are needed to represent any number: 0 or 1.

If we agree that when a light is ON it represents "1," and when it's OFF it stands for "0," then we have established an electronic way to represent any number. Fig. 10-1 shows a panel displaying the binary numeral 101, with the place values indicated above it.

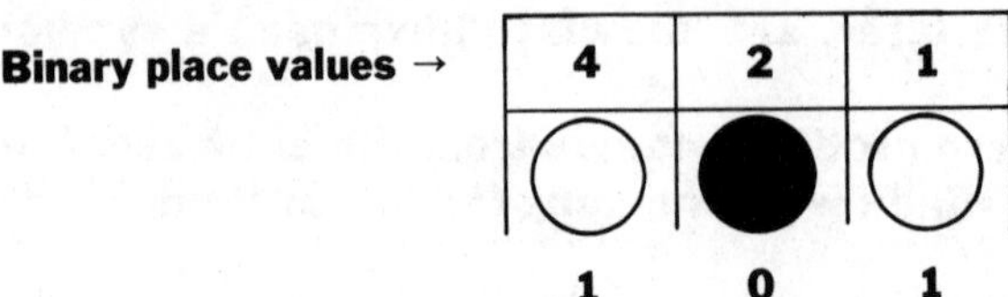

Figure 10-1

(continued)

101 in the binary system translates into 5 in the decimal system: 101 means 1 group of 4, *no* groups of 2, and one 1; that is,

4 + 0 + 1, or 5.

This means that if we see five planes in the sky, we can represent the number of planes—five—by 101 in binary language, or by 5 in decimal language.

Since electrical energy moves with the speed of light, we can understand why electronic computers solve problems with such fantastic speed.

READINESS FOR PLACE VALUE

Before your child can understand place value:

- He must be able to group objects by tens. Given a collection of counters, he should be able to break them up into groups of 10.
- He should be able to *read* and *write* the numerals 0, 1, 2, 3, 4, 5, 6, 7, 8, 9, 10.
- He should be able to count to 10.
- He should be able to recognize the written words *tens* and *ones*.

TEACHING PLACE VALUE

Getting Started

Place three counters in front of your child and ask him to tell you how many there are. Ask him to write the numeral that shows the number of counters.

Now place 14 counters in front of him and ask whether he knows the name of this number of counters. Can he write the numeral that represents this number of counters?

The essential idea in explaining *place value* is to tell your child that he will now learn a way to represent numbers of objects *greater* than 10. And that the method depends on his being able to separate a pile of objects into *groups of 10.*

Focus on the idea of forming as many sets of 10 as possible. Ten ones form a single group of 10; ten 10s form a single group of 100. These understandings must be rooted in experience with physical objects; they must also be tied to the verbal names and written symbols for these numbers.

If a child understands the concept of place value and can move easily from the objects to the spoken names and the written symbols, he will have acquired an essential and excellent foundation for later work with whole numbers.

Teaching Sequence

It is suggested that you teach place value in the following sequence.

A—Provide experiences with grouping objects in sets of 10.

1. Ask your child to separate a set of disks into groups of 10 and to then tell the number of tens and the number of single disks left over. In the example below

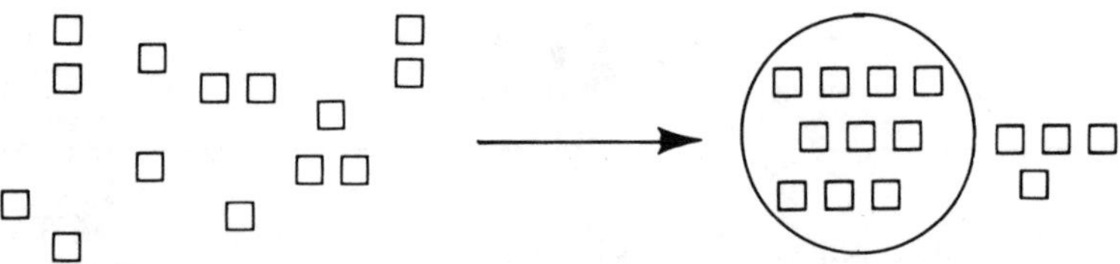

he should tell you that the pile contains 1 group of *10* and 4 *ones*. Repeat with different objects and different numbers.

NOTE: The essential questions to ask:

Are there enough disks left over to make another set of 10?
How many sets of 10 are there?
How many disks are left over?

Your child should think of a *group of 10* as a *unit*, just as individual disks are considered units.

2. Reverse the procedure. Name a grouping such as 3 tens 2 ones, and ask your child to set up this grouping with his disks.

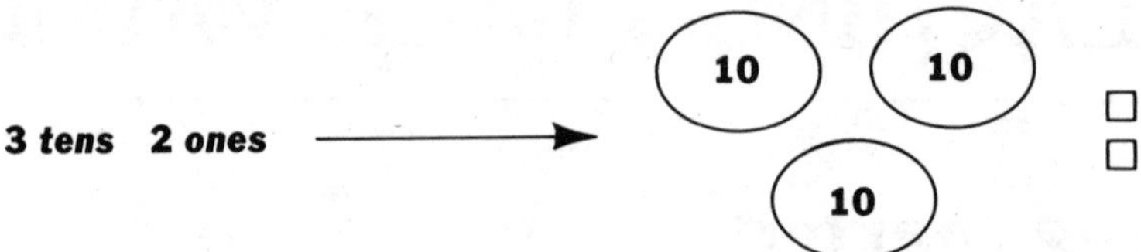

B—Help your child relate groups of 10 to the usual verbal names. Let him learn the names of multiples of 10: ten, twenty, thirty, . . . which he will need to read "2 tens 7 ones" as *twenty-seven*.

1. (a) Display a pile of 50 disks and ask your child to count them by ones.

(b) Now ask him to group the pile by tens and then count them by tens. As he points to each successive group, he counts: "ten, twenty, thirty, forty, fifty."

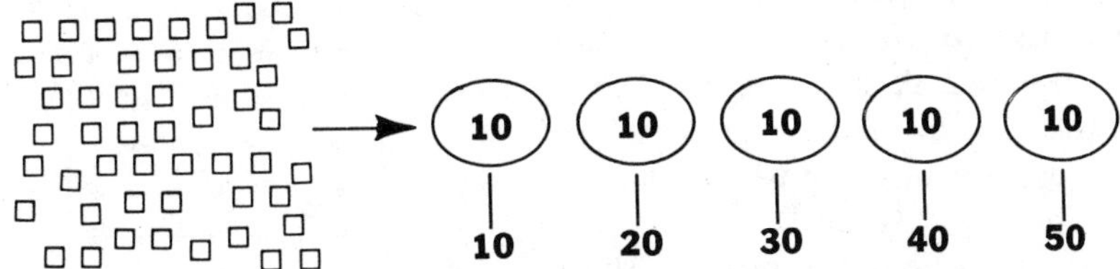

Make sure the child is convinced that counting by ones and counting by tens give the same result.

2. Because the child already knows that 2 tens is 20, and 3 tens is 30, he is ready for names like *thirty-five* from a grouping of 3 tens and 5 ones.
   (a) Arrange groupings of disks, like 4 tens 3 ones, and ask your child to give the verbal names for them:

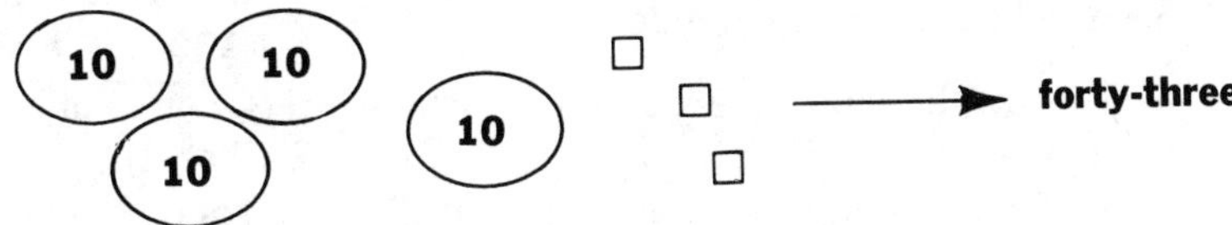

   (b) Say a number like *thirty-seven* and ask the child to represent this number with disks:

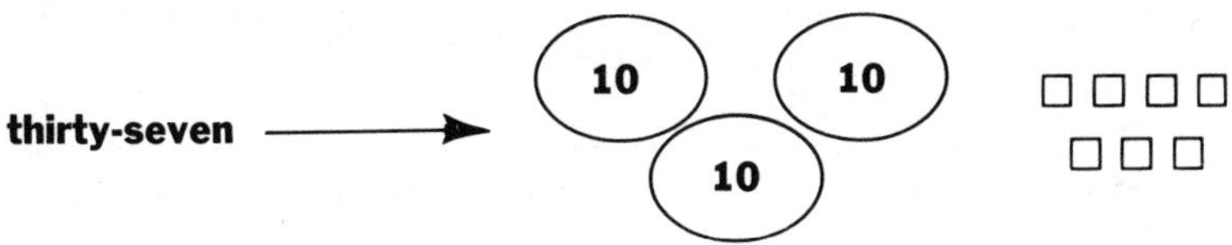

3. Now teach the **teens:**
   Eleven is a special name for "1 ten 1 one."
   Twelve is a special name for "1 ten 2 ones."
   The other names are all backward. For example, in *sixteen*, the part referring to "six" is named first and indicates 6 *ones*; "teen" indicates 1 ten.

C—Help your child write and interpret two-digit numbers and relate them to their verbal names.

1. Display a grouping of 2 tens 5 ones. Show how we record the groupings:

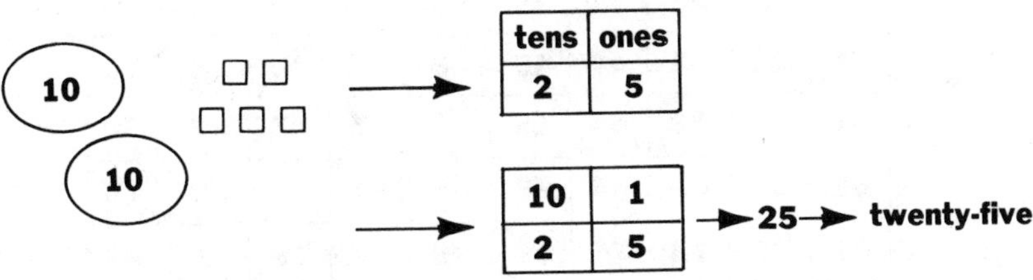

   Repeat with other numbers.
2. Ask the child to draw pictures of tens and ones to represent
   (a) 38 (b) 51 (c) 20 (d) 60

D—*When the child is ready,* move on to three-digit numbers. Use bird arrays to develop the concept.

A **bird array** is a rectangular arrangement of birds. A *10 × 10 bird array* consists of 10 rows, each row having 10 birds. A *1 × 10 bird array* consists of 1 row with 10 birds. A *1 × 1 array* consists of 1 row with 1 bird.

The 10 × 10 bird array on page 278 can be xeroxed to produce sheets of as many arrays as you wish. A sheet can be cut up into 10 strips, each a 1 × 10 array. A strip can be cut up into 10 single birds.

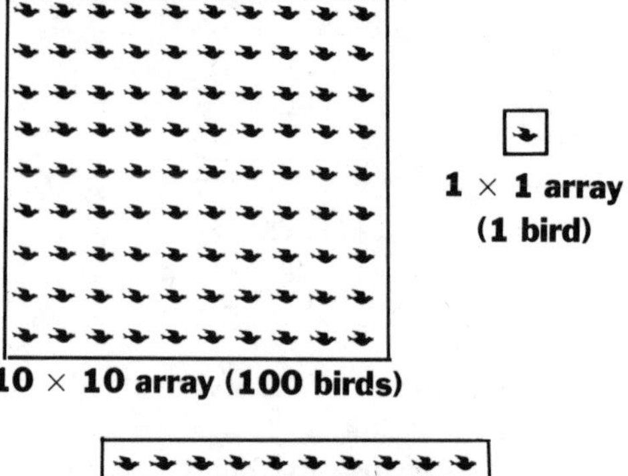

1. Using bird arrays, show your child how 10 single birds—*ones*—form a strip of 1 *ten*, and how 10 strips of ten form a sheet of a *hundred.* So our place values are now extended to three places:

100	10	1

2. Now follow the same procedures for teaching three-digit numbers as were used to teach two-digit numbers (A, B, C on pages 70–71).
3. (a) Call out numbers on a hundred chart (Fig. 10-2) and ask your child to cover them with disks.
   (b) Point to numbers on the chart and ask him to name them.
   (c) Make a hundred chart like the one in Fig. 10-3 and ask him to fill in the missing numbers.
4. Have your child fill in the missing numbers in:
   (a) ____ ____ 38 ____ ____ 41
   (b) 159 ____ ____ ____ 163 ____ ____.
5. Ask your child to write numbers *one less* and *one more* than the given number:
   (a) ____ 25 ____ (b) ____ 68 ____ (c) ____ 89 ____ (d) ____ 150 ____
6. Using bird arrays, ask your child to demonstrate numbers between:
   (a) 10 and 100 (b) 100 and 1000
   For each number, have him give you the verbal name and then write the numeral.

1	2	3	4	5	6	7	8	9	10
11	12	13	14	15	16	17	18	19	20
21	22	23	24	25	26	27	28	29	30
31	32	33	34	35	36	37	38	39	40
41	42	43	44	45	46	47	48	49	50
51	52	53	54	55	56	57	58	59	60
61	62	63	64	65	66	67	68	69	70
71	72	73	74	75	76	77	78	79	80
81	82	83	84	85	86	87	88	89	90
91	92	93	94	95	96	97	98	99	100

Figure 10-2

1				5					
	12							19	
		23				27			
				35			38		
91					96				

Figure 10-3

Other Activities for Teaching Place Value

1. Ask your child to obtain two numbers by spinning the hands on spinners A and B. Have him write them down next to each other to form a two-digit number. Let him then read the number and represent it with bird arrays. Repeat with many spins.

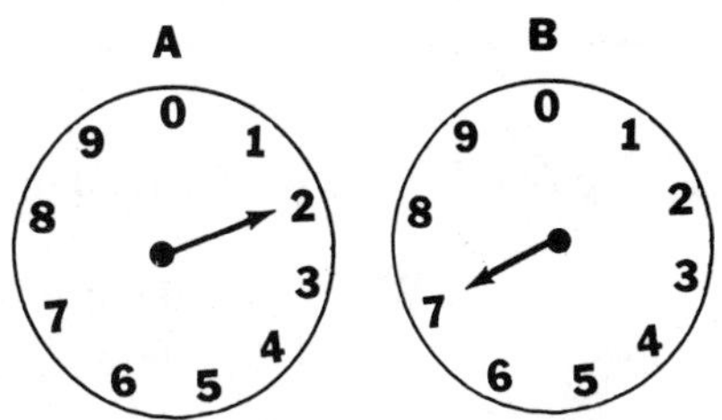

2. Place three spinners in front of child and ask him to follow the same procedure as in #1, but now form three-digit numbers.
3. Write any three digits on a piece of paper. Ask the child to form as many three-digit numbers as possible, using only the digits on the paper. For each number formed, ask him to:
   (a) demonstrate it with bird arrays
   (b) arrange all the numbers formed, from smallest to largest

HIGHLIGHTS OF PLACE VALUE

Upon completion of this chapter, your child should have acquired the following capabilities:

- Given a set of objects less than 100, he should be able to (1) group them by tens; (2) write the numeral showing the number of objects in the set; (3) give its verbal name.

- Given any number less than 100, he should be able to represent this number with a set of objects grouped in tens.
- Given any number less than 100, he should be able to identify the place value of each digit in the number.

The child should also have acquired the following understandings:

- In a place value system, the value of any digit depends upon the place it occupies in the number.
- The decimal system that we use in everyday life is a place value system.
- A number can be identified in three ways: by its (1) verbal name, (2) written symbol (numeral), (3) physical representation.
- Vocabulary:

place value	digit
decimal system	numeral

QUESTIONS ON THE CHAPTER

NOTE: If necessary, read these questions to your child.

1. Here is a pile of disks. (a) Group them in tens. (b) Write the numeral that shows the number of disks in the pile.
2. Gather groups of disks to show:
   (a) 2 tens 7 ones (b) 3 tens (c) 21 (d) 19 (e) 30
3. Read these numbers: 32; 87; 60. Tell how many tens and how many ones there are in each number.
4. Use bird arrays to show:
   (a) 54 (b) 87 (c) 99 (d) 152 (e) 675 (f) 800 (g) 702
5. Read the numbers in #4.
6. Read these numbers: 129; 682; 340; 501. Tell how many hundreds, tens, and ones there are in each number.
7. Make a table to show the place values of: 18; 97; 40; 175; 572; 908.
8. What's the place value of *3* in (a) 73 (b) 39 (c) 325 (d) 236?
9. True or false?
   (a) 10 ones is the same number as 1 ten.
   (b) 3 tens 2 ones is the same number as 2 tens 3 ones.
   (c) 10 tens is the same number as 1 hundred.
10. Count from 10 to 50 in ones.
11. Count in tens as high as you can.
12. Count in hundreds as high as you can.

13. Fill in the missing numbers.
    (a) 12 ____ ____ ____ ____17 ____ ____ ____ ____ 22.
    (b) ____ ____ 63 ____ ____ ____ ____ 68 ____ ____ ____ ____ .
    (c) 371 ____ ____ ____ ____ 376 ____ ____ ____ ____ 381.

14. What does **place value** mean?

15. How many different values can *2* have?

16. Why do we need a zero?

17. What is a **digit?**

18. How many different digits do we use in the decimal system?

19. What is a **numeral?**

20. What's the **decimal system?** Why is it called by that name?

21. Name the first three place values in the decimal system.

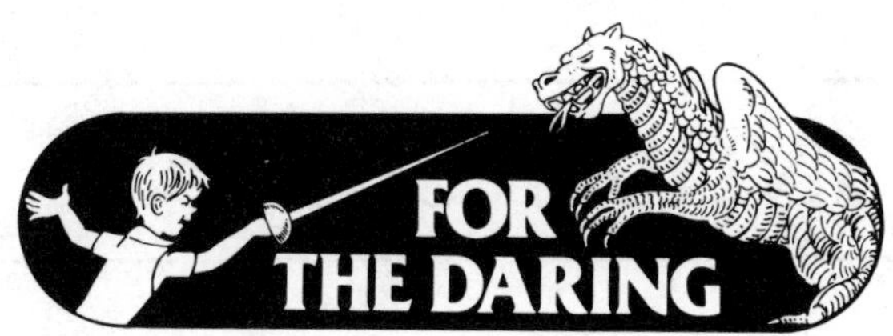

A Weighty Problem

What's the smallest number of weights needed for a balance scale to weigh any amount of candy from 1 pound to 15 pounds? What should be the weight of each? [Solution is on page 257.]

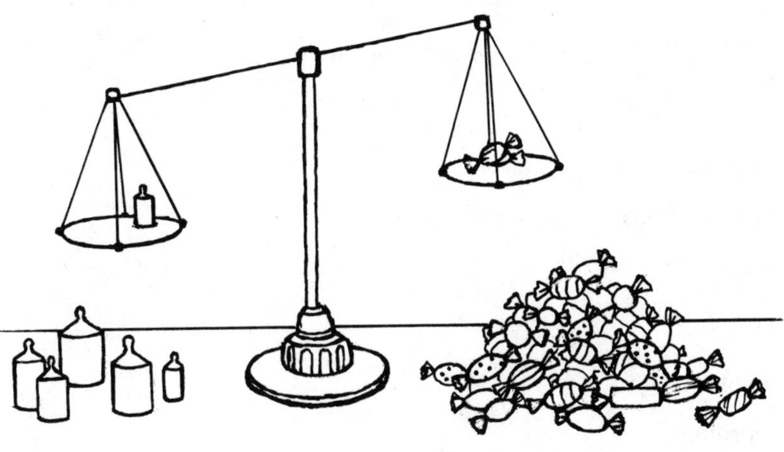

Chapter 11

Adding and Subtracting Whole Numbers

INTRODUCTION

When a child *joins* two disks and three disks to form a new set of five disks, he is performing *addition*. When he *removes* two disks from five disks to form a new set of three disks, he is performing *subtraction*. Once your child understands the meaning of these operations, he can move on to learning the basic addition and subtraction facts, and then to adding and subtracting larger numbers.

The two operations are combined in this chapter because there is an important relationship between them, and their teaching strategies are similar.

The **materials** needed for this chapter include:

Counters (such as checkers, disks, poker chips)

Bird arrays

Play money

Oak-tag strips (see page 83)

Beads

Dice

Spinners

Calculator

THE CHAPTER IN A NUTSHELL

In this chapter you will see how to

- Get your child ready for addition and subtraction
- Explain the meaning of addition and subtraction
- Have your child develop the addition table
- Explain addition and subtraction with larger numbers
- Use a variety of activities and games to reinforce the child's competency with these operations
- Identify and correct errors in his computations
- Provide interesting problems involving addition and subtraction

You will also be given

- A list of understandings and skills your child should be left with at the completion of the chapter
- Questions to test and reinforce the child's comprehension of addition and subtraction
- Insight into a famous young genius at work; a puzzling duet; and palindromes under **For the Curious**
- Magic squares and other puzzles under **For the Daring**

THE MEANING OF ADDITION AND SUBTRACTION

Addition is based on the joining of sets. If you start with two poker chips and then win three, how many chips do you have altogether? To find the answer, you join the two sets to form a new set:

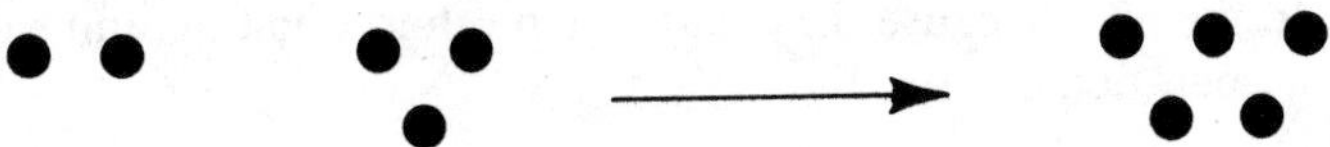

On the number line, addition means counting to the *right*. For instance, adding 2 and 3 means: Begin at 0. Move two spaces to the right. From there, move three more spaces to the right. The answer is the point where you end up, 5.

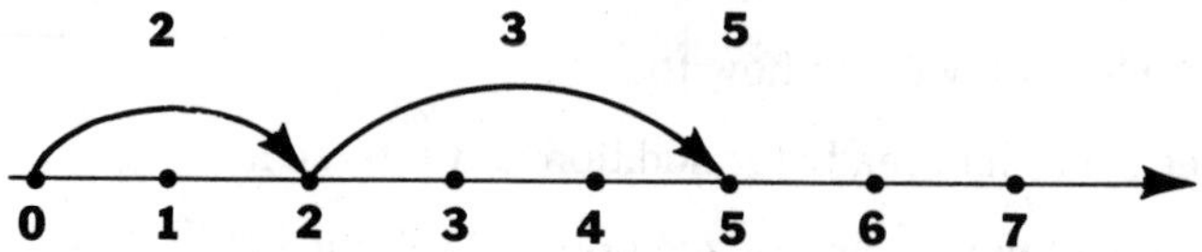

The numbers added are called *addends*. The answer to the addition is called the *sum*. In the example just given, 2 and 3 are the *addends*; 5 is their *sum*.

Subtraction is based on the removal of part of a set. If you start with six chips and lose two, how many do you have left?

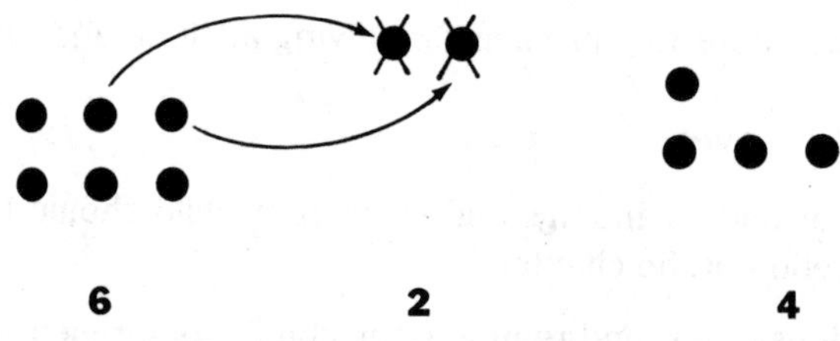

On the number line, subtraction means counting to the *left*. For example, 6 − 2 means: Starting at 0, count six spaces to the right. From here, count two spaces to the left, ending up at 4.

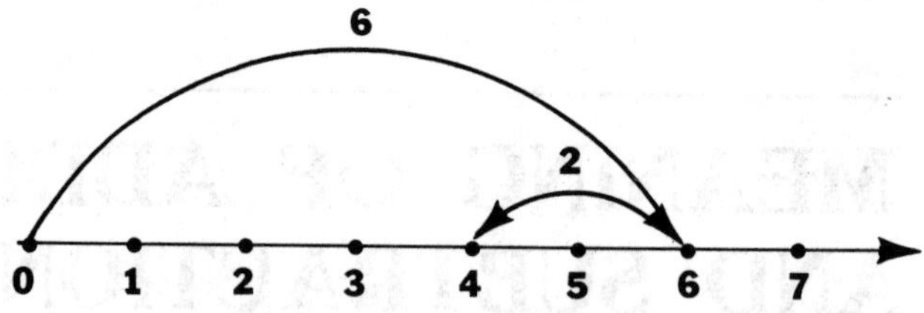

The answer to a subtraction is called the **difference.**

Notice that the two operations counteract each other: Addition joins sets; subtraction separates them. On the number line, addition means moving to the right; subtraction, to the left. Because they undo each other, addition and subtraction are called *inverse* operations.

NOTE: Because of this close connection between the two operations, we actually define subtraction in terms of addition. The subtraction 6 − 2 = ? asks: "What number must we *add* to 2 to give 6?" And we answer: 6 − 2 = 4 *because* 2 + 4 = 6.

You can see that in the subtraction 6 − 2 = ?, we are really looking for the missing addend in the addition 2 + ? = 6.

READINESS FOR ADDITION AND SUBTRACTION

Before your child can learn addition and subtraction, he must

- Be able to join and separate sets
- Be able to count
- Be able to write the numerals 0 through 9
- Understand place value

Specifically for subtraction, the child must first

- Understand the meaning of addition
- Know the basic addition facts

TEACHING ADDITION AND SUBTRACTION

Getting Started With Addition

1. With counters:
   (a) Display a set of 3 counters and another set of 2 counters.

   Ask the child to tell the number of counters in each set.
   (b) Have the child join the two sets.

   Ask him to tell the number of counters in the new set.
   (c) Say that joining 3 counters with 2 counters to produce a new set of 5 counters may be expressed as

$$3 + 2 = 5.$$

   Explain that the symbol +, read "plus," indicates *joining*. Another word for plus is **add.**

Explain that the symbol =, read "equals," means "is the same as."

We read "3 + 2 = 5" as "Three counters and two counters make five counters," or "Three *added* to two equals five," or "Three *plus* two equals five."

(d) Explain that:

We call "3 + 2 = 5" a **number sentence**—more specifically, an **addition sentence.**

The numbers added are called **addends.**
The answer is called the **sum.**

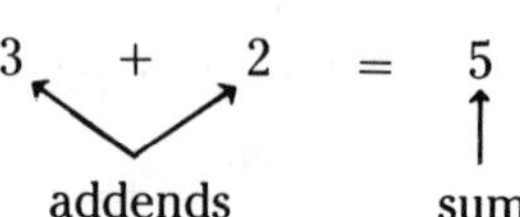

3 + 2 = 5 can also be written vertically:

$$\begin{array}{r} 3 \\ +\,2 \\ \hline 5 \end{array}$$

Repeat (a) through (d) with many examples.

(e) Display a set of 4 objects and another set of 5 objects. Ask the child to join the two sets and then write the number sentence that describes the action. [4 + 5 = 9]
Repeat with different set combinations and a variety of objects.

(f) Write 3 + 4 = □ and explain that the □ is called a *placeholder*, because it holds the place for the answer. We can also write 3 + 4 = ?

(g) Now write 2 + 4 = ? and ask child to demonstrate with counters the action described by this sentence. After joining a set of two objects with a set of four objects, he is to complete the sentence by filling in the sum: 2 + 4 = 6.
Repeat with a variety of sentences, including some in which one of the addends is zero, like 3 + 0 = ? and 0 + 2 = ?.

2. On the number line:
(a) Illustrate 3 + 4 = ? as follows:

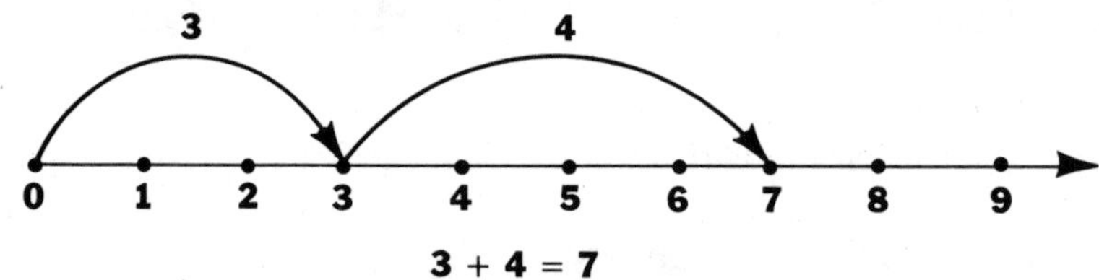

Stress that:

Addition on the number line means counting to the *right*.

Addends are represented by the *spaces between points*, not by the points.

The number line begins with 0, not with 1.

(b) Write several number sentences and ask the child to demonstrate them on the number line.

(c) Show several situations like the one below. Ask the child to write the addition sentence describing the picture.

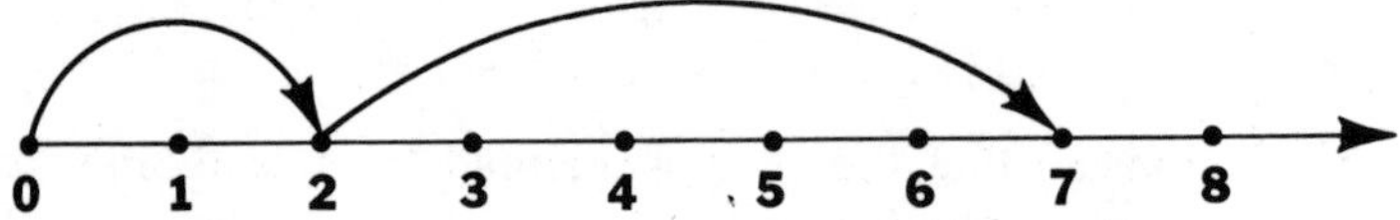

(d) Make up word problems like: Lisa had four cupcakes. Her brother gave her two more. How many did she have altogether? Ask the child to (1) write the number sentence describing the action in each problem; (2) demonstrate the problem on the number line

Getting Started With Subtraction

Introduce subtraction with materials and procedures similar to those used with addition.

1. With counters:

(a) Display 5 counters and ask how many there are. Then push 3 away and ask how many were removed. Ask how many counters are left.

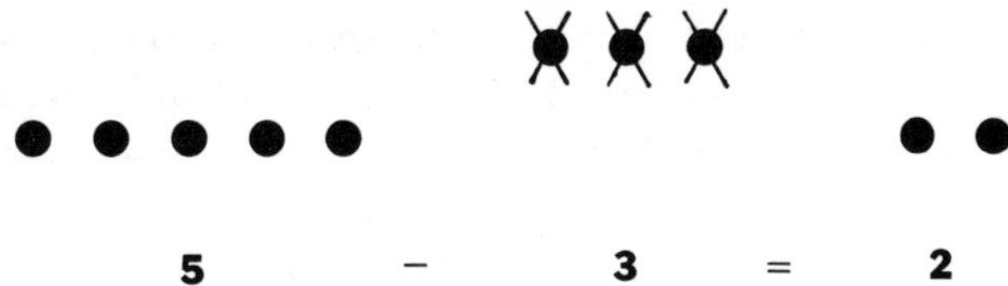

5 − 3 = 2

(b) Write the sentence 5 − 3 = 2, and ask your child what he thinks the sentence means.

Explain that the symbol −, read "minus," indicates separation or taking away. Another word for minus is *subtract*. Also, explain that:

The sentence 5 − 3 = 2 is read "5 minus 3 equals 2," or "3 taken away from 5 leaves 2," or "3 subtracted from 5 equals 2."

The answer to a subtraction is called the **difference.**

(c) Display a set of 7 counters. Ask the child to push away 3 counters and then write the number sentence that describes the action.

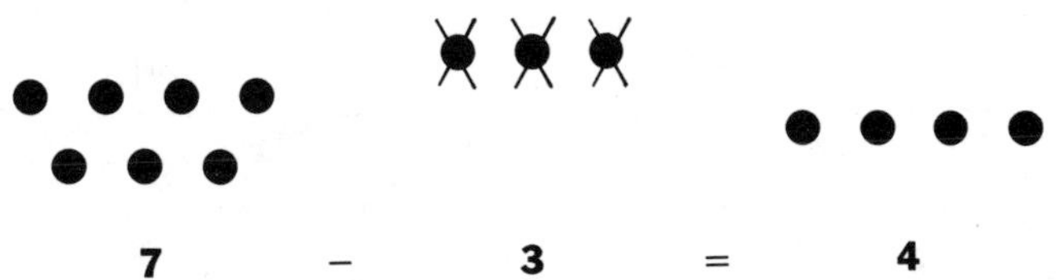

7 − 3 = 4

Repeat with other combinations.

(d) Write a number sentence like 6 − 2 = ?. Ask your child to display sets of counters that describe the action in the problem. Repeat with a variety of sentences like 3 − 0 = ?, 5 − 5 = ?, 6 − ? = 5, ? − 4 = 1.

(e) Ask the child to write number sentences for the following problems:

(1) Jack has three balloons and gives two to his sister. How many does he have left?

(2) Andy has four coupons. She needs seven to win a prize. How many more coupons does she need to win a prize?

(3) Laurie has five cupcakes and Fred has two. How many more cupcakes does Laurie have than Fred?

NOTE: Stress that in subtraction of counting numbers, the larger number is written first. Ask the child to demonstrate with counters why this must be so.

2. On the number line:
   (a) Ask: If we move on the number line four spaces to the *right*, how can we subtract one space from these four spaces?

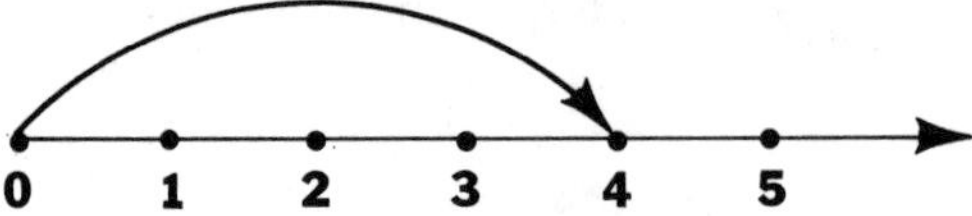

When the child understands that subtraction can be shown on the number line as counting to the *left*, the sentence 4 − 1 = ? can then be shown as follows: Starting at 0, count four spaces to the right. From here, count one space to the *left*, landing at 3:

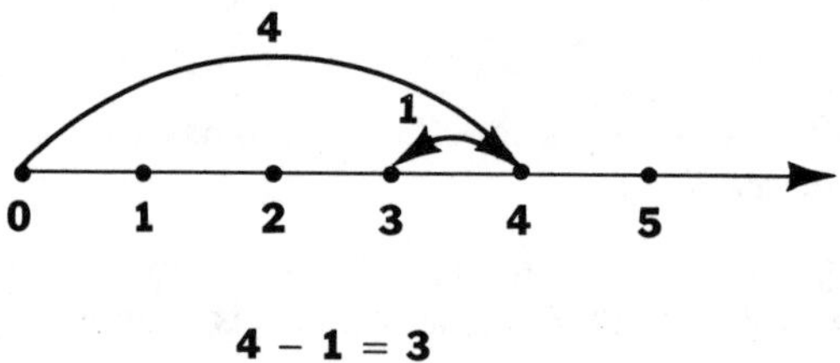

   (b) Write several subtraction sentences and ask the child to picture them on the number line.
   (c) Display several number lines, such as the one below. Ask the child to write the subtraction sentence describing the picture.

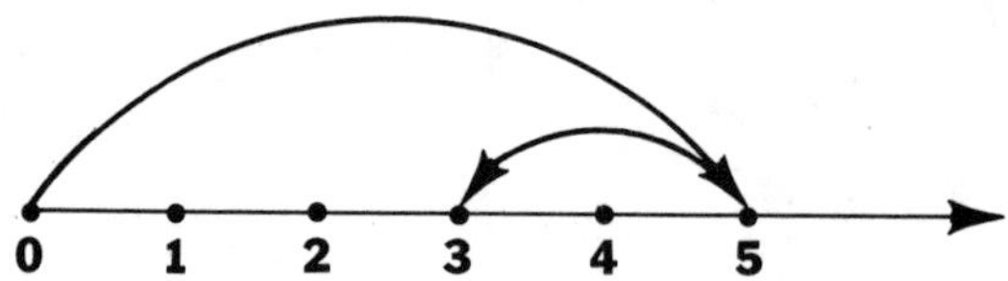

   (d) Make up word problems like: Jane had $6. She spent $2 for a birthday gift. How much does she have left? Ask the child to (1) write the subtraction sentence for the problem, (2) represent the problem on the number line.

Basic Addition and Subtraction Facts

A **basic addition fact** is the sum of any two numbers in the set {0, 1, 2, 3, 4, 5, 6, 7, 8, 9}. There are 100 basic addition facts (Fig. 11-1). Since every addition fact is associated with a corresponding subtraction fact, there are also 100 **basic subtraction facts.**

+	0	1	2	3	4	5	6	7	8	9
0	0	1	2	3	4	5	6	7	8	9
1	1	2	3	4	5	6	7	8	9	10
2	2	3	4	5	6	7	8	9	10	11
3	3	4	5	6	7	8	9	10	11	12
4	4	5	6	7	8	9	10	11	12	13
5	5	6	7	8	9	10	11	12	13	14
6	6	7	8	9	10	11	12	13	14	15
7	7	8	9	10	11	12	13	14	15	16
8	8	9	10	11	12	13	14	15	16	17
9	9	10	11	12	13	14	15	16	17	18

Figure 11-1
Addition Table

After your child understands the meaning of addition and subtraction and has learned how to use objects to find answers to simple addition and subtraction problems he is ready to develop the basic facts.

CAUTION: Do not move your child away from concrete materials into purely written work too rapidly. Also, do not have your child memorize the basic facts before he understands their meaning.

1. Prepare a set of oak-tag strips (or wood rods) of lengths 1 inch through 9 inches, with the length marked on each piece (Fig. 11-2).

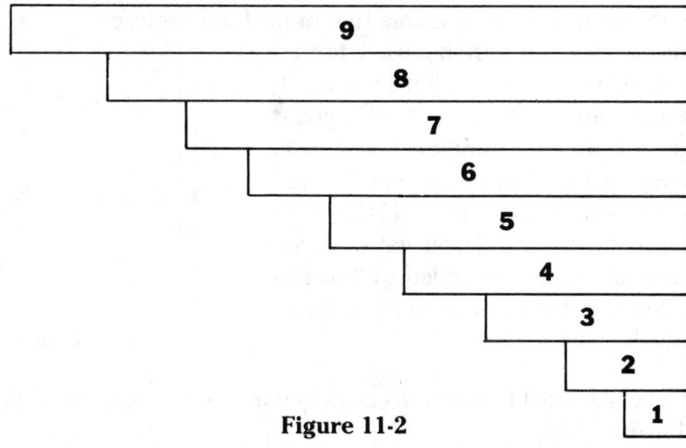

Figure 11-2

Your child can play with these strips and discover various relationships among their lengths (Fig. 11-3). He can observe, for instance, that

(a) A 2″ strip laid next to a 3″ strip is the same length as a 5″ strip.

(b) A 3″ strip laid next to a 2″ strip is the same length as a 5″ strip.

(c) If you take away a 2″ strip from a 5″ strip, you have a 3″ strip left.

(d) If you take away a 3″ strip from a 5″ strip, you have a 2″ strip left.

5	
3	2
2	3

Figure 11-3

The strips in Figure 11-3 show the following addition and subtraction facts about the number 5:

Addition	Subtraction
(a) 3 + 2 = 5	(a′) 5 − 2 = 3
(b) 2 + 3 = 5	(b′) 5 − 3 = 2

2. Use oak-tag strips to develop with your child the following basic addition and subtraction facts about the 4 "family":

Addition	Subtraction
4 + 0 = 4	4 − 0 = 4
3 + 1 = 4	4 − 1 = 3
2 + 2 = 4	4 − 2 = 2
1 + 3 = 4	4 − 3 = 1
0 + 4 = 4	4 − 4 = 0

Under this "family plan," all the basic facts surrounding one number are explored, and the child can see the connection between addition and subtraction.

Let your child now develop, through use of the oak-tag strips, the "family" relationships for each of the numbers 0 through 9.

3. As he develops the addition facts, let your child enter the results in a table like the one in Fig. 11-1. This addition table should be completed by the child and be available to him for future reference.
The same table can be used for the basic subtraction facts. For instance, to find the answer to 8 − 3 = ?, you're asked to find the number that must be added to 3 to make 8. So you locate 3 in the left column (Fig. 11-4), move across the row until you get to 8, and see what number is added to 3 to give 8. You find this number, in the top row, to be 5.

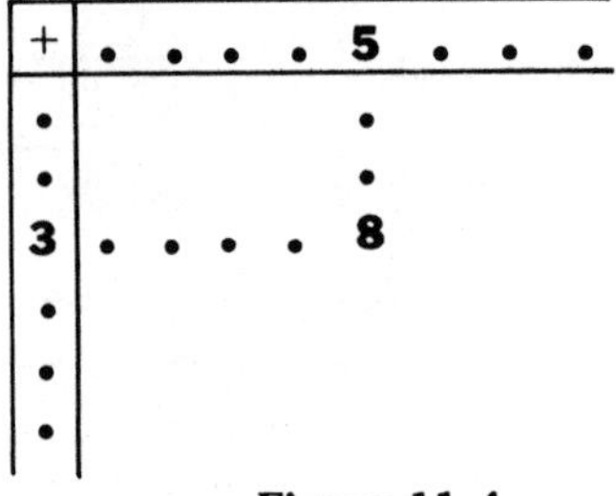

Figure 11-4

4. Your child should be able to draw several conclusions from the addition table (Fig. 11-1):

- The order in which numbers are added does not change the answer: 2 + 3 = 3 + 2.
- Zero added to any number equals that number: 0 + 5 = 5.
- The answer to any subtraction sentence can be gotten from the related addition sentence: 7 − 3 = 4 since 4 + 3 = 7.
- Subtracting zero from any number leaves the number unchanged: 4 − 0 = 4.
- Subtracting a number from itself leaves 0: 5 − 5 = 0.

Activities with Basic Facts

1. String eight beads (or clothespins) on a wire. Ask the child to separate the beads into any two parts, and to then write an addition sentence to describe the bead arrangement.

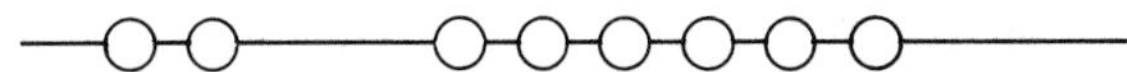

   Let him then work out with the beads the addition and subtraction facts for the 8 "family" the way he did with the oak-tag strips. Repeat with other numbers of beads.

2. Have the child roll a pair of dice. Ask him to give the sum (or difference) of the two numbers showing. (For larger sums, use two wood cubes with faces marked 4, 5, 6, 7, 8, 9.)

3. Have the child use two spinners, each marked 0 to 9, to get various number combinations. Ask the child to give the sum (or difference) of the numbers showing.

4. Write a number sentence such as $4 + 3 = 7$ or $9 - 4 = 5$. Ask the child to make up a verbal problem to fit each sentence.

5. Ask the child to determine two numbers you're thinking of, after being given clues about the numbers. For instance:

   Clue 1: The numbers are in the set {0,1, 2, 3, 4, 5, 6, 7, 8, 9} and their sum is 6. [From this clue, the child can conclude that the two numbers must be among the following pairs: (0,6), (1,5), (2,4), (3,3).]

   Clue 2: One of the numbers is 4 more than the other number. [From this clue, the child concludes that the numbers must be 1 and 5.]

6. Use a calculator to reinforce the child's competence with the basic facts.
   (a) The child enters, for example, $5 + 7$. But before pressing = for the answer, he calls out the answer and then checks it with the calculator. Repeat with many exercises.
   (b) The child enters "9" on the calculator. Ask: "What can you do to get 3?" Let him then press − and subtract the right number. Repeat with many combinations.

Adding Larger Numbers

Addition is a way to simplify counting. With small numbers, the child can add by manipulating objects and then counting them. But with large numbers, this becomes impractical. So we use a procedure, a sequence of steps called an *algorithm*, by which to add without having to count. The addition algorithm is the standard procedure by which we add.

Your child can use toy money—pennies, dimes, and dollars—to lead him to the addition algorithm. But you need not confine yourself to toy money. Vary your explanations with bird arrays, Popsicle sticks bundled into tens, and other materials.

Before your child can learn to add large numbers, he must know (1) the basic addition facts and (2) place value.

Make up problem situations involving store purchases by the child. These must be made with the fewest dimes and pennies possible. For instance, a 25¢ purchase must be paid for with 2 dimes and 5 pennies, rather than with 25 pennies or 1 dime and 15 pennies.

1. Suppose, you ask your child, he makes a 25¢ and a 12¢ purchase. How much would he pay altogether?
   (a) The child lays out the cost of each purchase:

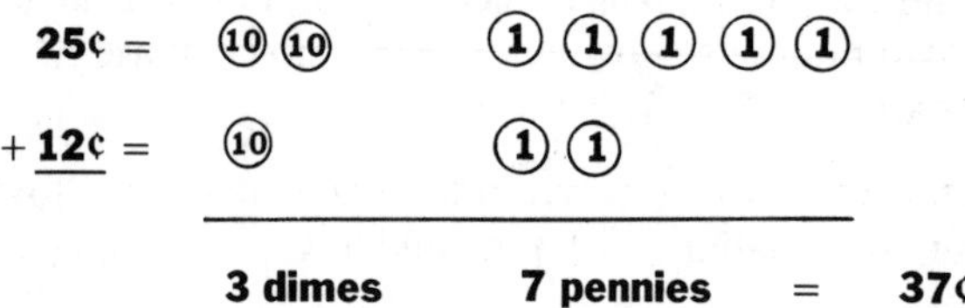

   (b) The next step is to add *ones* and *tens* without reference to money. Develop this step in three stages, reminding your child to write the ones in one column and the tens in the next column to the left:

(1)	(2)	(3)
25	25	25
+ 12	+12	+12
7 ones	7	37
3 tens	30	
37	37	

Note that stage (3), the shortest procedure, is the standard addition algorithm.

2. For purchases of 27¢ and 15¢, the child will have to trade 10 pennies for a dime. (This trading procedure is sometimes called *regrouping*.)
   (a)

27¢ = ⑩⑩ ① ① ① ① ① ① ①

+ **15¢** = ⑩ ① ① ① ① ①

3 dimes **12 pennies**

Since the child is to pay with the fewest possible coins, he trades 10 of the 12 pennies for 1 dime, and now has

3 dimes + 1 dime + 2 pennies
= 4 dimes + 2 pennies
= 42¢

(b) Now let the child do this addition without reference to money:

```
 (1)                 (2)        (3)

  27                 27         ¹27
 +15                +15        +15
  12 ones  ⟶        12          42
  3  tens  ⟶        30
  42                 42
```

The shortest procedure is (3), the addition algorithm.

Explain: In (3), in the ones column, $5 + 7 = 12$, we trade the 12 ones for 1 ten and 2 ones. So we write the "2" in the *ones* column, and "carry" the 1 ten to the *tens* column, writing a little "1" above the "2" in the tens column.

In the tens column, we now have 1 ten (that we carried) + 2 tens + 1 ten, or 4 tens altogether.

3. For purchases of $1.75 and $2.81, the child will need toy dollars, dimes, and pennies.
   (a) Using money,

$1.75 =	$	⑩⑩⑩⑩⑩⑩⑩	① ① ① ① ①
+ **$2.81** =	$ $	⑩⑩⑩⑩⑩⑩⑩⑩	①
	3 dollars	**15 dimes**	**6 pennies**

Trading 10 of the 15 dimes for 1 dollar, the child now has:

3 dollars + 1 dollar + 5 dimes + 6 pennies
= 4 dollars + 5 dimes + 6 pennies
= 456 pennies, or $4.56

(b) Now add 175 and 281 without reference to money.

```
 (1)                      (2)        (3)

  175                     175        ¹175
 +281                    +281       +281
    6 ones       ⟶         6         456
  15  tens       ⟶       150
  3   hundreds   ⟶       300
  456                     456
```

The shortest procedure is (3), the addition algorithm.

EXPLAIN: In the ones column in (3), we get $1 + 5 = 6$.
In the tens column, we get $8 + 7 = 15$ tens. We trade 10 of the 15 tens for 1 *hundred*, resulting in 1 hundred and 5 tens. So we write "5" in the tens column, and carry the "1" to the hundreds column. We show this "carry" by writing a little "1" above the 1 in the hundreds column.

In the hundreds column, we now have the 1 hundred, just carried, + 1 hundred + 2 hundreds, or 4 hundreds altogether.

4. Before your child works with addition requiring regrouping, provide him with practice exchanging 10 ones for 1 ten, and 10 tens for 1 hundred. Use money, bird arrays, and other objects that can be bundled conveniently into tens and hundreds.

   In subtraction, the child will need to do the opposite: exchange 1 hundred for 10 tens, and 1 ten for 10 ones. So let him practice trading in both directions.

   Remind your child, as needed, that:
   (1) When writing an addition example, you place ones under ones, tens under tens, and hundreds under hundreds.
   (2) You add each column, starting with the ones column.
   (3) If the sum of any column is 10 or more, you write down the last digit and "carry" the other digit to the next column.
   (4) You check the answer by adding the columns in the opposite direction.

What Price Genius

There is a story about the famous German mathematician Carl Gauss (1777–1855) who, as a precocious child, drove his teachers up a wall. On one occasion, hoping to keep the pest occupied for awhile, his teacher asked him to find the sum of all the counting numbers from 1 to 100. To his teacher's chagrin, Carl came up with the answer in a flash. This is how he did it.

He paired the numbers this way:

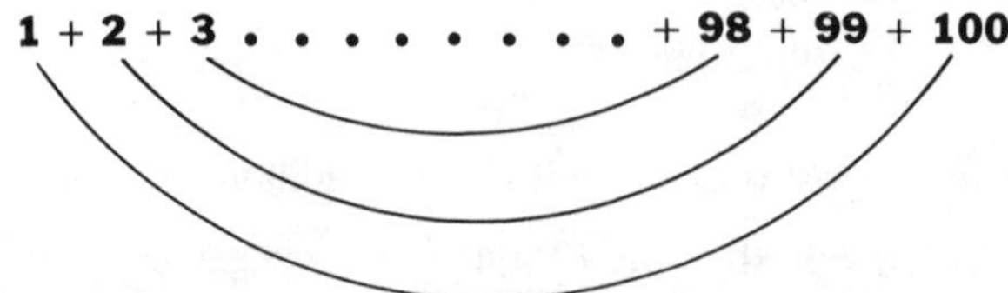

Then he reasoned that since each pair of numbers adds up to 101 (1 + 100, 2 + 99, 3 + 98, and so on), and since there are 50 such pairs, then the sum of the hundred pairs must be 50 × 101, or 5050.

(continued)

"How Did I Do It?"

	Example
Ask your child to write down any two-digit number.	**67**
[Now you write *166* on another piece of paper.]	
Ask the child to write under the 67 another two-digit number.	**43**
You write under the 43 *your* two-digit number.	**56**
Now ask the child to find the sum of all three numbers.	**166**

The child notes that the sum is the number you wrote on the piece of paper *before* you knew the other numbers. How did you do it?

***Explanation*: For each number the child writes, except the *first*, you respond with a number that makes the sum of the two numbers 99. For instance,**

If your child writes	**43**	**67**	**84**
You respond with	**56**	**32**	**15**
	99	**99**	**99**

Now think of 99 as 100 − 1. The given example then boils down to:

$$
\begin{array}{rl}
 & 67 \longrightarrow 67 \\
+ & \left.\begin{array}{r}43\\56\end{array}\right\} \rightarrow 100 - 1 \\
\hline
 & 166 \longrightarrow 167 - 1
\end{array}
$$

Therefore, you knew in advance that the sum would be 67 (the child's first number) + 100 − 1, or 166.

If you use two pairs of numbers, then you add to the child's first number *two* 99's, or 200 − 2. For example,

(continued)

Child's first number:	**78 ⟶ 78**
[Then you write on another piece of paper 276.]	
Child's second number:	**35**
You respond with	**64**
	⟶ 200 − 2
Child's third number:	**87**
You respond with	**12**
	276 ⟶ 278 − 2

Subtracting Larger Numbers

[Use materials and procedures described for addition.]

1. Shaun has 38¢ and spends 15¢ for a candy bar. How much does he have left?

   (a) Using money:

   38¢ = (10)(10)(10) (1)(1)(1)(1)(1)(1)(1)(1)

   −15¢ = (10) (1)(1)(1)(1)(1)

   2 dimes **3 pennies**

   = **23¢**

   (b) By the subtraction algorithm:

(1)		(2)	(3)
38		38	38
−15		−15	−15
3 ones	⟶	3	23
2 tens	⟶	20	
23		23	

   To check the answer, use the idea that subtraction is the inverse of addition, e.g., $5 - 2 = 3$ since $2 + 3 = 5$:

   $$\begin{array}{r} 38 \\ -15 \\ \hline 23 \end{array} \quad \text{is correct because} \quad \begin{array}{r} 15 \\ +23 \\ \hline 38 \end{array}$$

2. Linda has 52¢ and spends 15¢ for a balloon. How much does she have left?

(a) Using money:

52¢ =	(10)(10)(10)(10)(10)	(1) (1)
−15¢ =	(10)	(1) (1) (1) (1) (1)

Ask: Can you subtract 5 pennies from 2 pennies? How about exchanging one of the 5 dimes for 10 pennies?

The 52¢ will then consist of *4* dimes + 10 pennies + 2 pennies, or 4 dimes 12 pennies:

52¢ =	(10)(10)(10)(10)	(1) (1) (1) (1) (1) (1) (1) (1) (1) (1) (1) (1)
−15¢ =	(10)	(1) (1) (1) (1) (1)
	= **3 dimes**	**7 pennies**
	= **37¢**	

(b) By the algorithm:

$$\begin{array}{r} 52 \\ -\underline{15} \end{array}$$

Since we cannot subtract 5 ones from 2 ones, we exchange one of the 5 tens (in the top number) for 10 ones: 5 tens 2 ones = 4 tens + 10 ones + 2 ones = 4 tens 12 ones. We show this regrouping as

$$\begin{array}{r} 5\ 2 \\ -\underline{1\ 5} \end{array} \longrightarrow \begin{array}{r} {}^{4}\!\not{5}\ {}^{12}\!\not{2} \\ -\underline{1\ 5} \end{array}$$

and the child can now proceed to subtract ones from ones and tens from tens.

NOTE: Using superscripts helps the child understand the process of regrouping. The use of this "crutch" should be discouraged as soon as the child is ready for the more mature method.

(1)	(2)	(3)	(4)	*Check*
${}^{4}\!\not{5}\ {}^{12}\!\not{2}$	${}^{4}\!\not{5}\ {}^{12}\!\not{2}$	${}^{4}\!\not{5}\ {}^{12}\!\not{2}$	(without	15
− 1 5	− 1 5	− 1 5	the	+37
7 ones ⟶	7	3 7	"crutch")	52
3 tens ⟶	3 0			
3 7	3 7		52	
			−15	
			37	

3. Subtraction with three-digit numbers: 653 − 126

653
−126

Since we can't subtract 6 ones from 3 ones, we exchange one of the 5 tens in the top number for 10 ones:

5 tens 3 ones = 4 tens 13 ones

(1)		(2)	(3)	(4)	*Check*
4 13		4 13	4 13		
6 ~~5~~ ~~3~~		6 ~~5~~ ~~3~~	6 ~~5~~ ~~3~~	653	126
− 1 2 6		− 1 2 6	− 1 2 6	−126	+527
7 ones	⟶	7	5 2 7	527	653
2 tens	⟶	2 0			
5 hundreds	⟶	5 0 0			
5 2 7		5 2 7			

4. Subtraction with zeros:

	(1)	(2)	(3)	*Check*
		4 10		
(a)	50	~~5~~ ~~0~~	50	18
	−18	− 1 8	−18	+32
		3 2	32	50
		2 10		
(b)	305	~~3~~ ~~0~~ 5	305	132
	−132	− 1 3 2	−132	+173
		1 7 3	173	305
		5 9 13		
(c)	603	~~6~~ ~~0~~ ~~3~~	603	275
	−275	− 2 7 5	−275	+328
		3 2 8	328	603

Explain Step (2) in row (c): Think of *60* in *603* as 60 tens. 1 ten is then exchanged for 10 ones, resulting in 59 tens and 13 ones.

5. Provide your child with many examples like those in #1, 2, 3, and 4. Do not hurry the process because this is a point at which the child can get confused and lost. Work methodically and patiently with physical objects like play money and bird arrays, and only then should you move to the more formal procedure (the algorithm). Remind your child, as needed:
   (1) To place ones under ones, tens under tens, and hundreds under hundreds, when writing a subtraction
   (2) To subtract the bottom digit from the top digit, beginning with the ones place
   (3) To exchange tens for ones and hundreds for tens when the bottom digit is larger than the top digit
   (4) To check by adding the answer to the bottom number; the sum should be the same as the top number

FOR THE CURIOUS Logo

Palindromes

A *palindrome* is a word, sentence, or number that reads the same forward and backward. The words pop, level, and radar are examples of palindromes. So is the sentence "Able was I ere I saw Elba." So are the numbers 22, 505, and 11211.

Palindromes illustrate an intriguing property of whole numbers: Any whole number can be transformed to a palindrome by *reversing the digits and adding*. This procedure must sometimes be repeated several times, but it will eventually end with a palindrome.

Let's transform 52 to a palindrome:

52	starting number
25	digits reversed
77	sum (which is a palindrome)

Now let's transform 359 to a palindrome:

359	starting number
953	digits reversed
1312	sum
2131	digits reversed
3443	sum (which is a palindrome)

Besides being fun for the child, transforming given numbers to palindromes provides practice in addition.

Common Errors in Addition and Subtraction

When your child makes an error in computation, you must understand the reason for the error before you can help correct it. A good way to find it is to ask your child to do the problem "out loud," verbalizing each step and, thereby, revealing his thinking.

Below are some common addition and subtraction errors children make. The indicated reasons for these errors suggest the way to correct them.

1. Errors with basic facts:

(a) $$\begin{array}{r} 14 \\ \underline{+35} \\ 48 \end{array}$$

(b) $$\begin{array}{r} 57 \\ \underline{-24} \\ 23 \end{array}$$

2. Does not understand place value:

(a) $$\begin{array}{r} 35 \\ \underline{+47} \\ 712 \end{array}$$ (5 + 7 = 12)

(b) $$\begin{array}{r} 28 \\ \underline{+65} \\ 83 \end{array}$$ (5 + 8 = 13. The 3 was put down, but the 1 was not carried.)

(c) $$\begin{array}{r} 43 \\ \underline{-15} \\ 38 \end{array}$$ (4 tens 3 ones were not exchanged for 3 tens 13 ones.)

3. Subtracts smaller digit from larger digit:

$$\begin{array}{r} 63 \\ \underline{-17} \\ 54 \end{array}$$

4. Errors with zero:

(a) $$\begin{array}{r} 70 \\ \underline{-26} \\ 50 \end{array}$$

(b) $$\begin{array}{r} 50 \\ \underline{+13} \\ 60 \end{array}$$

(c) $$\begin{array}{r} 8 \\ \underline{+0} \\ 0 \end{array}$$

(d) $$\begin{array}{r} 7 \\ \underline{-0} \\ 0 \end{array}$$

5. Subtracts ones, adds tens:

$$\begin{array}{r} 69 \\ \underline{+26} \\ 83 \end{array}$$

6. Adds instead of subtracts:

$$\begin{array}{r} 56 \\ \underline{-23} \\ 79 \end{array}$$

7. Reverses digits:

$$\begin{array}{r} 7 \\ \underline{+8} \\ 51 \end{array}$$

8. Confuses 1 with 0:

(a) $$\begin{array}{r} 5 \\ \underline{+1} \\ 5 \end{array}$$

(b) $$\begin{array}{r} 5 \\ \underline{-1} \\ 5 \end{array}$$

9. Adds all the digits:

$$\begin{array}{r} 28 \\ \underline{+\ 3} \\ 13 \end{array}$$ (2 + 8 + 3 = 13)

Other Activities for Addition and Subtraction

1. How can you use the numbers 1, 2, 8 with addition and subtraction to get 9, if each number may be used only once? One answer is:

   $8 - 1 + 2 = 9$; another, $8 + 2 - 1 = 9$. Ask the child to use:

   (a) 3, 4, 6 to get 5 (b) 1, 5, 7, 9 to get 12

2. **Mental arithmetic:** Say to your child: "Think of a number from 1 to 20. Add 3. Subtract 5. Add 7. Now subtract the original number. I can read your mind: Your answer is . . . 5! How did I know?

   [*NOTE:* The original number $+ 3 - 5 + 7 -$ the number $= 5$ no matter what number you start with. Ask the child to make up his own number puzzle and try it on you.]

3. (a) Place the numbers 2, 3, 4, 5, 6 in the circles of the T in Fig. 11-5 so that the sum of the three numbers in each direction is the same.

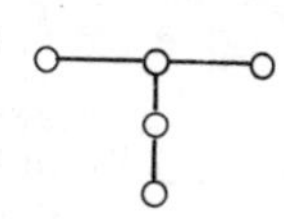

Figure 11-5

   (b) Do the same with the numbers 2, 3, 4, 5, 6, 7 in the circles of the A in Fig. 11-6.

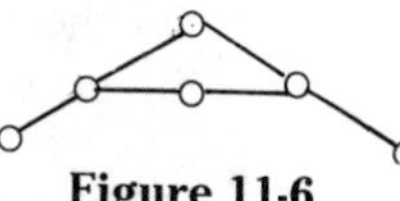

Figure 11-6

4. Fill in the missing digits in each example:

	(b)	(c)	(d)
2 ?	? 6	? 8	6 ?
+? 6	+3 ?	−3 ?	−? 7
8 3	9 5	1 6	3 8

5. **"Getting There First":** This is a game involving two players. The first player picks any number from 1 to 10. The second player then picks any number from 1 to 10 and adds it to the first player's number. The players then alternate, adding any number from 1 to 10 to the last sum. The play continues until one player reaches 100. This player is the winner.

   Win Strategy: Whoever reaches 89 first can be the winner. Reason: If A and B are the players and A reaches 89, then if B adds, say 10—making a sum of 99—A can win by adding 1. If B adds, say, 8—making a sum of 97—A can win by adding 3. So to win, A only has to add the difference between 11 and the number B just added.

   By the same reasoning, the winning sequence of numbers is 89, 78, 67, 56, 45, 34, 23, 12, 1. Therefore, whoever starts with 1 can force a win.

   NOTE: This game can be played with the use of a calculator. Players take turns adding their numbers to the sum shown on the display.

HIGHLIGHTS OF ADDITION AND SUBTRACTION

Upon completion of this chapter, your child should have acquired the following capabilities:

- Easy recall of the basic addition and subtraction facts
- Ability to add and subtract two- and three-digit numbers
- Ability to translate problem situations into number sentences
- Ability to solve word problems involving addition and subtraction

Your child should also have acquired the following understandings:

- Addition is based on joining sets; subtraction, on removal of part of a set.
- On the number line, addition means counting to the right; subtraction means counting to the left.
- Addition and subtraction are inverse operations; they undo each other.
- Subtraction is defined in terms of addition: $5 - 2 = 3$ because $2 + 3 = 5$.
- The order in which you add numbers does not change the answer.
- Zero added to or subtracted from any number leaves the number unchanged.
- In a subtraction example, you are looking for the missing addend in the corresponding addition sentence. In $7 - 3$, you are looking for the missing addend in $3 + ? = 7$.
- Vocabulary:

add	equals	minus
addend	number sentence	subtract
plus	placeholder	difference
sum		

QUESTIONS ON THE CHAPTER

1. $3 + ? = 7$
2. $7 = 4 + ?$
3. $? + ? = 4$
4. $4 + 3 + 2 = ?$
5. $30 + 20 = ?$
6. $40 + 6 = ?$
7. $25 + 3 = ?$
8. $25 + 10 = ?$
9. $25 + 8 = ?$
10. $13 - 3 = ?$
11. $70 - 40 = ?$
12. $35 - 5 = ?$
13. $35 - 2 = ?$
14. $45 - 10 = ?$
15. $30 - 2 = ?$
16. $\begin{array}{r} 20 \\ +\underline{15} \end{array}$
17. $\begin{array}{r} 25 \\ +\underline{12} \end{array}$
18. $\begin{array}{r} 25 \\ +\underline{17} \end{array}$
19. $\begin{array}{r} 45 \\ -\underline{12} \end{array}$
20. $\begin{array}{r} 70 \\ -\underline{42} \end{array}$
21. $\begin{array}{r} 35 \\ -\underline{7} \end{array}$
22. $\begin{array}{r} 45 \\ -\underline{17} \end{array}$
23. $\begin{array}{r} 208 \\ -\underline{62} \end{array}$
24. Find the following answers on the number line:
    (a) $9 + 4 = ?$
    (b) $12 - 5 = ?$
    (c) $8 + ? = 13$
    (d) $? - 4 = 7$
    (e) $? + 6 = 10$
25. (a) $\begin{array}{r} 35 \\ +\underline{27} \end{array}$ (b) $\begin{array}{r} 72 \\ +\underline{95} \end{array}$ (c) $\begin{array}{r} 305 \\ +\underline{279} \end{array}$ (d) $\begin{array}{r} 296 \\ +\underline{184} \end{array}$
26. (a) $\begin{array}{r} 48 \\ -\underline{23} \end{array}$ (b) $\begin{array}{r} 175 \\ -\underline{34} \end{array}$ (c) $\begin{array}{r} 52 \\ -\underline{17} \end{array}$ (d) $\begin{array}{r} 342 \\ -\underline{115} \end{array}$
    (e) $\begin{array}{r} 205 \\ -\underline{12} \end{array}$ (f) $\begin{array}{r} 625 \\ -\underline{239} \end{array}$

27. Discover the rule by which the sets of numbers below are written. Then fill in the blanks.
    (a) 3, 5, 7, 9, 11, ___, ___, ___.
    (b) 1, 2, 4, 7, 11, 16, ___, ___, ___.

28. Janet is 27 years old and her mother is 51. How much younger is Janet than her mother?

29. The construction of the Brooklyn Bridge was begun in 1869 and was finished in 1883. How many years did it take to build the bridge?

30. $24 - 9 = ? - 10$

31. Are the following statements true or false?
    (a) The distance between the points 1 and 2 on the number line is the same as the distance between the points 3 and 4.
    (b) The number line has no end.
    (c) Zero added to 5 is zero.

32. What number makes each sentence true?
    (a) $7 + ? = 9$ (b) $30 = 25 + ?$ (c) $99 + ? = 140$

33. What number is 5 more than 12?

34. In the following problem, the square and the triangle stand for different numbers. Find the numbers if

$$\square + \triangle = 13 \quad \text{and} \quad \square - \triangle = 3.$$

35. A man was 69 years old in 1982. When was he born?

36. Find the sum of 38, 81, and 152.

37. Find the difference between 180 and 75.

38. There are 27 children in first grade and 39 in second grade. How many children are there in the two grades?

39. A basketball team scored 112 points against 93 points by the other team. By how many points did the first team win?

40. On a telephone dial (Fig. 11-7), RADIO is a 22-point word:

    R A D I O
    $7 + 2 + 3 + 4 + 6 = 22$

    How many points do you get for:
    (a) your first name?
    (b) your last name?
    (c) Superman?

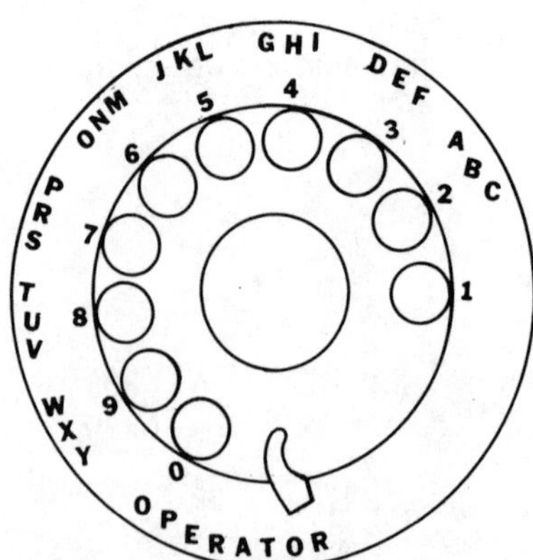

Figure 11-7

1. **Magic Squares: Can you arrange the numbers 1, 2, 3, 4, 5, 6, 7, 8, 9 in the squares in Fig. 11-8 so that the sum of the numbers in any one row, column, or diagonal is 15? All the numbers must be used.**

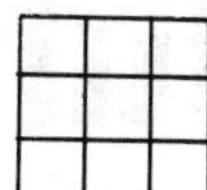

Figure 11-8

2. **Two U.S. coins total 55¢, but one of them is not a nickel. How is this possible?**

3. **A bottle and a cork cost $1.50 together. The bottle costs one dollar more than the cork. How much does each cost?**

4. **Two fathers and two sons have divided three apples among themselves, each receiving exactly one apple. How was this possible?**

5. **Arrange the numbers 3, 4, 5, 6, 7, 8, 9, 10, 11 in the squares in Fig. 11-9 so that the sum of the squares in any row, column, or diagonal is 21.**

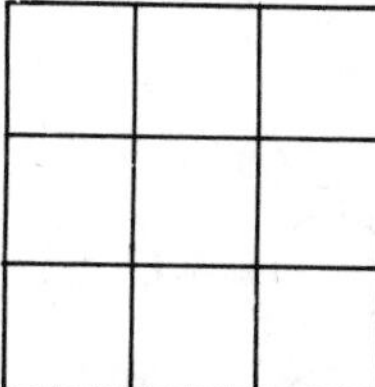

Figure 11-9

(The answers appear on page 258.)

Chapter 12

Multiplying Whole Numbers

INTRODUCTION

The advance made from addition to multiplication is like that made from automobiles to jets. Like the jet, an automobile can transport you, but more slowly. Addition can get you the same result as multiplication, but more slowly.

Addition is an efficient way to count; multiplication is an efficient way to add the same number many times.

A child's natural first encounter with multiplication is through an arrangement of cans on your kitchen table (Fig. 12-1):

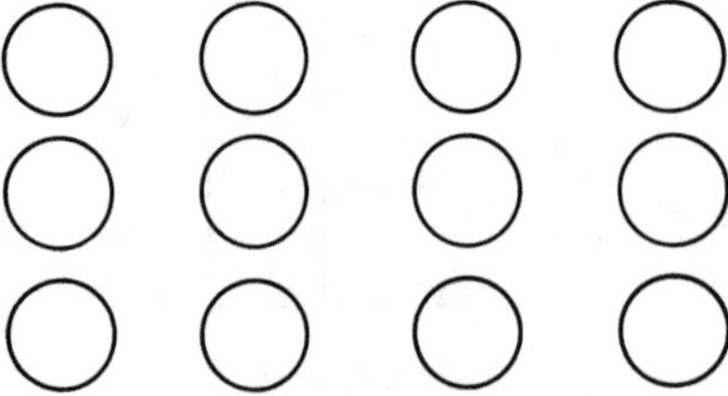

Figure 12-1

The child sees 3 rows with 4 cans in each row. Later, she sees it as "3 fours;" and then as "3 times 4."

One child, when first shown a row of cans, then two rows, and then three rows, exclaimed with disarming innocence, "Oh! Is this what multiplication is all about?"

The **materials** you will need for this chapter include:

Counters

Drinking straws

Toy money (pennies, dimes, and dollars)

A set of cards with dot arrays (see page 107)

A set of flash cards with the multiplication facts

A pegboard

Dice

Spinners

Two toy racing cars (see page 113)

THE CHAPTER IN A NUTSHELL

In this chapter you will see how to

- Get your child ready for multiplication
- Explain the meaning of multiplication:
  - as repeated addition
  - as the number of objects in an array
  - on the number line
- Explain the symbols and vocabulary used in multiplication
- Guide your child to construct her own multiplication table
- Teach multiplication with larger numbers
- Explain how to check multiplication
- Diagnose errors and show your child how to correct and avoid them
- Use a variety of activities and games to help strengthen your child's competence in multiplication, including the immediate recall of the basic multiplication facts
- Provide interesting word problems involving multiplication

You will also be given

- A list of understandings and skills your child should be left with at the completion of the chapter
- Questions to test and reinforce your child's comprehension and competence with multiplication; also, review questions on earlier material
- Insight into properties of multiplication; a "picture" of multiplication; and multiplication short-cuts under **For the Curious**
- Several puzzles under **For the Daring**

THE MEANING OF MULTIPLICATION

The meaning of multiplication is conveyed by

1. Looking upon it as a shorter and faster way to get the answer to a certain kind of addition problem—the problem of adding the *same* number many times. Instead of adding 13 + 13 + 13 + 13 + 13 + 13 + 13 and getting 91, we multiply 13 by 7 and use a procedure by which we quickly obtain the answer of 91.

2. Explaining multiplication in terms of the number of objects in a rectangular arrangement, called an **array.** Example: If, in a parking lot (Fig. 12-2), there are three rows, with five cars in each row, then multiplying 3 by 5 tells the number of cars in the parking lot.

Figure 12-2

3. Use of the number line. If you imagine the number line to be a subway track, then Fig. 12-3 shows a train making four stops, each stop covering three blocks. Here, 3 multiplied by 4 gives the distance traveled by the train in these four stops.

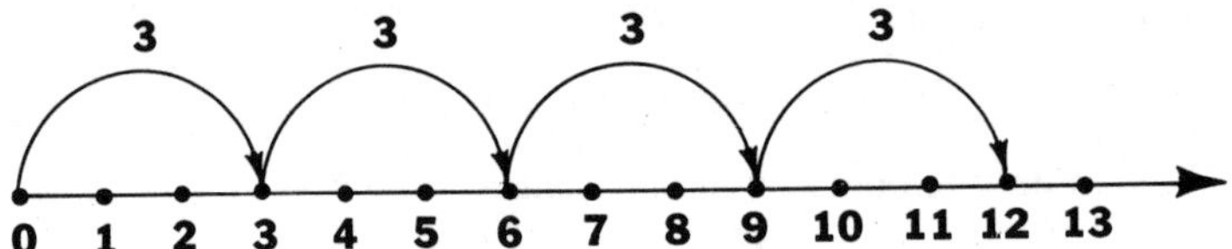

Figure 12-3

READINESS FOR MULTIPLICATION

Before a child can learn multiplication, she must

- Be able to add
- Have an understanding of place value (see Chapter 10)

You can further help your child become ready for multiplication with activities like the following:

1. Counting by twos, threes, fours, and fives (sometimes called "skip counting"). When a child counts by twos, calling out "2, 4, 6, 8, ...," she is actually reciting the two-table.

2. Making a 10 × 10 table (Fig. 12-4), and then asking the child to fill in every second number in blue, every third number in red, and so on. The child can then use the table as a check on her skip counting.

1	2	3	4	5	6	7	8	9	10
11									
21									
31									
41									
51									
61									
71									
81									
91									

Figure 12-4

3. Completing number patterns, such as
   (a) 1, 3, 5, ____, ____, ____.
   (b) 3, 6, 9, ____, ____, ____.
   (c) 10, 15, 20, ____, ____, ____.

4. Skip counting by twos, threes, etc., on the number line.

5. Providing your child with experience in
   (a) Marking off equal distances on the number line; for example: 2 threes; 4 twos; 3 fives
   (b) Adding equal addends; for example: 6 + 6 + 6 + 6; 8 + 8 + 8; 2 + 2 + 2 + 2 + 2 + 2.

TEACHING MULTIPLICATION

Getting Started

For an overview of what multiplication is all about, demonstrate to your child three views of the operation: as repeated addition; as the number of objects in an array; and as distance on the number line.

1. As repeated addition

Lay out on the table two containers, each with three disks (Fig. 12-5):

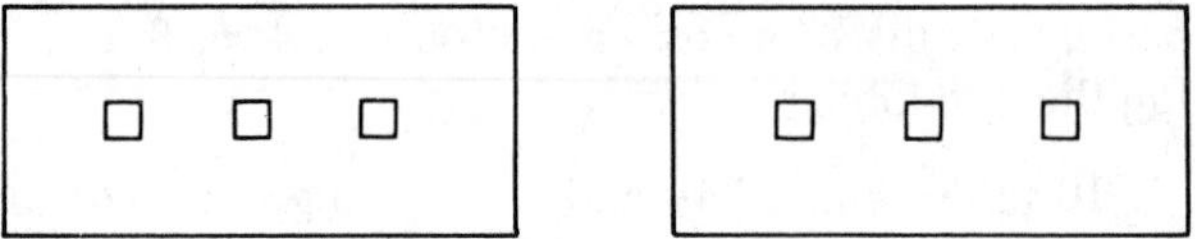

Figure 12-5

Ask the child how many disks there are altogether in the two containers. To find the answer, she can count the disks and get 6, or she can add 3 disks and 3 disks and get the same answer: $3 + 3 = 6$.

Explain that another way to write $3 + 3 = 6$ is to write $2 \times 3 = 6$, which is read "2 threes are 6," or "2 times 3 equals 6." The symbol × is called the **times,** or **multiplication,** sign.

We call $2 \times 3 = 6$ a **multiplication sentence.**

The numbers 2 and 3 that are multiplied are called **factors.**

The answer, 6, is called the **product.**

$$\underset{\substack{\uparrow \\ \text{factor}}}{2} \times \underset{\substack{\uparrow \\ \text{factor}}}{3} = \underset{\substack{\uparrow \\ \text{product}}}{6}$$

In the same way,

$2 + 2 + 2 + 2 + 2 = 10$ can be written as $5 \times 2 = 10$, and $3 + 3 + 3 + 3 + 3 + 3 + 3 = 21$ can be written as $7 \times 3 = 21$.

Show that a multiplication sentence may be written horizontally or vertically:

$$5 \times 2 = 10, \text{ or } \begin{array}{r} 5 \\ \times\ \underline{2} \\ 10 \end{array} \qquad 7 \times 3 = 21, \text{ or } \begin{array}{r} 7 \\ \times\ \underline{3} \\ 21 \end{array}$$

Repeat these explanations with various problem situations like:

(1) How many pennies are there in three nickels?

(2) How many wheels are there on five cars?

For reinforcement, ask the child to answer the following questions:

(1) Find the number of pennies in four nickels by addition. Then write the multiplication sentence for the problem.

(2) In the multiplication sentence $3 \times 5 = 15$, what do we call
(a) the 3? (b) the 15? (c) the 5?

Let your child conclude that multiplication is a shorter and faster way to get the answer to a *repeated addition* problem—a problem involving addition of the *same* number many times. Tell her that after she learns the multiplication procedure she will have a much shorter and faster way to find the answer to a problem like 72×45 than to add 45 seventy-two times!

2. As the number of objects in an array

You can also explain multiplication in terms of a rectangular arrangement of objects. For instance (Fig. 12-6), arrange the following array of counters on the table:

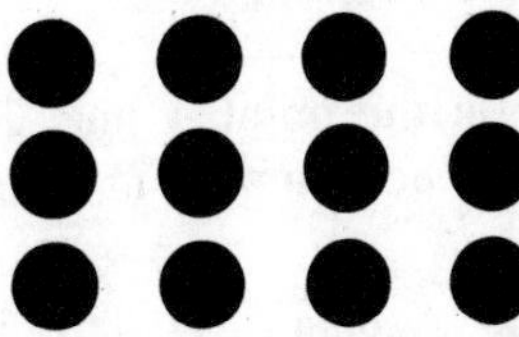

Figure 12-6

Explain that an arrangement of objects in rows (the rows are horizontal) and columns (vertical), with the same number of objects in each row, is called an *array*.

Help your child describe the array as "3 fours"; then as "3 times 4"; and then as "3 × 4" or "3 by 4."

Ask how the child can determine the number of counters in the array. She should understand that counting by ones to 12; counting by fours: 4, 8, 12; and counting by threes: 3, 6, 9, 12—yield the same result, 12.

Present various arrays and ask your child how many rows, columns, and objects there are in each array.

Explain that the counters in Fig. 12-6 are arranged in a "3 by 4" or "3 × 4" array. Since the number of objects in the array is 12, we write the multiplication sentence: 3 × 4 = 12. Explain further that

The first factor, 3, indicates the number of rows.

The second factor, 4, tells the number of counters in each row.

The product, 12, tells the number of counters in the entire array.

For reinforcement, ask the child to do the following exercises:

(1) Form an array with three rows, each row containing five counters.
   (a) How many counters are there in this array?
   (b) Write the multiplication sentence for this array.

(2) Form a 2 × 4 array; a 3 × 3 array; a 5 × 2 array.
   (a) Tell how many objects there are in each array.
   (b) Write the multiplication sentence for each array.

(3) Use arrays to complete the following multiplication sentences:
   (a) 2 × 3 = ?   (c) 5 × 5 = ?
   (b) 4 × 5 = ?   (d) 6 × 2 = ?

(4) Form a 3 × 6 array on a pegboard. Then write the multiplication sentence describing the array.

3. As the distance on the number line

Let the child imagine a grasshopper leaping along the number line. The number line in Fig. 12-7 illustrates the multiplication sentence 3 × 4 = 12: The grasshopper takes 3 hops, each covering 4 spaces, landing on point 12.

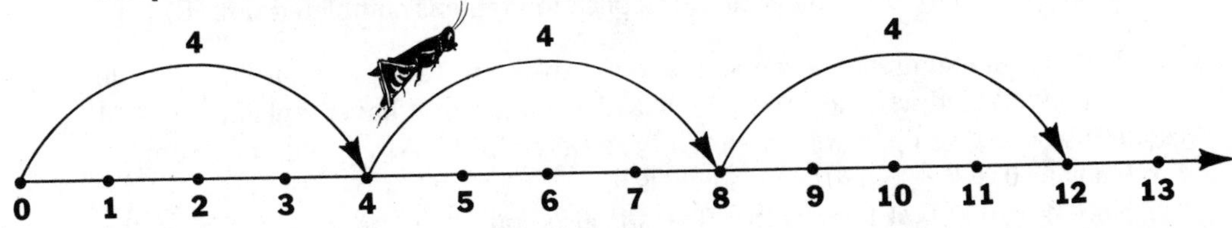

Figure 12-7

The factor 3 tells the number of hops; the factor 4 tells the number of spaces in each hop; the product 12 gives the number of spaces the grasshopper leaped altogether.

After several more illustrations, ask your child to

(1) Find the product 2×3 on the number line. Then write the number sentence and explain what each number in the sentence means on the number line.

(2) Write the multiplication sentence illustrated in Fig. 12-8.

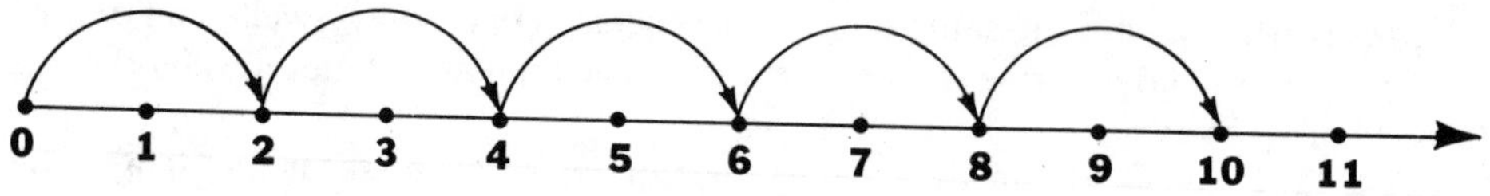

Figure 12-8

The Basic Multiplication Facts

A **basic multiplication fact** is the product of any pair of numbers in the set {0, 1, 2, 3, 4, 5, 6, 7, 8, 9}. The traditional way of learning these facts is by building a *multiplication table.*

The child herself can develop the table by using drinking straws or Popsicle sticks. After drawing (or being given) a table like the one shown in Fig. 12-9, the child can be guided to develop the multiplication table in the following way:

To construct, say, the two-table, tie drinking straws in sets of two, and ask the child how many straws there are in one set.

When she answers correctly, let her write the multiplication sentence $1 \times 2 = 2$, and then read it: "1 two is 2, or 1 times 2 is 2." Then have the child enter the product in the table.

×	0	1	2	3	4	5	6	7	8	9
0										
1										
2										
3										
4										
5										
6										
7										
8										
9										

Figure 12-9

Repeat this procedure with two sets of two straws, three sets of two straws, four sets of two straws, and continue through nine sets of two straws. When the activity is completed, all 10 multiplication sentences in the two-table will have been recorded, verbalized, and the products entered in the multiplication table (Fig. 12-10). Follow this procedure until the entire multiplication table is completed (Fig. 12-11).

NOTE: Something should be said about explaining the zero-table. If the child does not understand why, for instance, $4 \times 0 = 0$ after you have explained it as "4 sets of 0 straws give 0 straws," then fall back on addition. As an addition example, 4×0 means $0 + 0 + 0 + 0$, which equals 0.

Since $4 \times 0 = 0$, it follows that 0×4 is also equal to 0, because reversing the order of the factors does not change the answer.

×	0	1	2	3	4	5	6	7	8	9
0										
1										
2	0	2	4	6	8	10	12	14	16	18
3										
4										
5										
6										
7										
8										
9										

Figure 12-10

×	0	1	2	3	4	5	6	7	8	9
0	0	0	0	0	0	0	0	0	0	0
1	0	1	2	3	4	5	6	7	8	9
2	0	2	4	6	8	10	12	14	16	18
3	0	3	6	9	12	15	18	21	24	27
4	0	4	8	12	16	20	24	28	32	36
5	0	5	10	15	20	25	30	35	40	45
6	0	6	12	18	24	30	36	42	48	54
7	0	7	14	21	28	35	42	49	56	63
8	0	8	16	24	32	40	48	56	64	72
9	0	9	18	27	36	45	54	63	72	81

Figure 12-11 Multiplication Table

Another good way to develop the multiplication table with your child is to use the following nine arrays of dots on a set of nine cards:

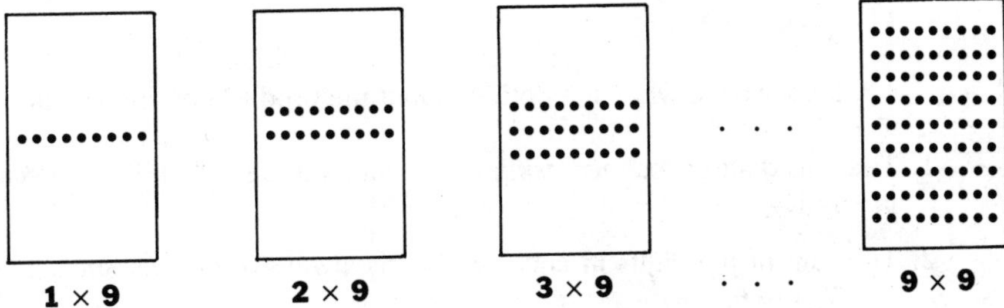

Figure 12-12

Each card can be used to develop a part of the multiplication table. For instance, to develop the three-table, uncover successive columns of the 3 × 9 array by using a blank cover card (Fig. 12-13). The child writes the multiplication sentence for each array as it is uncovered, and enters the product in the multiplication table.

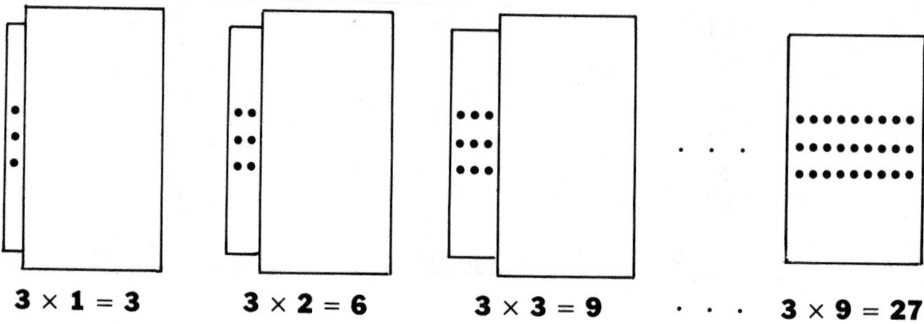

Figure 12-13

ABOUT THE MULTIPLICATION FACTS

From a close inspection of the multiplication table, your child can be guided to the following conclusions:

- There are 100 multiplication facts.
- You can count by ones, twos, threes, ... nines across rows and columns.
- When any number is multiplied by 0, the product is always 0. Unless the child understands the reasoning behind this idea, confusion can result, with the child stating that 4 × 0 = 4 (see Note on page 106).
- Whenever any number is multiplied by 1, the product is that number.
- The product of any two numbers is the same regardless of the order of the factors. For instance, 2 × 3 = 3 × 2 and 7 × 5 = 5 × 7. (*NOTE:* This fact is known as the "commutative" property of multiplication. Because of this property, every *row* of numbers has a corresponding *column* with the same numbers.)
- The numbers in the diagonal of the multiplication table, from the upper left corner to the lower right corner, are *square* numbers; that is, they are products of numbers multiplied by themselves. For instance, 1 (1 × 1); 4 (2 × 2); 9 (3 × 3).

Ask your child what she notices about the products in the 9 column.

1. The tens digit keeps *increasing* by 1, while the ones digit keeps *decreasing* by 1.
2. The sum of the digits in each product is always 9. For instance, 1 + 8; 2 + 7; 3 + 6.

   Can the child explain why this happens?

1. Because adding 9 is the same as *adding 10* and *subtracting 1*.
2. Starting with the first product 9, adding 1 and subtracting 1 with each succeeding product leaves the sum of the digits unchanged; that is, the sum remains 9.

What conclusion can the child draw from the table about:

The product of two *even* numbers?
(The product is always even.)

The product of two *odd* numbers?
(The product is always odd.)

The product of *an even number and an odd number*?
(The product is always even.)

When your child understands how each product in the table is obtained, she should commit the basic multiplication facts to memory.

Multiplication by Powers of 10

Before learning to multiply larger numbers, the child must understand one more idea: How to multiply a single digit by 10, 100, and 1000; for instance 2 × 10, 2 × 100, 2 × 1000.

An easy way to grasp this idea is to use the concept of place value:

2 × 10 means 2 tens, or 20.
2 × 100 means 2 hundreds, or 200.
2 × 1000 means 2 thousands, or 2000.
2 × 30 means 2 times 3 tens, or 6 tens, or 60.

Less To Memorize

If a child understands the ideas just gleaned from the multiplication table, she will be left with only 36 basic facts to memorize instead of 100, because

28 facts are eliminated by the commutative property. If a child knows that 2 × 3 = 6, then she also knows that 3 × 2 = 6; and if she knows that 7 × 5 = 35, then she also knows that 5 × 7 = 35.

(continued)

19 facts are eliminated if the child knows that whenever any number is multiplied by 0, the product is 0.
17 facts are eliminated if the child knows that whenever any number is multiplied by 1, the product is always that number.

This leaves 36 out of the 100 multiplication facts that need to be memorized.

Do 3 × 4 and 4 × 3 Mean the Same Thing?

Although 3 × 4 and 4 × 3 produce the same answer, they do not *mean* the same thing. Think of women and skirts:
3 × 4 would represent 3 women, each having 4 skirts.
4 × 3 would represent 4 women, each having 3 skirts.
In both cases there are 12 skirts, but each presents a different situation.

Proof Positive

By using an array of dots on a card, a child can easily be convinced that the order of the factors in a multiplication does not change the product (the Commutative Property). For instance, a 2 × 3 array, when rotated through a 90° angle, becomes a 3 × 2 array as shown in Fig. 12-14.

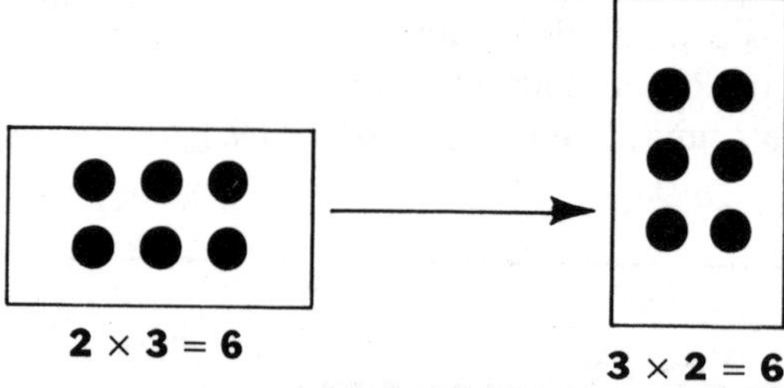

Figure 12-14

Since each array contains six dots, the child sees that 2 × 3 = 3 × 2. A 3 × 2 array is just another view of a 2 × 3 array.

Activities for Mastering the Multiplication Facts

1. Prepare a set of flash cards, each with a different multiplication fact. Write the product on the back of each card (Fig. 12-15). (Such cards are available commercially.) These cards can be used in two ways:

(a) As you flash each multiplication fact, the child reads it and then gives the product. For Fig. 12-15 she would read, "3 times 4 equals 12." Use the flash cards often enough for the responses to become automatic.

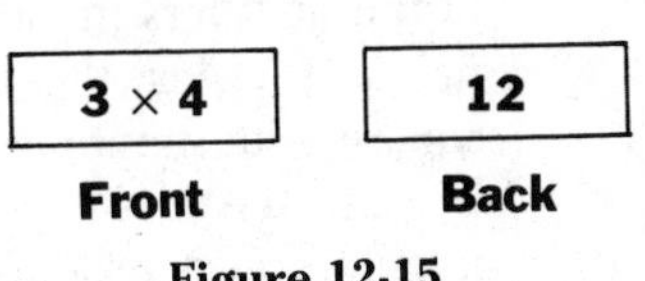

Figure 12-15

(b) Flash the *product* and ask the child to give you all the combinations of factors that give this product. For instance, flash a 12, and the child should be able to say: $3 \times 4 = 12$, $6 \times 2 = 12$, and $12 \times 1 = 12$.

2. Make a partially completed multiplication table and ask your child to complete it.

3. Make up examples like the following for the child to answer:
   (a) $5 \times ? = 15$ (b) $? \times 4 = 24$ (c) $6 \times 9 = ?$
   (d) $12 = ? \times 4$ (e) $18 = 6 \times ?$ (f) $? = 4 \times 8$
   (g) $3 \times 6 = ? \times 2$ (h) $2 \times 3 \times 5 = ?$
   NOTE: Let the child multiply the three numbers in (h) in different orders; that is, $2 \times 3 \times 5$, $2 \times 5 \times 3$, $3 \times 5 \times 2$, and see that each time she gets the same answer.
   (i) Name all the combinations of two factors whose product is 24.
   (j) Write a multiplication sentence that means the same as: $3 + 3 + 3 + 3 = 12$.
   (k) Write an addition sentence that means the same as $4 \times 5 = 20$.

4. Make up word problems, like those below, for the child to solve.
   (a) If you live four blocks from school, and Ron lives three times as far, how many blocks from school does Ron live?
   (b) Judy scored four times as many points as Irene in a basketball game. If Irene scored seven points, how many points did Judy score?
   (c) There are eight rows of cars in a parking lot, and seven cars in each row. How many cars are there in the parking lot?
   (d) Show three ways to find the number of dots in Fig. 12-16 without counting them. How many dots are there in the array? Write the multiplication sentence for this array.

   Figure 12-16

   (e) There are eight steps in the stairway leading from the living room to my bedroom. One day I walked up and down the steps three times. How many steps did I walk altogether?

5. Ask the child to form a 4×5 array on a pegboard, and then write the corresponding multiplication sentence. Repeat with other arrays.

6. Say: "I'm thinking of two numbers. One is a product; the other is a factor. The product is 36 and the factor is 9. What is the other factor?"

7. Say: "I'm thinking of two numbers whose product is 24 and whose difference is 10. What are the numbers?

8. Say: "I have a mystery number. If you multiply my mystery number by 5 and then add 10, you get 40. What's my mystery number?"

9. In the magic square shown in Fig. 12-17, the sum of the numbers in each row, column, and diagonal is 15. Ask the child to construct another square with numbers *two times* those in Fig. 12-17. Let the child determine whether the new numbers also produce a magic square; that is, whether the sum of the numbers in each row, column, and diagonal of the new square is again the same.

6	7	2
1	5	9
8	3	4

Figure 12-17

10. Ask your child to construct squares in which the numbers are three, four, five, ... nine times the numbers in the square in Fig. 12-17, and determine in each case whether it is a magic square.

11. Ask your child to draw a conclusion about magic squares. [If you multiply each number in a magic square by any number, you still get a magic square. The sum of each row, column, and diagonal in the new magic square will be 15 times the number by which you multiplied each number. For example, if you multiply the numbers in Fig. 12-17 by 2, the new sum will be 15 (the old sum) × 2, or 30.]

Games for Mastering the Multiplication Facts

Games are special kinds of activities involving rules, competition, and winners. They are an excellent and natural way to motivate an activity and sustain interest. They are especially useful in providing a fresh and stimulating setting for needed practice, such as that needed to learn the multiplication facts.

The games presented here can be played by just you and your child. If you can enlarge the circle of players, it will be simple enough to adapt the games to the larger number. The rules set down are only suggestions, and can be modified as the situation dictates. Have prizes ready for the winners.

1. The child rolls a pair of dice and calls out the product of the two numbers showing. Ten correct answers in a set amount of time wins.

2. Write on a sheet of paper a set of numbers from 0 to 9, in random order. Then call out a multiplier by which each number on the paper is to be multiplied.

   For instance, write the numbers 2, 5, 9, 8, 0, 7, 3, 6, 1, 4. When you call out "multiply by four," the child writes the product of 4 and each number on the sheet of paper as quickly as possible. If she gets eight or more correct answers in a set amount of time, she is a winner.

3. The child rotates a spinner (Fig. 12-18) and writes down the number showing. She spins it a second time and writes down the second number showing. She then finds the product of the two numbers. Every correct product earns a point, and every wrong answer loses a point. Ten points earned in a set amount of time wins.

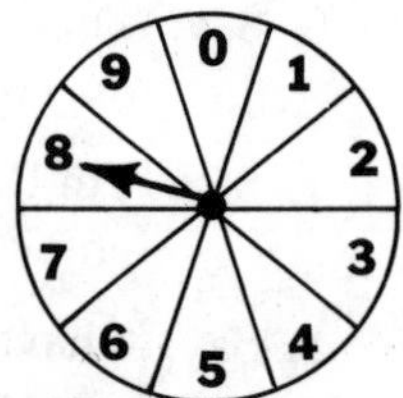

Figure 12-18

4. This is an auto racing game you can call "The (name of child) 200." Materials needed are (a) a spinner, (b) a track to accommodate as many players as desired, marked off with numbers 1–200 in equal spaces (Fig. 12-19), (c) a toy racing car for each player.

Player A	**1**	**2**	**3**	**4**	**. . .**	**200**
Player B	**1**	**2**	**3**	**4**	**. . .**	**200**

Figure 12-19

Game Action: A player obtains two numbers by spinning twice (as in #3 above) and calling out their product. A correct product advances the player's car that many spaces on her track. An incorrect product moves her car two spaces *back*. The play alternates between the players, the winner being the one whose car reaches the end of the track first.

NOTE: (1) Instead of spinning twice with one spinner, you can use two spinners and obtain one number on each.

(2) This game can also be played with *addition*. Instead of finding the product of the two numbers, the child finds their *sum*. (You may then wish to reduce the goal from 200 to 100.)

Multiplication with Larger Numbers

Before learning to multiply numbers with more than a single digit, the child must

- Know the basic multiplication facts
- Know how to multiply a single digit by 10; 100; 1000
- Understand the function of zero as a placeholder
- Understand the concept of place value

Example 1: If a candy bar costs 23¢, how much will three candy bars cost?

Ask the child to set up the problem with dimes and pennies, reasoning that if one bar requires three dimes and two pennies, then three bars will require three times that many dimes and pennies:

1 candy bar	**3 candy bars**	
(10)(10) (1) (1) (1)	(10)(10)	(1) (1) (1)
	(10)(10)	(1) (1) (1)
	(10) (10)	(1) (1) (1)
	6 dimes	**9 pennies** = **69¢**

After working through a number of such problems with money, the child is ready to think of dimes as *tens* and pennies as *ones*. She can then represent 23¢ × 3, the cost of three candy bars, as

(a)	2 tens + 3 ones × 3 6 tens + 9 ones = 69	=	(b)	20 + 3 × 3 60 + 9 = 69

The 9 and 60 in (b) are called **partial products**, which are added to get the final answer to the multiplication.

Another way to write (b) is

(c)
```
   23
 ×  3
    9  (3 × 3)
   60  (3 × 20)
   69
```
=

(d)
```
   23
 ×  3
   69
```

(d) is a shorter and more compact way to write (c), and is the standard way we perform such multiplications. It is the *multiplication algorithm* for finding the product of a two-digit number and a one-digit number, where no "carrying" is involved.

MULTIPLICATION VOCABULARY

Besides **factor** and **product**, there are other words used in multiplication you should know. (But don't push your child too fast to learn them.)

The **multiplicand** is the number you multiply.

The **multiplier** is the number by which you multiply.

A **partial product** is the product obtained by multiplying the multiplicand by any digit in the multiplier.

The **product** is the answer in multiplication.

A **factor** is any one of the numbers being multiplied. Both the multiplicand and the multiplier are factors of the product.

23	⟵	multiplicand
× 3	⟵	multiplier
9	⟵	partial product
60	⟵	partial product
69	⟵	product

Example 2: For a product like 24 × 3, requiring "carrying" (or "regrouping"), the child starts with coins again.

Let a bar of candy now cost 24¢. The child is asked to find the cost of three bars. She proceeds as follows:

1 candy bar	**3 candy bars**	
(10)(10) (1) (1) (1) (1)	(10)(10)	(1) (1) (1) (1)
	(10)(10)	(1) (1) (1) (1)
	(10)(10)	(1) (1) (1) (1)

6 dimes   **12 pennies**
= **6 dimes** + **1 dime** + **2 pennies**
= **7 dimes** + **2 pennies**
= **72¢**

To find 24 × 3 without reference to money, we have

	(a)		(b)		(c)
24	= 2 tens + 4 ones		24		[1]24
× 3	× 3	=	× 3	=	× 3
	6 tens + 12 ones		12 (3 × 4)		72
	= 6 tens + 1 ten + 2 ones		60 (3 × 20)		
	= 7 tens + 2 ones		72		
	= 72				

Notice that (c) is a shorter way of writing (b). Instead of writing the partial products 12 and 60 separately as in (b), they are combined into a single product in (c). The 72 is the sum of 12 and 60.

With more experience, the child can "carry" the 1 in (c) in her head, and just write 24 without the superscript.

$$\begin{array}{r} 24 \\ \times\ 3 \\ \hline 72 \end{array}$$

Example 3: 12 × 17. Here, we first multiply the 12 by the 7 ones, then by the 1 ten, and then add all the partial products:

	(a)	(b)	(c)
	12	12	12
	×17	×17	×17
(7 × 2) ⟶	14 }	→ 84	84
(7 × 10) ⟶	70 }	→ 120	12
(10 × 2) ⟶	20 }	204	204
(10 × 10) ⟶	100 }		
	204		

Make clear to the child that the 84 in (b) is the sum of 14 and 70 in (a), and that the 120 in (b) is the sum of 20 and 100 in (a). The procedure in (c) is the same as in (b), except that the 0 was not entered for the 120. The 0 is not needed since the 2 is already entered in the tens place and the 1 in the hundreds place.

The procedure in (c) is the multiplication algorithm, the standard way we multiply a two-digit number by a two-digit number.

A Picture of Multiplication

An interesting way to picture the "guts" in the multiplication 12 × 17 as developed in Example 3 is by means of a 12 × 17 array (Fig. 12-20).

(continued)

Since 17 can be written as 10 + 7, and
12 can be written as 10 + 2,

we subdivide the array in Fig. 12-20 into the four pieces A, B, C, D shown in Fig. 12-21, such that:

A is a 7 × 2 array whose product is 14,
B is a 7 × 10 array whose product is 70,
C is a 10 × 2 array whose product is 20,
D is a 10 × 10 array whose product is 100

Fig. 12-21 pictures the development in Example 3:

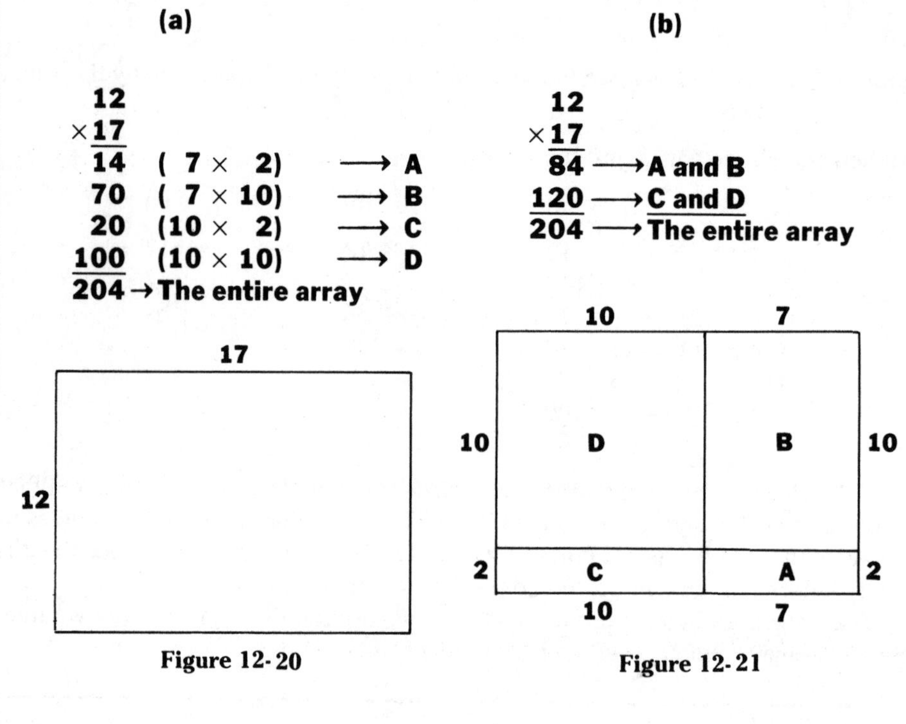

Figure 12-20

Figure 12-21

Checking Multiplication

A child can check a multiplication answer in several ways:

By *estimating* the answer and then seeing whether the result of his computation is reasonable. The idea in estimation is to round off the factors to convenient multiples of 10, 100, etc., so that computation with the resulting numbers can be done mentally. For instance,

Computation	*Estimate*
42	40
× 29	× 30
378	1200
84	
1218	

By *redoing* the example.

By *reversing the order of the factors* and then multiplying. For example,

(a)	(b)
14	12
× 12	× 14
28	48
14	12
168	168

Through reversing the factors, different multiplication combinations are introduced, increasing the chances of a more accurate check.

Reminder: After your child understands the rationale behind the multiplication algorithm, it's well to remind her, from time to time, of the steps in the procedure:

- In writing down the example, place ones under ones, tens under tens, and hundreds under hundreds.
- If the multiplier is a one-digit number, multiply each figure in the multiplicand by this digit, starting from the right.
- If the multiplier has more than one digit, find the partial products. Then add the partial products.
- Check your answer by (1) estimating the answer, (2) redoing the example, or (3) interchanging the multiplier and multiplicand, and multiplying again.

Common Multiplication Errors

Children often make certain types of errors in their multiplication. Some are systematic—they show a pattern of incorrect responses; others are random—with no discernible pattern. Many are careless errors: the child knows how to perform the computation correctly, yet makes mistakes because of distractions, boredom, or a lapse of attention. Studies have shown the most common multiplication errors to involve zeros, carrying, and the multiplication facts of tables over five.

A pattern of errors with the multiplication facts points to the need for more practice through games and other activities suggested earlier. The child's response to any combination in the table should be automatic.

A pattern revealing lack of understanding of "place value" requires review of the concept through use of physical materials as described earlier. Frequent mistakes in addition will be reduced by a review of the operation and extended practice through games and other activities previously suggested.

Forgetting the multiplication process might be dealt with by the "Reminder" on page 117. Lots of practice with multiplication, including word problems, will help fix the process in mind. Errors in "carrying" can be explained through the use of toy money.

Following are major causes of multiplication errors.

- Not knowing the basic multiplication facts:

$$(1)\ \begin{array}{r} 16 \\ \times\ \underline{3} \\ 46 \end{array} \qquad (2)\ \begin{array}{r} 24 \\ \times\ \underline{7} \\ 172 \end{array}$$

- Not understanding place value:

$$(3)\ \begin{array}{r} 13 \\ \times\ \underline{7} \\ 721 \end{array} \qquad (4)\ \begin{array}{r} 25 \\ \times\underline{38} \\ 200 \\ \underline{75} \\ 275 \end{array}$$

In example (4) the child forgot that the 75 is really 750; the 5 should appear in the tens column and the 7 in the hundreds column.

- Errors in addition:

$$(5)\ \begin{array}{r} 37 \\ \times\ \underline{46} \\ 222 \\ \underline{108} \\ 1262 \end{array}$$

- Forgetting the multiplication process:

$$(6)\ \begin{array}{r} 21 \\ \times\underline{43} \\ 83 \end{array}$$

The child multiplied the *ones by the ones*, and the *tens by the tens* (3×1 and 4×2), and forgot to multiply the *ones by the tens* and the *tens by the ones* (3×20 and 40×1).

- "Carrying" mistakes:

$$(7)\ \begin{array}{r} {}^{2}64 \\ \times\ \underline{3} \\ 201 \end{array}$$

When the child multiplied 4 by 3 and got 12, she wrote down the 1 and carried the 2.

$$(8)\ \begin{array}{r} {}^{1}74 \\ \times\ \underline{3} \\ 242 \end{array}$$

Instead of adding the 1 (that's carried) to the product of 3×7, she added the 1 to the 7 and then wrote the product of 3×8.

- Mistakes involving zero:

$$(9)\ \begin{array}{r} 20 \\ \times\ \underline{6} \\ 12 \end{array}$$

Neglected to multiply the 0 by 6.

$$(10)\ \begin{array}{r} 30 \\ \times\ \underline{70} \\ 210 \end{array}$$

Failed to use both zeros.

Multiplication Shortcuts

Multiplying by 11

Let's see what happens when we multiply 43 by 11. Note that the answer is obtained by adding the two digits in the multiplicand (4 + 3 = 7) and then placing their sum, 7, between the two digits, 4 and 3.

$$\begin{array}{r} 43 \\ \times\ \underline{11} \\ 43 \\ \underline{43} \\ 473 \end{array}$$

Therefore, to multiply a two-digit number by 11, *first add the two digits and then place their sum in the middle*. For example,

$43 \times 11 = 473$ $\qquad$ $23 \times 11 = 253$
$72 \times 11 = 792$ $\qquad$ $54 \times 11 = 594$

Now let's see what happens if the sum of the digits is 10 or more, as in 85×11. We again add the two digits in the multiplicand: $8 + 5 = 13$. We write the 3, and then add the 1 we carried to the 8, getting 9.

$$\begin{array}{r} 85 \\ \times\ \underline{11} \\ 85 \\ \underline{85} \\ 935 \end{array}$$

Therefore, if the sum of the digits is 10 or more, *we add the digits and then add the 1 we are carrying to the first digit.* For example,

$85 \times 11 = 935$ $\qquad$ $76 \times 11 = 836$
$59 \times 11 = 649$ $\qquad$ $98 \times 11 = 1078$

Multiplying by 9 or by 99

To multiply any number by 9, first multiply it by 10, then subtract the given number. For instance,

$$9 \times 425 = \begin{array}{r} 4250 \\ -\ \underline{425} \\ 3825 \end{array} \leftarrow (10 \times 425)$$

Similarly, to multiply any number by 99, first multiply it by 100, then subtract the given number. For example,

$$99 \times 687 = \begin{array}{r} 68700 \\ -\ \underline{687} \\ 68013 \end{array} \leftarrow (100 \times 687)$$

HIGHLIGHTS OF MULTIPLICATION

The study of multiplication should leave your child with the following understandings:

- Multiplication can be thought of as:
  1. a shorter and quicker way to add the same number many times, or
  2. finding the number of objects in an array, or
  3. moving from left to right on the number line in equal groups of spaces.
- The order in which two numbers are multiplied does not affect the result.
- Any number multiplied by 1 equals that same number.
- Any number multiplied by 0 equals 0.
- Knowing the properties of multiplication helps memorization of the basic facts.

Also, your child should be able to

- Recall with ease the basic multiplication facts
- Multiply two- and three-digit numbers by two-digit numbers
- Solve word problems involving multiplication
- Explain and use the following vocabulary:

multiply	product	array
times	multiplier	
factor	multiplicand	

QUESTIONS ON THE CHAPTER

Multiply and check your answers in Questions 1 through 17.

1. 32 × 3
2. 23 × 4
3. 35 × 6
4. 146 × 5
5. 134 × 6
6. 105 × 6
7. 403 × 5
8. 305 × 4
9. 36 × 21
10. 42 × 27
11. 135 × 40
12. 216 × 60
13. 162 × 15
14. 143 × 51
15. 34 × 45
16. 23 × 187
17. 12 × 27 × 34

18. Find the product of 65 and 23 and check your answer.

19. (a) $2 \times 3 = ? \times 6$ (b) $4 \times 6 = 8 \times ?$

20. What two numbers have a sum of 14 and a product of 45?

21. How many different arrays (or rectangles) can you make with
    (a) 12 objects   (b) 24 objects   (c) 5 objects?

22. An auditorium has 37 rows, each with 26 seats. How many seats are there in the auditorium?

23. Danny earns $14 a week delivering newspapers. How much will he earn after 25 weeks?

24. If a car can go 29 miles on a gallon of gas, how many miles can it go on 18 gallons?

25. If a plane flies 325 miles an hour, how far will it fly in 16 hours?

26. In the George School there are 4 classes with 27 children in each class, and 9 classes with 23 children in each class. How many children are there in the school?

27. Are the following statements true or false?
    (a) The product of two odd numbers is always an odd number.
    (b) The product of an odd and an even number is always an even number.
    (c) The product of two even numbers is always an even number.

28. Round to the nearest *ten:*   (a) 23   (b) 87   (c) 36

REVIEW

1. Find the sum of 24, 105, and 36.
2. Find the difference between 205 and 17.
3. What number is 15 more than 37?

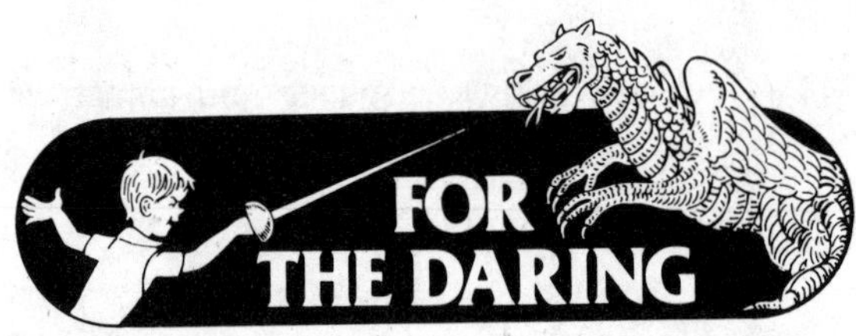

1. **The number of bacteria in a bottle doubles every minute. If the bottle is filled with bacteria in one hour, how long will it take for the bottle to be half full?**
2. **You meet two people. One always lies; the other always tells the truth.**

   "Are you the truth teller?" you ask the first person.

   "Looge," he replies.

   "He says 'yes,'" says the other. "But him big liar."

   Which of the two is telling the truth?
3. **How can you measure out exactly 2 pounds of flour if you have only an 8-pound can and a 5-pound can?**

(The answers appear on page 258.)

Chapter 13

Dividing Whole Numbers

INTRODUCTION

Think back to when you were a child trying to learn long division. Unless you were one of the lucky few, you were given a mechanical procedure to be memorized for grinding out answers, with little attempt to explain why the procedure worked. For a division like 13)351, you might have been given the following instructions:

27	1. Estimate how many times 13 goes into 35.
13)351	2. Multiply 13 by 2.
26	3. Subtract 26 from 35.
91	4. Estimate how many times 13 goes into 91, then multiply 13 by 7.
91	
0	

Does all this evoke joyless memories?

Why estimate 13 into 35? Why multiply? Why subtract? Why not add? Unless you were one of a very few, you never really understood what question a division like 13)351 asks. And you never understood why going through the routine of estimate, multiply, subtract, . . . estimate, multiply, subtract, . . . produced the correct answer. These questions will be answered in this chapter.

Of the four operations in arithmetic, division is the hardest for children to learn because it is the most complex. It requires extensive use of multiplication and subtraction, estimating, dealing with remainders, and working from left to right in contrast with the right-to-left direction used in the other operations.

Much of the difficulty can be overcome if you introduce division through real-life situations and activities with concrete or pictorial materials. This foundation will provide the child with the rationale for the more abstract procedures in division to be learned later.

Materials used in this chapter include:

Counters (Popsicle sticks, disks)

A set of five cards, each containing two dots (page 128)

Pennies

Division flash cards (page 134)

A metric ruler

Bird arrays

Two sets of cards: one set marked 1 through 10, the other marked 1 through 20 (page 135)

Two spinners

A set of 21 cards marked 0 through 20 (page 135)

THE CHAPTER IN A NUTSHELL

In this chapter you will see how to

- Get your child ready for division
- Introduce division meaningfully through physical and pictorial materials
- Teach division with larger numbers
- Justify the steps in the division procedure
- Provide motivating activities to reinforce your child's competency with division
- Explain how to check division
- Diagnose common errors in division
- Provide interesting word problems involving division

You will also be given

- A list of understandings and skills your child should be left with at the completion of the chapter
- Questions to test and reinforce your child's comprehension and competence with division; also, review questions on earlier material
- Easy ways to determine whether a number is divisible by, say, 6, 8, or 9—under **For the Curious**
- A variety of puzzles under **For the Daring**

THE MEANING OF DIVISION

The meaning of division can be explained in several ways: as the inverse of multiplication; as repeated subtraction; on the number line; and in terms of an array.

Division as the Inverse of Multiplication

The division $6 \div 2 = ?$ asks: By what number do I multiply 2 to get 6? Answer: Since $2 \times 3 = 6$, the number we are looking for is 3. So we say $6 \div 2 = 3$ because $2 \times 3 = 6$. Similarly, $6 \div 3 = 2$ because $3 \times 2 = 6$.

The three statements $2 \times 3 = 6$, $6 \div 2 = 3$, and $6 \div 3 = 2$ are three different ways of saying the same thing. In the same way, *every* multiplication fact is associated with two corresponding division facts. Thus, $5 \times 6 = 30$ is associated with $30 \div 6 = 5$ and $30 \div 5 = 6$; $3 \times 8 = 24$ is associated with $24 \div 3 = 8$ and $24 \div 8 = 3$.

Division and multiplication are *inverse* operations because one operation undoes the other. For instance, if you multiply a number by 5, then divide by 5, you're back to the original number. Likewise, if you first divide a number by 3, then multiply by 3, you're back to the original number.

Division is the inverse of multiplication in still another sense. The multiplication sentence $3 \times 4 = ?$ asks: Given two factors (3 and 4), find their product. The division sentence $12 \div 4 = ?$ asks: Given the product (12) and one of the factors (4), find the other factor.

In the division $12 \div 4 = 3$, 12 is called the *dividend*; 4, the *divisor*; and 3, the *quotient*.

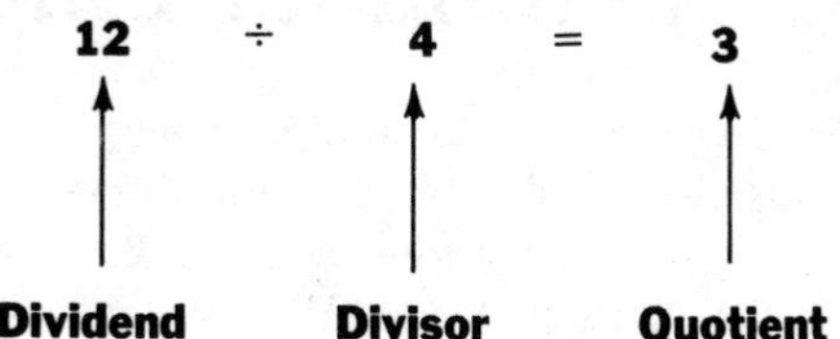

Division as Repeated Subtraction

We can also think of division as repeated subtraction. This approach is so easy for children to understand that we'll use it to explain the entire process of division.

In this view, $12 \div 4 = ?$ asks: How many fours are contained in 12? Worded differently, how many fours can be subtracted from 12?

To explain the problem with concrete objects, give the child 12 counters, let him remove 4 at a time, and then count the number of fours he was able to remove (Fig. 13-1):

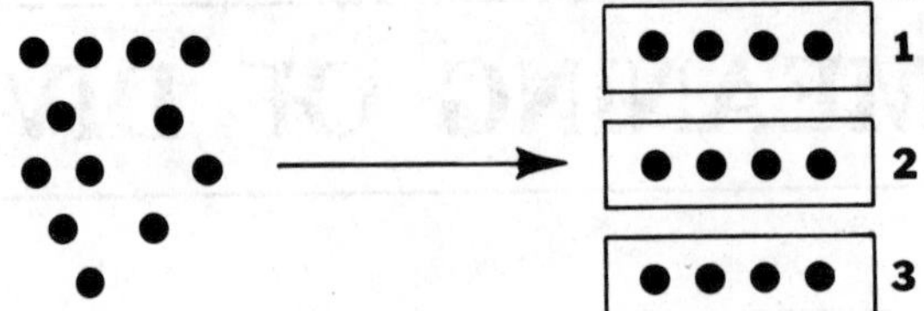

Figure 13-1

The child can prove the answer by recombining the three groups and seeing that he is back to the 12 counters he started with.

In written form, what the child did is described this way:

<table>
<tr>
<td>

$$\begin{array}{r} 4\overline{)12} \\ -\ \underline{4*} \\ 8 \\ -\ \underline{4*} \\ 4 \\ -\ \underline{4*} \\ 0 \end{array}$$

</td>
<td>

By counting, the child notes that 4 has been subtracted 3 times; that is, there are exactly 3 fours in 12. We express this conclusion with the *division sentence*

$$12 \div 4 = 3.$$

To prove the answer, we multiply it by 4 and see whether the result is 12. Since $3 \times 4 = 12$, the answer is correct.

</td>
</tr>
</table>

In this example, the remainder is 0; but not all divisions have a remainder of 0. In the division $7 \div 2 = ?$, we find that after subtracting 3 twos, a *remainder* of 1 is left:

<table>
<tr>
<td>

$$\begin{array}{r} 2\overline{)7} \\ -\underline{2*} \\ 5 \\ -\underline{2*} \\ 3 \\ -\underline{2*} \\ 1 \end{array}$$

</td>
<td>

To check the answer, we multiply 2 by 3 and add the remainder. We look to see whether we get back the 7:

$$(2 \times 3) + 1 = 6 + 1 = 7$$

</td>
</tr>
</table>

Division on the Number Line

On the number line, division means counting spaces to the left in equal groups. For example, $6 \div 2$ means: Start at 6. Then count to the left in groups of 2 until you reach 0 (Fig. 13-2). Note that there are three groups of 2: That is, $6 \div 2 = 3$.

In Fig. 13-3 we see the division $9 \div 4$. Here we get two groups of 4, with a remainder of 1.

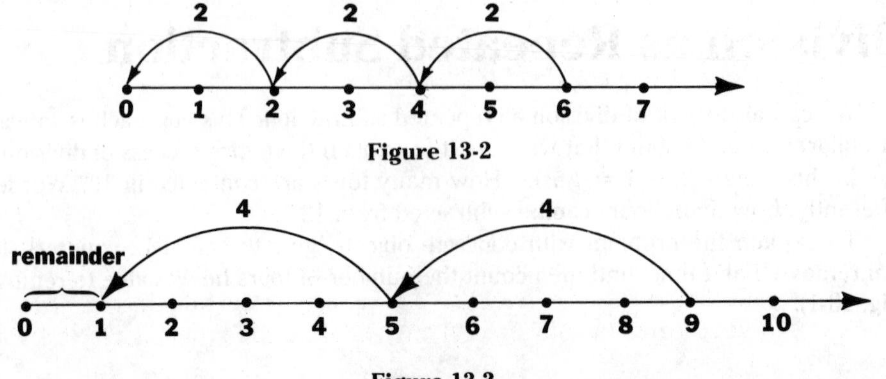

Figure 13-2

Figure 13-3

Division in Terms of an Array

Division helps us determine the number of rows or columns in an array. For instance, if you know that the array in Fig. 13-4 contains 12 dots and has 3 rows, how can you determine the number of columns in the array?

We ask and answer this question with the division sentence

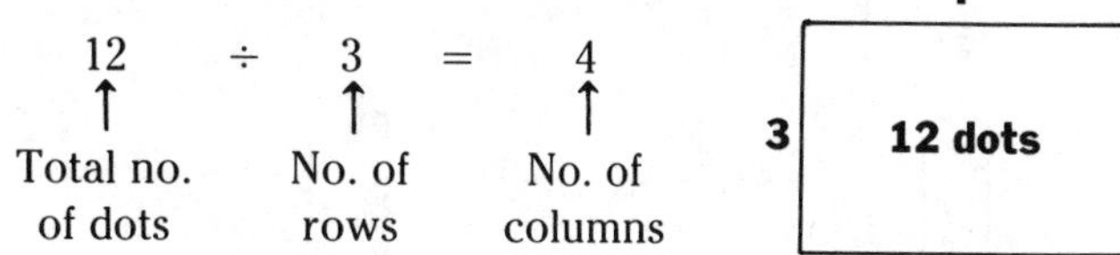

Figure 13-4

If you know that the array in Fig. 13-5 contains 12 dots and has 4 columns, how can you determine the number of rows? We ask and answer this question with the division sentence

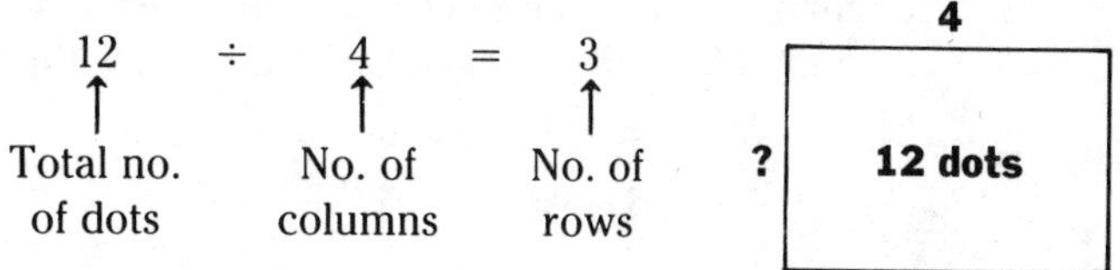

Figure 13-5

READINESS FOR DIVISION

Before your child can learn division, he must

- Know how to multiply and subtract
- Know the multiplication facts well enough to find the answers to sentences of the type: $? \times 3 = 21$ or $5 \times ? = 30$
- Know how to multiply by powers of 10. The child should know that annexing a zero to a number multiplies that number by 10; annexing 2 zeros multiplies the number by 100; annexing 3 zeros multiplies the number by 1000

You can further prepare your child for division by helping him develop the concept of repeated subtraction with activities like the following:

1. Have the child separate a set of Popsicle sticks into groups of two and put a rubber band around each group. Place all the sticks in an open box, and ask the child to remove the groups one at a time. As each group is removed, let the child count backward to indicate the number of Popsicle sticks left in the box. Then ask him to tell you the number of groups he removed from the box. Repeat with different size groups and with different objects.

2. Display a set of five cards, each containing two dots (Fig. 13-6). Remove the cards, one by one, and ask the child at each step to:
   (a) write down the subtraction involved, and
   (b) tell how many dots are left

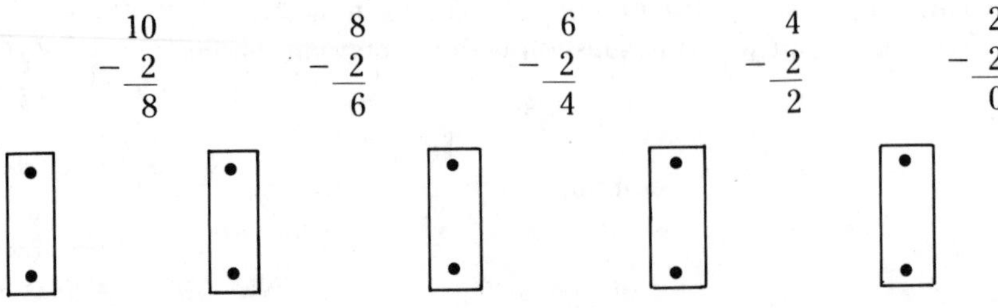

Figure 13-6

3. Place 12 pennies on the table and ask the child to determine the number of children who can receive 3 pennies each. As he sets aside a group of 3 pennies for a different child, let him show the subtraction at each stage:

$$\begin{array}{r}12\\-\ 3\\\hline 9\end{array}\qquad\begin{array}{r}9\\-\ 3\\\hline 6\end{array}\qquad\begin{array}{r}6\\-\ 3\\\hline 3\end{array}\qquad\begin{array}{r}3\\-\ 3\\\hline 0\end{array}$$

   Let your child see that the number of children receiving pennies is the same as the number of subtractions performed.

4. Provide practice finding the missing factor in problems like these:
   (a) $5 \times 3 = ?$ (b) $5 \times ? = 15$ (c) $? \times 3 = 15$

TEACHING DIVISION

Getting Started

1. Use concrete materials to develop the basic division facts. Place 12 disks on the table and ask the child to separate them into sets of 3 (Fig. 13-7). Then ask how many sets of 3 there are.

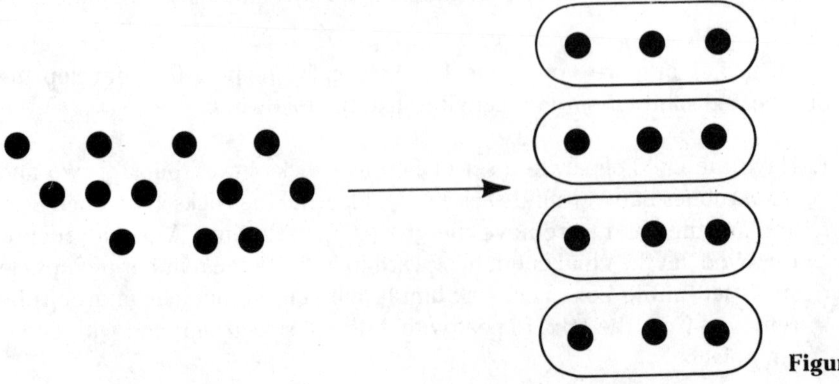

Figure 13-7

After doing several such problems, write down the division sentence $12 \div 3 = 4$ and explain that this sentence tells that we started with 12 disks, which were then separated into groups of 3, and found that there are 4 such groups in the set of 12.

Explain that the symbol $\div$ is read "divided by." Show that the division $12 \div 3 = 4$ is also written $3\overline{)12}$ with quotient 4.

$$\begin{array}{r} 4 \\ 3\overline{)12} \end{array}$$

Later the child will learn that another way to write this division is $\frac{12}{3} = 4$.

Repeat this exercise with many different sets of objects, asking the child to write the division sentence for each problem and then to interpret it.

2. Explain that the division sentence $6 \div 2 = ?$ asks you to find the number that, multiplied by 2, gives 6. Since $2 \times 3 = 6$, the number we are looking for is 3.

   Now ask for the answer to the division $6 \div 3 = ?$ Let the child use the same reasoning to conclude that $6 \div 3 = 2$ because $3 \times 2 = 6$.

   Provide many examples to relate division facts to their corresponding multiplication facts. For instance, $8 \div 2 = 4$ because $2 \times 4 = 8$,
   $5 \div 1 = 5$ because $1 \times 5 = 5$.

3. *Division by zero is not possible.* This important idea needs to be explained very carefully. Consider the division $3 \div 0 = ?$ This question asks us to find a number which when multiplied by 0 gives 3. Since *no* number multiplied by 0 equals 3, the division is not possible. For the same reason, no other number can be divided by 0.

   Now consider the division $0 \div 2 = ?$. Here we are looking for a number such that $2 \times$ the number $= 0$. The number must be 0 since only $2 \times 0 = 0$. We can show by the same reasoning that 0 divided by *any* number equals 0.

   What about $0 \div 0$? This division, too, is not possible (or is "undefined") for the following reason: $0 \div 0$ asks us to find a number which when multiplied by 0 equals 0. But *any* number times 0 equals 0. Since the result of a division (as with the other operations) must be *one* and only one number, having too many numbers makes the operation unacceptable.

   These ideas are subtle and should be explained very carefully to your child only when he is ready for them.

4. After the child feels comfortable with a division sentence and its interpretation, introduce the division vocabulary:

   The **dividend** is the number you divide.
   The **divisor** is the number by which you divide.
   The **quotient** is the answer to a division.

$$\underset{\text{dividend}}{\underset{\uparrow}{15}} \div \underset{\text{divisor}}{\underset{\uparrow}{3}} = \underset{\text{quotient}}{\underset{\uparrow}{5}}$$

NOTE: It's possible to use the multiplication table (page 107) for the division facts if the table is used without zero as a factor. Since division by 0 is not allowed, only 90 division facts are possible.

5. Below are sample problems involving the basic division facts. In each case have your child:
   (a) work through the problem with counters or other physical materials
   (b) write the division sentence and then prove the answer by multiplication
      (1) If I have 12 balloons and separate them into bunches of 4 balloons, how many bunches do I have?
      (2) Michael's mother bought 21 candy bars for a party. If she gave each child at the party 3 bars, how many children were at the party?
      (3) Linda's class of 24 children is going on a trip. They will be driven 4 in a car. How many cars will be needed?
      (4) If you want to pack 48 oranges in bags that hold 6 oranges, how many bags will you need?
6. Explain division as repeated subtraction (as on page 125).
7. Demonstrate division on the number line (as on page 126).
8. Explain division in terms of an array (as on page 127).

Dividing Larger Numbers

We shall now use the repeated subtraction method to explain division with larger numbers by the standard division algorithm.

1. Consider the division $48 \div 3$. Think of this problem as you having 48 cupcakes and wishing to package them in boxes of 3. How many boxes will you need?

   One way to find out is to subtract threes repeatedly from 48, one at a time, until you have fewer than 3 cupcakes left. But this would be quite tedious and time-consuming. A better and more efficient way is to subtract *multiples* of 3 instead of single threes.

   But *what* multiple of 3 shall we subtract? Let's guess. Try 10 threes and check how good an estimate this is: Since $10 \times 3 = 30$ cupcakes—and we have 48 cupcakes to package—we will certainly need at least 10 boxes. When we subtract these 30 from the 48, we have 18 cupcakes left to package:

   (a)
$$\begin{array}{r} 10 \\ 3\overline{)48} \\ -\underline{30} \\ 18 \end{array}$$

   Since $6 \times 3 = 18$, 6 more boxes will take care of all 48 cupcakes:

   (b)
$$\begin{array}{r} 6 \\ 3\overline{)18} \\ -\underline{18} \\ 0 \end{array}$$

We now have 10 boxes plus 6 boxes, or 16 boxes of threes altogether. Steps (a) and (b) can be combined into one step:

```
        16                             16
(c)   3)48                   (d)     3)48
       -30   (10 × 3)                 -3
        18              ------>        18
       -18   ( 6 × 3)                 -18
         0   (16 × 3)                   0
```

(Note that (d) is a more compact form of (c). Writing each partial quotient in (c) at the side keeps the child aware of what has been accomplished at each step and what still needs to be done.)

We check the answer by multiplying the quotient by the divisor and seeing whether we get the dividend: $16 \times 3 = 48$.

2. As your child divides larger numbers, let him pay careful attention to place-value ideas and follow a well-planned sequence of steps to find the answer. In estimating multiples of the divisor to subtract, we use ones, tens, hundreds, and so on, to conform to the place values we ordinarily use in arithmetic.

Example 1:

```
       86 r1                          86 r1
     4)345                          4)345
      -320   (80 × 4)  ------>       -32
        25                             25
       -24   ( 6 × 4)                 -24
         1   (86 × 4)                   1
```

Steps: (1) Estimate the answer from this table:

$$\begin{aligned} 1 \times 4 &= 4 \\ \left\{\begin{aligned} 10 \times 4 &= 40 \\ 100 \times 4 &= 400 \end{aligned}\right\} \end{aligned}$$

Since the dividend 345 lies between 40 and 400 in the table, the quotient must lie somewhere between 10×4 and 100×4. The closest multiple of 10 that fits is 80 since $80 \times 4 = 320$. So we write 8 in the tens place. Subtracting 80 fours, or 320, from 345 leaves a remainder of 25.

(2) The closest multiple of 4 that can be subtracted from 25 is 6. So we write 6 in the ones place. Subtracting 6 fours, or 24, from 25 leaves a remainder of 1.

(3) Show the remainder 1, and write r1 next to the quotient.

(4) To check the answer, multiply the divisor by the quotient, add the remainder, and see whether the result is the dividend: $(86 \times 4) + 1 = 344 + 1 = 345$.

NOTE: Division is essentially a process of making estimates of the quotient. In step (1) we start with an estimate. In step (2), we improve that estimate by adding a correction to it. This process is repeated until we are left with a remainder smaller than the divisor.

When estimating a quotient, the child needs to know that estimates may be either too large or too small. He needs to know what to do in either case—try a smaller or a larger number.

Example 2:

```
    536 r3                                    536 r3
  5)2683                                    5)2683
   -2500    (500 × 5) ———————————→           -25
     183                                      18
    -150    ( 30 × 5)                        -15
      33                                      33
     -30    (  6 × 5)                        -30
       3    (536 × 5)                          3
```

Steps: (1) Estimate the answer from this table:

$$\begin{aligned} 1 \times 5 &= 5 \\ 10 \times 5 &= 50 \\ 100 \times 5 &= 500 \\ 1000 \times 5 &= 5000 \end{aligned}$$

Since the dividend 2683 lies between 500 and 5000 in the table, the quotient must lie somewhere between 100 × 5 and 1000 × 5. The closest multiple of 100 that fits is 500. So we write 5 in the hundreds place. Subtracting 500 fives, or 2500, from 2683 leaves a remainder of 183.

(2) Estimating from the table the number of fives contained in 183, we get 30 fives. So we write 3 in the tens place. Subtracting 30 fives, or 150, from 183 leaves a remainder of 33.

(3) Estimating the number of fives contained in 33, we get 6 fives. So we write 6 in the ones place. Subtracting 6 fives, or 30, from 33 leaves a remainder of 3.

(4) Show the remainder by writing r3 next to the quotient.

(5) To check the answer, multiply the quotient by the divisor and add the remainder, and then see whether the result is the same as the dividend:

$$\begin{aligned} (236 \times 5) + 3 &= 2680 + 3 \\ &= 2683 \end{aligned}$$

Example 3:

```
      54 r16                                  54 r16
  37)2014                                 37)2014
    -1850    (50 × 37)                       -185
      164    ———————————————————→             164
     -148    ( 4 × 37)                       -148
       16    (54 × 37)                         16
```

Steps: (1) Estimate the answer from this table:

$$\begin{aligned} 1 \times 37 &= 37 \\ 10 \times 37 &= 370 \\ 100 \times 37 &= 3700 \end{aligned}$$

Since the dividend 2014 lies between 370 and 3700 in the table, the quotient must lie somewhere between 10 × 37 and 100 × 37. The closest multiple of 10 that fits is 50. So we write 5 in the tens place. Subtracting 50 thirty-sevens, or 1850, from 2014 leaves a remainder of 164.

(2) Estimating the number of thirty-sevens contained in 164, we get 4 thirty-sevens. So we write 4 in the ones place. Subtracting 4 thirty-sevens, or 148, from 164, leaves a remainder of 16.

(3) Show this remainder by writing r16 next to the quotient.

(4) Check: Does (54 × 37) + 16 = 2014?

NOTE: Particular attention needs to be given to divisions that have zeros in the quotient, such as

(a) $3\overline{)1524}$ with quotient 508 (b) $5\overline{)2100}$ with quotient 420

Sometimes children omit the zeros and get for (a) an answer of 58; and for (b), 42.

Checking Division

We have already seen one way to check division: Multiply the divisor by the quotient, add the remainder (if any), and see whether you get back the dividend. We can also check division by using the quotient as a divisor and then dividing. The new quotient should be the same as the former divisor. For instance, to check $27\overline{)1053}$ with quotient 39, we divide $39\overline{)1053}$. If 39 is the correct answer, then 1053 ÷ 39 should give 27.

A third way to check is to redo the original division.

Finding an Average

Experience with *averages* is commonplace in the lives of children—average test scores, batting averages, etc. Since the concept of average is tied to division, this is a good time for your child to get a clearer picture of what it means and how to find it.

Explain that we often seek a single number that can best represent a set of data—for instance, a single number that best describes how well a child did on several tests, or a single number that tells how well a baseball player is batting.

There are several ways to find such a number, but the most common is to find the average. The average of a set of scores is the *sum of all the scores, divided by the number of scores*. For instance, if in five weekly tests a child received 80, 75, 92, 65, 98, the average score is

$$(80 + 75 + 92 + 65 + 98) \div 5 = 410 \div 5 = 82$$

Properties of Division

From the basic division facts, your child should be able to conclude that

- The order in which two numbers are divided does affect the answer. For instance, 6 ÷ 2 does not give the same answer as 2 ÷ 6. 6 ÷ 2 = 3, but 2 ÷ 6 does not equal *any* whole number, since there is no whole number which when multiplied by 6 gives 2.
- When a number is divided by 1, the result is that number. For instance, 3 ÷ 1 = 3 since 1 × 3 = 3.
- When a number is divided by itself, the result is 1. For instance, 5 ÷ 5 = 1 since 1 × 5 = 5.
- Dividing a number by 0 is not possible (see page 129).
- Zero divided by any number equals zero. For instance, 0 ÷ 7 = 0 since 7 × 0 = 0.

Activities Involving Division

1. Use division flash cards to reinforce your child's mastery of the basic division facts. Such cards are available commercially. You can also make them (on 3×5 cards) from the multiplication table on page 107.

2. Have your child draw a 12-inch line segment and divide it into three equal parts.
   (a) Let him tell you how many inches there are in each segment.
   (b) Then ask him to write a division sentence describing what he did.

3. Have the child use a centimeter (cm) ruler to cut a piece of string 58 cm long.
   (a) Ask him to cut from it as many pieces as possible that are 7 cm long.
   (b) Let him then tell you how many pieces he cut and how many centimeters are left.
   (c) Ask him to write a division sentence that describes what he did.

4. Have the child take 48 disks and separate them into 4 equal piles.
   (a) Let the child tell you how many disks there are in each pile.
   (b) Ask him to write a division sentence describing what he did.
   (c) Let him then tell you which number in the sentence is the quotient; the dividend; the divisor.

5. The child can use bird arrays to find the answers to the following divisions:

   (a) $7\overline{)22}$ (b) $45\overline{)180}$

   Let him check the answers.

6. Ask the child to find the mistakes in each division below, and then correct the errors.

   (a)
   $$\begin{array}{r} 13\ \ r7 \\ 4\overline{)509} \\ \underline{4} \\ 19 \\ \underline{12} \\ 7 \end{array}$$

   (b)
   $$\begin{array}{r} 6 \\ 7\overline{)420} \\ \underline{42} \end{array}$$

7. Let the child use a calculator to work through each step in the following divisions:

   (a) $26\overline{)8164}$ (b) $357\overline{)44982}$

Games Involving Division

1. Draw a model football field on a sheet of paper. Players alternate giving the answer to a division fact from a flash card. For every correct answer, the player moves one space on the football field. The winner is the first to score a touchdown.

2. Prepare two sets of cards, one set marked 1–10 and the other, 1–20. Shuffle each deck separately and lay each on the table face down. The child lifts the top card from each set and divides the larger number by the smaller one, dropping all remainders. Ten correct answers in a set amount of time wins.

3. Write on a sheet of paper a random set of two- and three-digit numbers. Then call out a divisor by which each number is to be divided. If your child gets eight or more correct answers in a set amount of time, he is a winner.

4. For this game you will need (a) two spinners, and (b) a set of 21 cards marked 0–20. Game action:
   (1) One card is turned face up; remaining cards are placed in a stack face down.
   (2) The player gets two numbers—one on each spinner.
   (3) The object is for the player to add, subtract, multiply, or divide the two numbers so as to obtain the number showing on the face-up card. For instance: If spinners show 5 and 2 and the number showing on the card is 10, then the player says "$5 \times 2 = 10$" and wins the 10 card. The player takes the card and turns the next card in the stack face up.
   (4) The play should alternate; the winner is the player with the most cards.

Common Errors in Division

The following are major causes for errors in division.

(1) (error in a multiplication fact)

```
   8
3)29
  27
   2
```

(2) (ignored the remainder)

```
    5
12)65
   60
```

(3) (did not complete the division)

```
     1
13)175
   13
    45
```

(4) (disregarded the tens place in the dividend because of the 0)

```
   14
2)208
```

(5) (left a remainder greater than the divisor)

```
  18 r6
4)78
  4
  38
  32
   6
```

(6) (wrote first digit in quotient in extreme right position instead of left position)

```
   25
3)158
  15
    8
    6
    2
```

(7) (confused dividend with divisor in part of the division)

```
  132
3)316
```

(8) (used only first digit of divisor to find quotient)

```
   423
23)846
```

(9) (divided remainder from first subtraction [2] by the divisor before bringing down the 5)

```
  105
5)75
  5
  25
  25
```

NOTE: Checking each division answer by estimation and multiplication will help your child eliminate many of these errors.

Divisibility Tests

Can you tell—without actually dividing—whether the number 1492 is divisible by 6? Maybe you can. But how about 56940?

We often want an easy way to tell whether a given number is divisible by another number. By "divisible" we mean that the division leaves no remainder.

We'll now show ways to determine whether a number is divisible by 2, 3, 4, 5, 6, 8, 9, and 10. To prove that these methods always work is fairly simple, but beyond the scope of this book.

1. **A number is divisible by 2 if its last digit is even.**

   ***Example*: 258 is divisible by 2 because the last digit, 8, is even.**

2. **A number is divisible by 3 if the sum of its digits is divisible by 3.**

   ***Example*: 92352 is divisible by 3 since 9 + 2 + 3 + 5 + 2 = 21, and 21 is divisible by 3.**

3. **A number is divisible by 4 if the number formed by its last two digits is divisible by 4.**

   ***Example*: 568912 is divisible by 4 since the number formed by its last two digits, 12, is divisible by 4.**

4. **A number is divisible by 5 if its last digit is 0 or 5.**

   ***Example*: 930 and 12965 are divisible by 5.**

5. **A number is divisible by 6 if (a) it is an even number, and (b) the sum of its digits is divisible by 3.**

   ***Example*: 5754 is divisible by 6 because**
   (a) it's an even number
   (b) 5 + 7 + 5 + 4 = 21, and 21 is divisible by 3

6. **A number is divisible by 8 if the number formed by its last three digits is divisible by 8.**

   ***Example*: 5209824 is divisible by 8 because the number formed by its last three digits, 824, is divisible by 8.**

7. **A number is divisible by 9 if the sum of its digits is divisible by 9.**

(continued)

Example: 3081654 is divisible by 9 since 3 + 0 + 8 + 1 + 6 + 5 + 4 = 27, and 27 is divisible by 9.

8. A number is divisible by 10 if its last digit is 0.

Example: 2590 is divisible by 10 since its last digit is 0.

HIGHLIGHTS OF DIVISION

The study of division should leave your child with the following understandings:

- Division is defined in terms of multiplication. For instance, $8 \div 2 = ?$ asks: By what number do I multiply 2 to get 8?
- Division is the inverse of multiplication. For instance, in the multiplication $3 \times 5 = ?$ we are asked to find the product of two factors, 3 and 5. In the division $15 \div 3 = ?$ we are given the product, 15, and one factor, 3, and asked to find the missing factor.
- Each multiplication fact is associated with two corresponding division facts. For instance, $5 \times 2 = 10$ is associated with $10 \div 2 = 5$ and $10 \div 5 = 2$.
- Division can be seen as a repeated subtraction process. For instance, $21 \div 3$ asks: How many threes can we subtract from 21 before we have a remainder less than 3?
- On the number line, division means counting to the left in equal groups (Fig. 13-8).

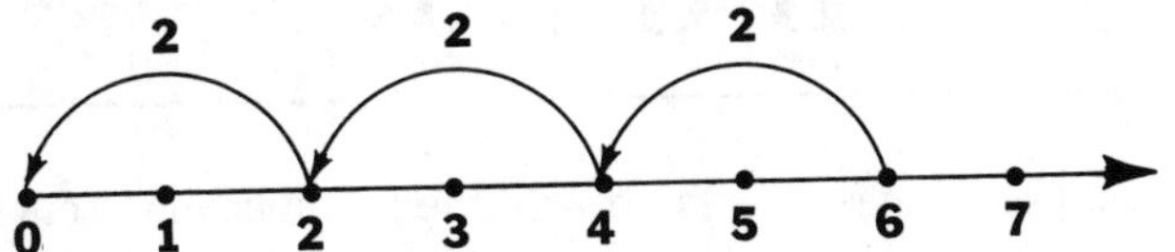

Figure 13-8

- In an array, division means
  1. Finding the number of rows if you know the number of columns and the total number of objects in the array (Fig. 13-9), or
  2. Finding the number of columns if you know the number of rows and the total number of objects in the array (Fig. 13-10).

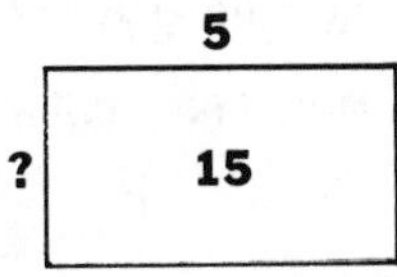

Figure 13-9

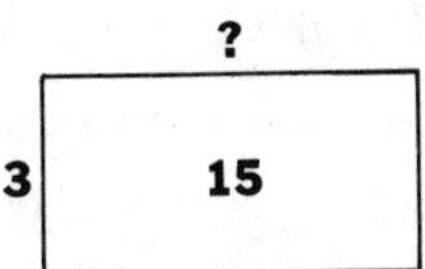

Figure 13-10

- When a number is divided by 1, the result is that number.
- When a number is divided by itself, the result is 1.

  You cannot always divide a whole number by another whole number and get a whole number for an answer. For instance, $2 \div 5$ does not result in a whole number.
- Division by 0 is not possible. (See page 129.)
- 0 divided by any number is 0. (See page 129.)
- The order in which two numbers are divided affects the answer. For instance, $8 \div 2$ is not equal to $2 \div 8$.
- An average is a single number that describes a set of data.

Also, at the completion of this chapter, your child should be able to

- Demonstrate division with concrete materials
- Recall with ease the basic division facts
- Compute the quotient of two whole numbers when the divisor is less than 100
- Solve word problems involving division
- Find the average of a set of numbers
- Explain and use the following vocabulary:

divide	quotient	average
dividend	remainder	
divisor	estimate	

QUESTIONS ON THE CHAPTER

1. Write two division sentences associated with the multiplication sentence $3 \times 9 = 27$.
2. Write a multiplication sentence associated with the division sentence $72 \div 8 = 9$.
3. Make up a word problem to fit each of these divisions: (a) $12 \div 5$ (b) $30\overline{)360}$
4. Show $17 \div 3$ on the number line.
   (a) What does the 17 represent? (b) the 3? (c) the answer?
5. An array containing 36 cans of soup has 4 rows. How many columns are there?
6. (a) $9 \div 1 = ?$ (b) $35 \div 35 = ?$ (c) $0 \div 129 = ?$ (d) $4 \div 9 = ?$
7. Divide and check your answers.
   (a) $3\overline{)69}$ (b) $2\overline{)58}$ (c) $2\overline{)426}$ (d) $4\overline{)724}$ (e) $6\overline{)534}$
8. Divide and check your answers. (a) $5\overline{)134}$ (b) $3\overline{)2561}$

9. Estimate the answers, then divide and check. See how good your estimates were.
   (a) $17\overline{)68}$ (b) $23\overline{)184}$ (c) $32\overline{)864}$ (d) $65\overline{)3079}$

10. Ms. Watson earns $25,000 a year. What is her monthly salary?

11. Mr. Jordan drove 364 miles in 7 hours. (a) How many miles an hour did he travel? (b) If the car averages 17 miles on a gallon of gas, how many gallons were needed to make the trip?

12. Larry's mother bought a washing machine for $385. If she paid $50 in cash and arranged to pay off the balance in 8 equal monthly payments, what is the amount of each payment?

13. Knowing that $18 \times 23 = 414$, find the answers to the following problems *mentally:*
    (a) $414 \div 23$ (b) 23×18 (c) $414 \div 18$ (d) 23×9 (e) 230×18

14. The square and the triangle in the sentence below stand for different numbers. What number does each shape stand for?
    $$(\square \div 3) + (2 \times \triangle) = 14$$

15. What is Jake's average in his arithmetic tests if he received the following marks: 85, 92, 73, 60, 100?

16. The temperature readings for a week in Philadelphia were: Monday, 62°; Tuesday, 70°; Wednesday, 63°; Thursday, 60°; Friday, 57°; Saturday, 61°; Sunday, 55°. What was the average temperature for that week?

17. A basketball player scores the following numbers of points in four games: 20, 15, 12, 24. How many points does he have to score in the fifth game to give him an average score of 18 points per game?

REVIEW

1. **Mental arithmetic:** See whether you can follow these instructions.
   (a) Begin with 16. Divide by 2. Add 5. Subtract 3. Multiply by 5. What's your answer? [50]
   (b) Begin with 30. Subtract 10. Divide by 4. Multiply by 3. Add 7. What's your answer? [22]

2. **Cross-number puzzle**

Across	*Down*
(a) 28×2	(a) $250 \div 5$
(b) $90 \div 5$	(b) $(12 + 18) \div 2$
(c) $(3 + 4) \times 5$	(c) $106 - 67$
(e) $46 - 17$	(d) $4 \times 4 \times 4$
(f) $12 + 6 + 16$	(e) $162 \div 6$
(g) $(5 \times 10) - 3$	(f) $(2 \times 20) - 2$
(h) $666 \div 37$	(g) $360 \div 8$
(i) 5 fours	(h) $100 \div 10$
(k) $540 \div 15$	(i) $(5 + 8) \times 2$
(l) 9×8	(j) $4140 \div 45$

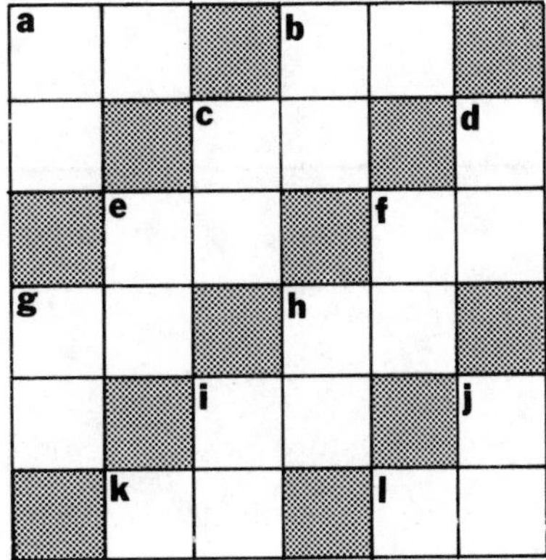

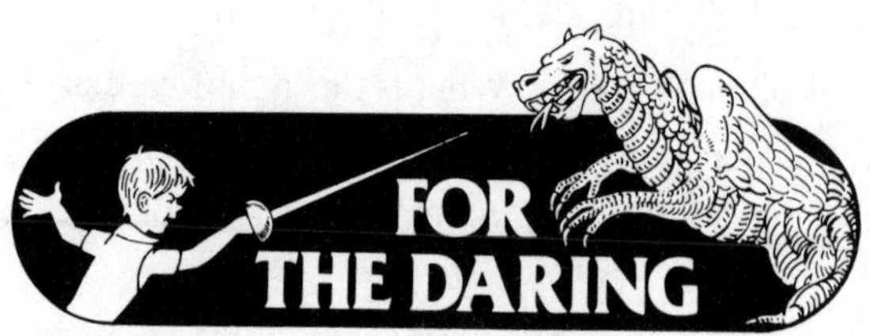

1. **Use 20 counters to make 4 piles of counters so that the first pile contains 4 more counters than the second pile, the second pile contains 1 counter less than the third pile, and the fourth pile contains twice as many counters as the second pile.**
2. **Which is greater: six dozen dozen or a half-dozen dozen?**
3. **Write a mathematical sequence that equals 20, using four 9's and no other numbers.**
4. **Insert operational signs (+, −, ×, ÷) between the threes to make a true statement:**

   3 3 3 3 = 2

5. ***Break the Code*: Each letter stands for a digit. The O stands for zero. Figure out what number each letter represents.**

```
    9AB
 D)R8CR
   RP
    16
    MA
      M2
      MR
       0
```

(The answers appear on page 258.)

Chapter 14
Fractions

INTRODUCTION

Among the mathematical concepts encountered by young children, those relating to fractions are most bothersome and least understood. Often, the root of the problem is a lack of adequate experience with physical materials. With heavy reliance on manipulation of objects, a child can learn much about fractions—what they mean, how to compare them, how to rename them, and how to compute with them—before being given a single formal rule. Only after this spadework can a child grasp the formal rules.

In his study of children's understanding of fractions, Jean Piaget, the famous Swiss psychologist, showed a child a circular clay slab and two dolls. He asked her to imagine the slab to be a "cake" the dolls will eat, but that each doll must be given the same amount. The child was then given a wooden knife for "cutting the cake." The experiment was later repeated with another "cake" and three dolls.

Piaget found that children up to about four years of age could not easily divide the cake in halves, nor could they relate the parts to the whole cake. When the child held half a cake, she could not conclude that the other part was also half the cake. But at the age of six or seven, she could divide the cake into halves and into thirds. Division into fifths or sixths could not be accomplished until the age of seven to nine.

Why do we need fractions? They were created to eliminate two major defects in the earlier system of whole numbers: (1) *parts* of things, like half a cake or three quarters of a dozen, cannot be represented by whole numbers; (2) division of whole numbers is not always possible. You can't divide 2 by 3 and get a whole number answer. Fractions, sometimes called *rational numbers*, correct these shortcomings. (But you still can't divide by 0.)

Materials used in this chapter include: A fraction chart, Fraction strips (see page 146)

A set of cards containing basic fractions (see page 164)

Three spinners (see page 164)

Several paper strips about 12 inches long

THE CHAPTER IN A NUTSHELL

In this chapter you will see how to

- Prepare your child for the study of fractions
- Use fraction strips to explain informally the basic concepts related to fractions
- Explain how to add, subtract, multiply, and divide fractions
- Recognize and correct common errors with fractions
- Use a variety of activities and games to motivate the study of fractions

You will also be given

- A list of understandings and skills your child should be left with at the completion of the chapter
- Questions to test and reinforce your child's comprehension of fractions; also, review questions on earlier material
- A variety of insights into the history of fractions; "seeing" $\frac{1}{3} \times \frac{1}{2}$; the reasons for some of the rules used with fractions; and an easy way to compare fractions—under **For the Curious**
- Several challenging problems and puzzles—under **For the Daring**

WHAT'S A FRACTION?

A fraction has several meanings, but the key idea in all of them is that a fraction is *part of something*. Consider the fraction $\frac{1}{2}$:

1. If we divide a rectangle into two equal parts and shade one of the parts, then the shaded part is one of two equal parts. The unshaded part is also one of two equal parts. The two parts together make up the whole rectangle (Fig. 14-1).

Figure 14-1

We say that each part is "one half" of the whole rectangle, and denote this fact by the symbol $\frac{1}{2}$. We call $\frac{1}{2}$ a *fraction* in which the 2 indicates the number of equal parts into which the whole rectangle was subdivided; the 1 indicates the number of parts that were shaded.

To create the fraction $\frac{1}{2}$, we started with a *unit*—the rectangle—and divided it into two pieces, each piece the same size. The number of pieces into which the unit was subdivided, 2, is called the *denominator*; the number of shaded pieces, 1, is called the *numerator*.

2. If a set of six objects is divided into two equal groups, there will be three objects in each group. Each group of three objects is one half of the original set of six objects. In this case, the original set of six objects is the *unit*.

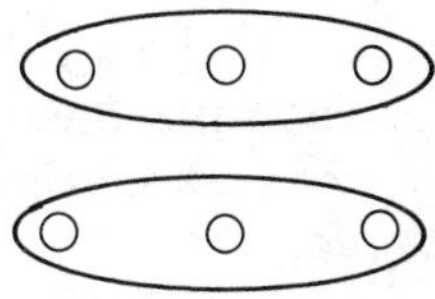

Figure 14-2

The 2 in the fraction indicates the number of equal groups into which the original set of 6 was divided, and the 1 indicates the number of groups in which we are interested.

In both instances—the rectangle and the set of six objects—the fraction concept involves (1) a unit, and (2) dividing the unit into equal pieces.

3. A fraction is also used to indicate division. The fraction $\frac{3}{4}$ means "3 divided by 4"; that is, $3 \div 4$, or $4\overline{)3}$. While the solution to $10 \div 2$ is the whole number 5, there is no whole number answer to the division $3 \div 4$. So the fraction $\frac{3}{4}$ provides the answer.

Random Notes on Fractions

A *fraction* is a number that can be expressed in the form $\frac{a}{b}$, where *a* and *b* are integers. However, *b* can't be 0. (Why?)

This definition suggests the existence of negative as well as positive fractions, like $^{-}\frac{2}{3}$ or $-\frac{3}{4}$. This, in fact, is the case; the set of positive fractions, 0, and the set of negative fractions together make up the *rational number* system. In this book, however, we will not deal with negative fractions.

(continued)

The word *fraction* is derived from the Latin word *fracto* ("to break"). A fraction was first thought of as a broken number.

The early Greeks were the first to use two integers to write fractions. Some Greek writers wrote the two numerals side by side, e.g., 2, 3. Some wrote the numerator above the denominator, e.g., $\frac{2}{3}$. Others wrote the denominator above the numerator, e.g., $\frac{3}{2}$.

The Egyptians had an interesting way of representing fractions between 0 and 1. They represented all such fractions as a *sum of fractions with a numerator of 1*. For example,

$$\frac{1}{2} = \frac{1}{4} + \frac{1}{4}$$

$$\frac{1}{3} = \frac{1}{4} + \frac{1}{12}$$

$$\frac{2}{5} = \frac{1}{3} + \frac{1}{15}$$

About 100 years ago, an English mathematician proved that it is possible to express *any* fraction between 0 and 1 as a sum of fractions, each with a numerator of 1.

READINESS FOR FRACTIONS

Developing Meaning of a Fraction

1. Fraction concepts should be developed slowly, with emphasis on "hands-on" experience with a variety of concrete materials of different shapes. For instance, develop the concept of $\frac{1}{2}$ by giving the child several paper cutouts (Fig. 14-3) and asking him to fold each shape in as many ways as possible to make two equal size parts. In each case the child sees that the two parts make the whole cutout.

   Now proceed with the explanation of "fraction" as developed on page 142.

   Repeat this activity with $\frac{1}{4}$: let the child see that the four fourths make up the whole cutout.

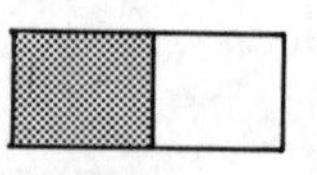
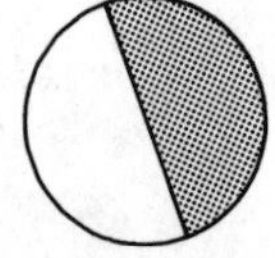
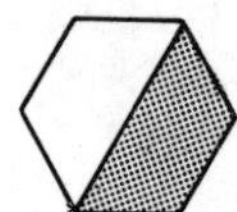

Figure 14-3

2. To understand the concept of a fraction, the child needs to ask four key questions:

   What is the unit?

   How many pieces are in the unit?

   Are the pieces the same size?

   In how many pieces am I interested? For instance, how many pieces are shaded?

   The number of equal size pieces in the unit identifies the fraction as *halves*, *thirds*, *fourths*, and so on.

3. Show your child a strip of paper folded in thirds, two thirds shaded red and one third blue, and ask:
   (a) What is the unit? (b) How many pieces are in the unit? (c) Are all the pieces the same size? (d) What part of the unit is each piece? (e) How many pieces are shaded red? (f) What part of the unit is shaded red? (g) Write the *fraction* that tells what part of the unit is red. (h) What part of the unit is shaded blue? (i) Write the *fraction* that tells what part of the unit is blue.

   Repeat with other fractions.

4. Show your child a strip of paper folded in fourths. Shade all four fourths. Ask questions like those in 3. Let the child see that $\frac{4}{4}$ and 1 are different ways of naming the unit.

5. A common misconception among children is that a fraction must be less than a unit. To help correct this misconception, give your child two paper strips to serve as units. Then ask:
   (a) Show me $\frac{1}{2}$ a unit.
   (b) Show me $\frac{3}{2}$.
   (c) What does $\frac{3}{2}$ mean?
   (d) Is $\frac{3}{2}$ less than the unit, the same as the unit, or greater than the unit?

   Repeat with other fractions equal to or greater than the unit.

Fractions on the Number Line

The number line is also useful for developing the concept of a fraction. Mark a point *P* midway between 0 and 1 (Fig. 14-4) and ask the child to compare the distance from 0 to P with the distance from 0 to 1. The child will see that the distance from 0 to P is one half the distance from 0 to 1. Here, the *unit* is the distance from 0 to 1, and the equal size pieces are the two segments 0 to P and P to 1.

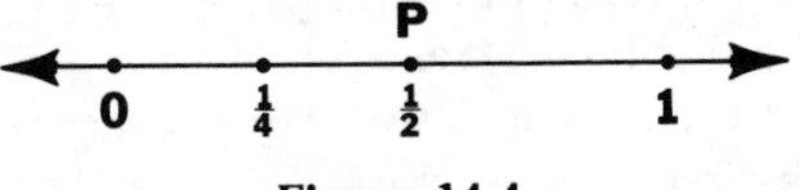

Figure 14-4

Repeat this explanation with fourths and other fractions.

A Fraction as Part of a Set

To explain the concept of a fraction as *part of a set of objects*, follow the development on page 143.

Fraction Strips

A most useful aid for building fraction concepts consists of two parts: (1) a *fraction chart* (Fig. 14-5), and (2) another identical chart cut up into *fraction strips* of halves, thirds, fourths, and so on, showing the relative sizes of the most common fractions. Using different colors for different fractional sizes can be helpful.

The strips serve as physical representations of fractions; the chart is used for overall comparisons. Both the chart and strips are available commercially or can be made by you out of any stiff material such as oak tag.

With these strips the child can compare the sizes of fractions, rename them, and perform the basic operations on them—without any of the formal rules taught later. Formal rules without lots of prior experience with physical objects (like fraction strips) are useless because the child isn't ready for them. Rules not rooted in physical experience inevitably lead to errors like $\frac{1}{4} + \frac{1}{4} = \frac{1}{8}$, and $\frac{1}{3} > \frac{1}{2}$. [> means "is greater than." < means "is less than."]

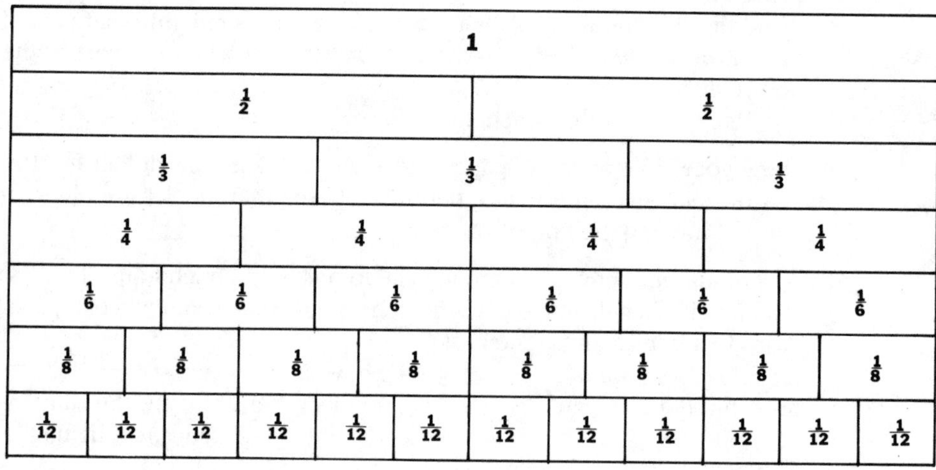

Figure 14-5

DEVELOPING KEY CONCEPTS WITH FRACTION STRIPS

1. Verifying relationships

Let your child verify the following (and similar) relationships by lining up fraction strips directly under the unit 1:

If a unit is divided into two equal parts, then each part is $\frac{1}{2}$ of the unit, and $\frac{2}{2} = 1$ (Fig. 14-6).

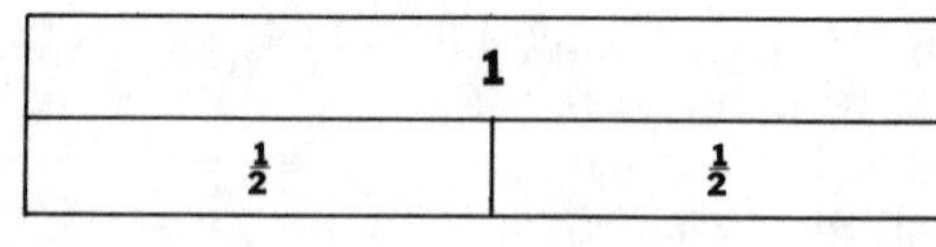

Figure 14-6

If a unit is divided into three equal parts, then each part is $\frac{1}{3}$ of the unit, and $\frac{3}{3} = 1$ (Fig. 14-7).

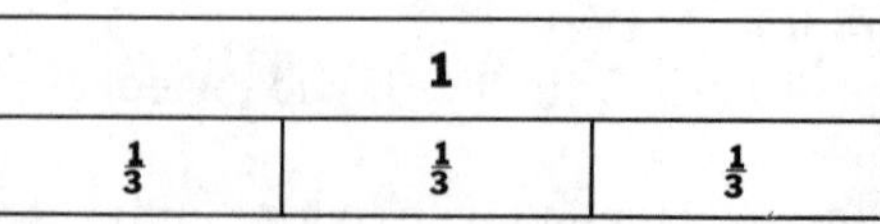

Figure 14-7

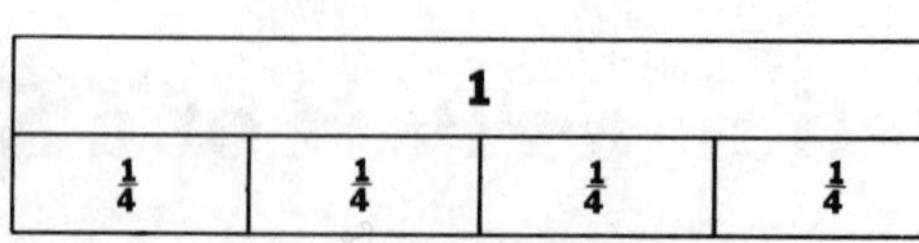

Figure 14-8

If a unit is divided into four equal parts, then each part is $\frac{1}{4}$ of the unit, and $\frac{4}{4} = 1$ (Fig. 14-8).

2. Comparing fractions

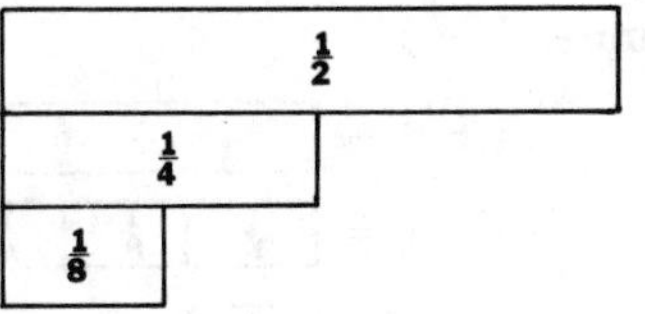

Figure 14-9

To tell which fraction strip is longer, the child places one strip directly under the other and compares (Fig. 14-9). After observing that the order, from largest to smallest, is $\frac{1}{2}$, $\frac{1}{4}$, $\frac{1}{8}$, guide the child to the conclusion that the *larger fraction* is the one with the *smaller denominator*, and that the *smaller fraction* is the one with the *larger denominator*. Then ask why this is so.

Later, ask the child to use the same method to compare fractions like $\frac{2}{3}$ and $\frac{3}{4}$, $\frac{3}{8}$ and $\frac{5}{6}$.

3. Equivalent fractions

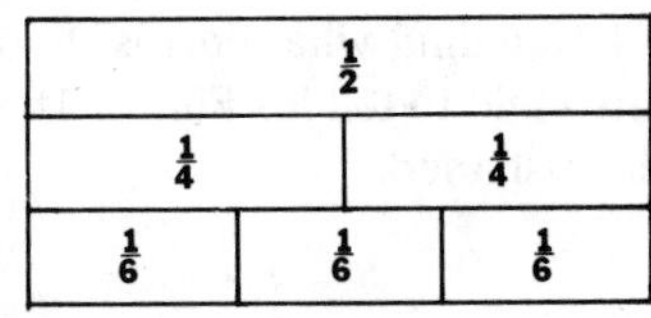

Figure 14-10

Ask your child to lay out, directly under each other, the following fraction strips and compare their lengths: $\frac{1}{2}$, two $\frac{1}{4}$'s, three $\frac{1}{6}$'s (Fig. 14-10).

Since $\frac{1}{2}$, $\frac{2}{4}$, and $\frac{3}{6}$ are the same length, we say that they are *equivalent fractions;* that is,

$$\frac{1}{2} = \frac{2}{4} = \frac{3}{6}$$

Ask the child to tell you what other fractions are equivalent to $\frac{1}{2}$, and then to demonstrate any answer with strips.

4. Adding two fractions with the same denominator

Example: $\frac{1}{4} + \frac{2}{4}$

Have your child lay out the $\frac{1}{4}$ strips, end to end, (Fig. 14-11) and find that $\frac{1}{4} + \frac{2}{4} = \frac{3}{4}$. Repeat with other fractions having the same denominators.

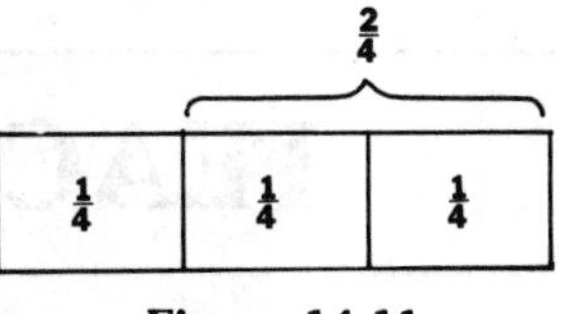

Figure 14-11

5. Subtracting two fractions with the same denominator

Example: $\frac{4}{5} - \frac{1}{5}$

Have the child lay out four $\frac{1}{5}$'s and then remove one $\frac{1}{5}$ (Fig. 14-12), finding that $\frac{4}{5} - \frac{1}{5} = \frac{3}{5}$.

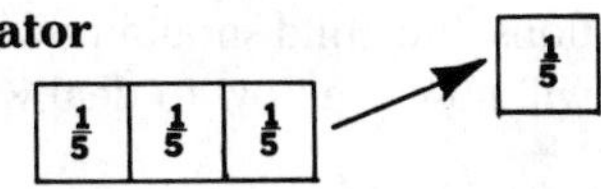

Figure 14-12

6. Adding two fractions with different denominators

Example: $\frac{1}{2} + \frac{1}{4}$

The child should:

(a) Exchange strips so that the two fractions have the same denominator. Looking at the fraction chart, she sees that $\frac{1}{2}$ is equivalent to $\frac{2}{4}$.

(b) By laying the strips end to end, the child gets:
$\frac{1}{2} + \frac{1}{4} = \frac{2}{4} + \frac{1}{4} = \frac{3}{4}$ (Fig. 14-13)

$\frac{1}{2} + \frac{1}{4} =$ [$\frac{1}{2}$ | $\frac{1}{4}$]

$=$ [$\frac{1}{4}$ | $\frac{1}{4}$ | $\frac{1}{4}$] $= \frac{3}{4}$

Figure 14-13

7. Subtracting two fractions with different denominators (Fig. 14-14)

Example: $\frac{1}{2} - \frac{3}{8}$

$\frac{1}{2} - \frac{3}{8} = \boxed{\frac{1}{2}} - \boxed{\frac{1}{8}}\boxed{\frac{1}{8}}\boxed{\frac{1}{8}}$

$\frac{1}{2} - \frac{3}{8} = \boxed{\frac{1}{8}}\boxed{\frac{1}{8}}\boxed{\frac{1}{8}}\boxed{\frac{1}{8}} - \boxed{\frac{1}{8}}\boxed{\frac{1}{8}}\boxed{\frac{1}{8}}$

$\frac{1}{2} - \frac{3}{8} = \boxed{\frac{1}{8}} = \frac{1}{8}$

Figure 14-14

8. Other activities.

(a) To help the child compare the sizes of fractions, write several fractions, in random order, on a piece of paper. Ask your child to arrange them in ascending order (from smallest to largest). When in doubt, let her use fraction strips.

(b) Ask the child to use fraction strips to verify (1) $\frac{3}{5} + \frac{1}{5} = \frac{4}{5}$, (2) $\frac{3}{4} + \frac{1}{8} = \frac{7}{8}$, (3) $\frac{5}{8} - \frac{1}{4} = \frac{3}{8}$.

(c) Ask the child to write the fraction next to each row and under each column in Fig. 14-15 telling what part is shaded.

(d) Ask the child to tell for Fig. 14-16 what part of each shape is shaded, and what part is unshaded.

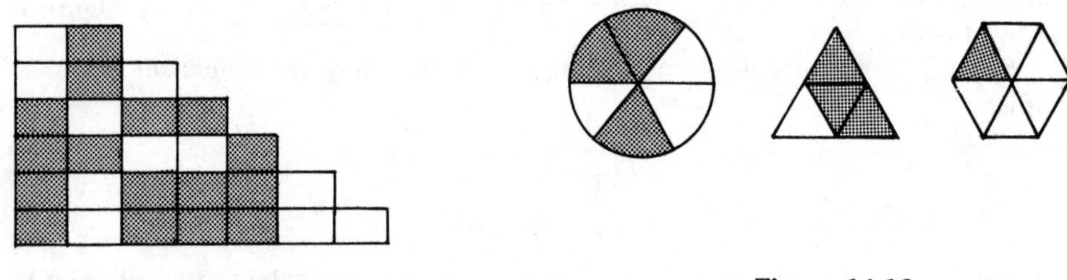

Figure 14-15

Figure 14-16

TEACHING FRACTIONS

Having acquired an intuitive understanding of the basic concepts related to fractions, the child should now be ready for a more formal and abstract presentation. He will now see how to deal with fractions as numbers rather than as physical objects.

Renaming Fractions

From the fraction chart, the child already knows that $\frac{1}{2}$, $\frac{2}{4}$, and $\frac{3}{6}$ are *equivalent* fractions because they have the same value:

$$\frac{1}{2} = \frac{2}{4} = \frac{3}{6}$$

Now ask him to figure out a way to obtain equivalent fractions *without* looking at the fraction chart.

The child should notice that from $\frac{1}{2}$ he can obtain $\frac{2}{4}$ by multiplying *both* the numerator and denominator by 2:

$$\frac{1}{2} = \frac{1 \times 2}{2 \times 2} = \frac{2}{4}$$

Similarly, the child can obtain $\frac{3}{6}$ from $\frac{1}{2}$ by multiplying *both* the numerator and denominator by 3:

$$\frac{1}{2} = \frac{1 \times 3}{2 \times 3} = \frac{3}{6}$$

Since there are infinitely many numbers by which to multiply the numerator and denominator of $\frac{1}{2}$, $\frac{1}{2}$ has infinitely many equivalent fractions:

$$\frac{1}{2} = \frac{2}{4} = \frac{3}{6} = \frac{4}{8} = \frac{5}{10} = \frac{6}{12} \cdots$$

Not just $\frac{1}{2}$, but *every* fraction has infinitely many equivalent fractions for the same reason. For instance,

$$\frac{3}{4} = \frac{6}{8} = \frac{9}{12} = \frac{12}{16} \cdots$$

and

$$\frac{2}{3} = \frac{4}{6} = \frac{6}{9} = \frac{8}{12} \cdots$$

The point to stress is that multiplying the numerator and denominator by the *same* number is the same as multiplying the fraction by 1, which does not change its value.

When we multiply the numerator and denominator of a fraction by the same number, we rename the fraction in "higher terms." $\frac{1}{2}$ renamed in higher terms becomes $\frac{2}{4}$, $\frac{3}{6}$, etc. To rename a fraction in "lower terms," we do the opposite: we *divide* the numerator and denominator by the same number. For instance, to reduce $\frac{2}{4}$ to lower terms, divide the numerator and denominator by 2:

$$\frac{2}{4} = \frac{2 \div 2}{4 \div 2} = \frac{1}{2}$$

We say that a fraction is in its "lowest terms" (or in its "simplest form") when its numerator and denominator cannot be divided by any number other than 1. For instance, the fractions $\frac{2}{3}$, $\frac{3}{5}$, and $\frac{9}{11}$ are in their lowest terms.

To reduce a fraction to its lowest terms, the child must determine the *largest* number by which *both* numerator and denominator may be divided, and then carry through the division. For instance, to reduce $\frac{24}{30}$ to its lowest terms, we look for the largest number by which both 24 and 30 can be divided. Though you can divide by 2 or 3, neither is the *largest* number; 6 is:

$$\frac{24}{30} = \frac{24 \div 6}{30 \div 6} = \frac{4}{5}$$

So $\frac{24}{30}$, reduced to lowest terms, is $\frac{4}{5}$.

Here the point to stress is that dividing the numerator and denominator by the *same* number is the same as dividing the fraction by 1, which does not change its value.

The examples that follow illustrate what has been said about renaming fractions.

Example 1: Change $\frac{1}{3}$ to sixths.

$$\frac{1}{3} = \frac{?}{6}$$

Question: By what number do you multiply 3 to get 6?

Since the denominator 3 must be multiplied by 2 to get 6, the numerator 1 must also be multiplied by 2:

$$\frac{1}{3} = \frac{1 \times 2}{3 \times 2} = \frac{2}{6}\text{; so } \frac{1}{3} = \frac{2}{6}$$

Example 2: Change $\frac{2}{5}$ to fifteenths.

$$\frac{2}{5} = \frac{?}{15}$$

Question: By what number do you multiply 5 to get 15?
Since the denominator 5 must be multiplied by 3 to get 15, the numerator 2 must also be multiplied by 3:

$$\frac{2}{5} = \frac{2 \times 3}{5 \times 3} = \frac{6}{15}\text{; so } \frac{2}{5} = \frac{6}{15}$$

Example 3: Change $\frac{5}{8}$ to sixteenths.

$$\frac{5}{8} = \frac{?}{16}$$

Question: By what number do you multiply 8 to get 16?
Since the denominator 8 must be multiplied by 2 to get 16, the numerator 5 must also be multiplied by 2:

$$\frac{5}{8} = \frac{5 \times 2}{8 \times 2} = \frac{10}{16}\text{; so } \frac{5}{8} = \frac{10}{16}$$

Example 4: Reduce $\frac{6}{9}$ to lowest terms.
Question: What is the *largest* number that divides evenly into 6 and 9? Asked another way, what is the *highest common factor* of 6 and 9?
Since 3 is the highest common factor, we divide the numerator and denominator by 3:

$$\frac{6}{9} = \frac{6 \div 3}{9 \div 3} = \frac{2}{3}$$

Example 5: Reduce $\frac{12}{16}$ to lowest terms.
Question: What is the highest common factor of 12 and 16?
Since 4 is the highest common factor, we divide by 4:

$$\frac{12}{16} = \frac{12 \div 4}{16 \div 4} = \frac{3}{4}$$

EXERCISES

1. Write each of the following as a fraction.

   (a) $4\overline{)2}$ (b) $3 \div 5$ (c) 7 divided by 10 (d) two thirds (e) three eighths

2. Fill in the missing numbers.

   (a) $\frac{2}{3} = \frac{?}{6}$ (b) $\frac{1}{4} = \frac{?}{12}$ (c) $\frac{1}{5} = \frac{?}{25}$ (d) $\frac{2}{3} = \frac{?}{15}$ (e) $\frac{3}{4} = \frac{?}{20}$

3. Reduce to lowest terms.

   (a) $\frac{3}{9}$ (b) $\frac{4}{8}$ (c) $\frac{4}{10}$ (d) $\frac{5}{15}$ (e) $\frac{4}{14}$ (f) $\frac{10}{25}$ (g) $\frac{12}{18}$

Adding Fractions

Addition is usually developed in three stages:

(a) addition of fractions with the same denominator
(b) addition of fractions with different denominators
(c) addition of mixed numbers

1. Addition of fractions with the same denominator
From his experience with fraction strips, the child already knows that

$$\frac{1}{3} + \frac{1}{3} = \frac{2}{3} \qquad \frac{2}{4} + \frac{1}{4} = \frac{3}{4} \qquad \frac{2}{8} + \frac{5}{8} = \frac{7}{8}$$

He now has to see that when he added fractions with the same denominator, in effect he *added the numerators and placed their sum over the common denominator:*

$$\frac{1}{3} + \frac{1}{3} = \frac{1+1}{3} = \frac{2}{3} \qquad \frac{2}{4} + \frac{1}{4} = \frac{2+1}{4} = \frac{3}{4} \qquad \frac{2}{8} + \frac{5}{8} = \frac{2+5}{8} = \frac{7}{8}$$

NOTE: ▪ To reduce errors such as $\frac{1}{3} + \frac{1}{3} = \frac{2}{6}$, have the child initially express the denominator as a *word* rather than as a number. Let him write $\frac{1}{3} + \frac{1}{3}$ as

1 third + 1 third 2 thirds	and $\frac{2}{4} + \frac{1}{4}$ as	2 fourths + 1 fourth 3 fourths

This makes it easier for the child to think of the denominator as a unit of measure, like quarts or inches.

- The child should learn to write solutions both horizontally and vertically:

$$\frac{1}{5} + \frac{2}{5} = \frac{3}{5} \quad \text{and} \quad \begin{array}{r} \frac{1}{5} \\ +\frac{2}{5} \\ \hline \frac{3}{5} \end{array}$$

EXERCISES

1. (a) $\frac{1}{5} + \frac{3}{5}$ (b) $\frac{2}{7} + \frac{4}{7}$ (c) $\frac{2}{9} + \frac{5}{9}$ (d) $\frac{1}{3} + \frac{2}{3}$ (e) $\frac{3}{4} + \frac{2}{4}$
2. A boy spends $\frac{1}{6}$ of his allowance on going to a movie and $\frac{2}{6}$ on a birthday present for his sister. What part of his allowance did he spend?
3. On a family trip, Mr. Daley drove $\frac{2}{7}$ of the distance the first day and $\frac{3}{7}$ of the distance the next day. What part of the trip was covered in the first two days?

2. Addition of fractions with different denominators
To add such fractions, your child must first know how to

- Rename fractions with equivalent fractions
- Add fractions with the same denominator

The child then converts an addition problem with *different* denominators to an addition problem with the *same* denominator.

Therefore, to add fractions with different denominators, *rename the fractions with equivalent fractions having the same denominator, and then add.*

Example 1: $\frac{1}{2} + \frac{1}{4}$

$$\begin{array}{r} \frac{1}{2} = \frac{2}{4} \\ + \frac{1}{4} = \frac{1}{4} \\ \hline \frac{3}{4} \end{array}$$

Example 2: $\frac{1}{2} + \frac{1}{3}$

$$\begin{array}{r} \frac{1}{2} = \frac{3}{6} \\ + \frac{1}{3} = \frac{2}{6} \\ \hline \frac{5}{6} \end{array}$$

Example 3: $\frac{2}{3} + \frac{3}{10}$

$$\begin{array}{r} \frac{2}{3} = \frac{20}{30} \\ + \frac{3}{10} = \frac{9}{30} \\ \hline \frac{29}{30} \end{array}$$

To rename two fractions with equivalent fractions having a common denominator, we usually look for the *smallest*, or *least*, common denominator so as to give the answer in its simplest form, that is, in its *lowest terms.*

For instance, in Example 1 we changed the $\frac{1}{2}$ and $\frac{1}{4}$ to fourths and got $\frac{1}{2} + \frac{1}{4} = \frac{3}{4}$, an answer expressed in its lowest terms. But we could have chosen to change the $\frac{1}{2}$ and $\frac{1}{4}$ to eighths:

$$\begin{array}{r} \frac{1}{2} = \frac{4}{8} \\ + \frac{1}{4} = \frac{2}{8} \\ \hline \frac{6}{8} \end{array}$$

Though $\frac{6}{8}$ is a perfectly good answer, it's not expressed in lowest terms because 8 is not the *least* common denominator of $\frac{1}{2}$ and $\frac{1}{4}$; 4 is.

Finding the Least Common Denominator (LCD)

A systematic way to find the least common denominator is to list the multiples of each denominator and then select the least common multiple. Let's use this method to find the least common denominator for $\frac{1}{2}$ and $\frac{1}{3}$:

The multiples of 2 are:	2, 4, (6), 8, 10, (12), 14, 16, (18), ...
The multiples of 3 are:	3, (6), 9, (12), 15, (18) ...

The common multiples are 6, 12, 18, ... But the *least* of these is 6. So we say that the *least common denominator* for $\frac{1}{2}$ and $\frac{1}{3}$ is 6.

Example 4: $\frac{1}{2} + \frac{1}{3}$

Since the LCD for these two fractions is 6, we change $\frac{1}{2}$ and $\frac{1}{3}$ to sixths and then add:

$$\begin{array}{r} \frac{1}{2} = \frac{3}{6} \\ + \frac{1}{3} = \frac{2}{6} \\ \hline \frac{5}{6} \end{array}$$

Example 5: $\frac{3}{4} + \frac{1}{5}$

The multiples of 4 are:	4, 8, 12, 16, (20), 24, ...
The multiples of 5 are:	5, 10, 15, (20), 25, 30, ...

Since the LCD is 20, we change $\frac{3}{4}$ and $\frac{1}{5}$ to *twentieths*:

$$\begin{array}{r} \frac{3}{4} = \frac{15}{20} \\ + \frac{1}{5} = \frac{4}{20} \\ \hline \frac{19}{20} \end{array}$$

Example 6: $\frac{1}{6} + \frac{4}{9}$

The multiples of 6 are: 6, 12, (18), 24, . . .
The multiples of 9 are: 9, (18), 27, 36, . . .

Since the LCD is 18, we proceed as follows:

$$\begin{array}{rcl} & \frac{1}{6} = & \frac{3}{18} \\ + & \frac{4}{9} = & \frac{8}{18} \\ \hline & & \frac{11}{18} \end{array}$$

EXERCISES

1. (a) $\frac{1}{2} + \frac{1}{5}$ (b) $\frac{2}{3} + \frac{1}{4}$ (c) $\frac{2}{5} + \frac{3}{10}$ (d) $\frac{1}{4} + \frac{1}{8} + \frac{1}{3}$

2. A woman completes $\frac{1}{3}$ of a job on Monday and $\frac{2}{5}$ of the job on Tuesday. What part of the job did she complete at the end of the two days?

3. Addition of mixed numbers

A *mixed* number is a number containing a whole number and a fraction. An example is $1\frac{1}{2}$, which is a short way of writing $1 + \frac{1}{2}$.

With fraction strips, the addition $1\frac{1}{2} + 2\frac{1}{4}$ looks like this:

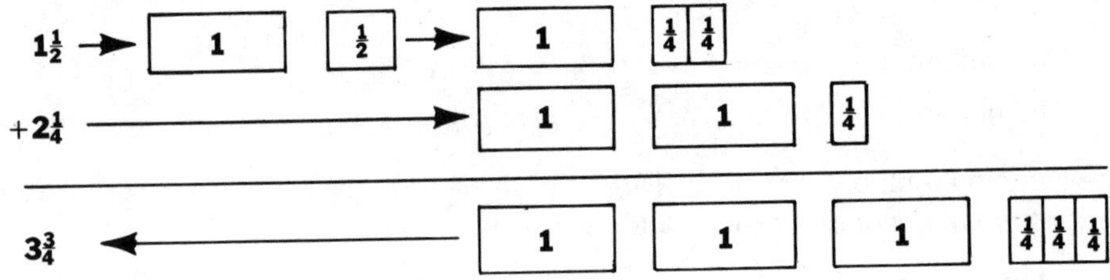

Figure 14-17

Without fraction strips, the addition looks like this:

$$\begin{array}{rcl} & 1\frac{1}{2} = & 1\frac{2}{4} \\ + & 2\frac{1}{4} = & 2\frac{1}{4} \\ \hline & & 3\frac{3}{4} \end{array}$$

The child should note that the fractions are added first, then the whole numbers.

Another way to add mixed numbers is to first express them as *improper fractions*. Then you add them the way you would any other fractions.

An *improper fraction* is a fraction whose numerator is equal to or larger than the denominator, like $\frac{2}{2}$ or $\frac{5}{3}$. The value of an improper fraction is 1 or greater than 1.

Changing Mixed Numbers to Improper Fractions

How do you change $1\frac{1}{2}$ dollars to half-dollars? Exchange the 1 dollar for 2 halves. These 2 halves, plus the 1 half you started with, give you altogether 3 halves; that is, $1\frac{1}{2} = \frac{3}{2}$. What we've done, step by step, to change the mixed number $1\frac{1}{2}$ to the improper fraction $\frac{3}{2}$, is this:

$$1\frac{1}{2} = 1 + \frac{1}{2} = \frac{2}{2} + \frac{1}{2} = \frac{3}{2}$$

Following the same procedure, we'll now rename $2\frac{3}{4}$ as an improper fraction:

$$2\frac{3}{4} = 2 + \frac{3}{4} = \frac{8}{4} + \frac{3}{4} = \frac{11}{4}$$

Using the money analogy, you started with 2 dollars and 3 quarters. To change the entire amount to quarters, change the 2 dollars to $\frac{8}{4}$, and add to that the original $\frac{3}{4}$, giving a total of $\frac{11}{4}$.

Similarly,

$$1\frac{3}{5} = 1 + \frac{3}{5} = \frac{5}{5} + \frac{3}{5} = \frac{8}{5},$$

and

$$3\frac{1}{8} = 3 + \frac{1}{8} = \frac{24}{8} + \frac{1}{8} = \frac{25}{8}.$$

To change the improper fraction $\frac{3}{2}$ back to a mixed number, we think of $\frac{3}{2}$ as meaning $2\overline{)3}$. Carrying through the division, we get

$$\begin{array}{r} 1 \text{ r1} = 1\frac{1}{2} \\ 2\overline{)3} \\ \underline{2} \\ 1 \end{array} \qquad \text{Similarly, } \frac{11}{4} = \begin{array}{r} 2 \text{ r3} = 2\frac{3}{4} \\ 4\overline{)11} \\ \underline{8} \\ 3 \end{array}$$

To add $1\frac{1}{2}$ and $2\frac{1}{4}$ by changing them to improper fractions, we write

$$\begin{array}{r} 1\frac{1}{2} = \frac{3}{2} = \frac{6}{4} \\ + 2\frac{1}{4} = \frac{9}{4} = \frac{9}{4} \\ \hline \frac{15}{4} = 3\frac{3}{4}. \end{array} \qquad \text{Similarly, } \begin{array}{r} 2\frac{1}{6} = \frac{13}{6} = \frac{13}{6} \\ + 3\frac{2}{3} = \frac{11}{3} = \frac{22}{6} \\ \hline \frac{35}{6} = 5\frac{5}{6} \end{array}$$

EXERCISES

1. (a) $1\frac{1}{4} + 2\frac{1}{2}$ (b) $2\frac{1}{3} + 1\frac{1}{2}$ (c) $3\frac{1}{8} + 2\frac{1}{4}$
2. Change the following mixed numbers to improper fractions.

   (a) $2\frac{1}{4}$ (b) $1\frac{3}{5}$ (c) $4\frac{2}{3}$ (d) $3\frac{1}{7}$ (e) $8\frac{4}{9}$ (f) $5\frac{3}{10}$
3. Change the following improper fractions to mixed numbers.

   (a) $\frac{8}{3}$ (b) $\frac{11}{2}$ (c) $\frac{13}{4}$ (d) $\frac{10}{5}$ (e) $\frac{9}{8}$

4. Sue bought $1\frac{1}{4}$ yards of blue cloth and $3\frac{1}{2}$ yards of red cloth. How many yards of cloth did she buy altogether?
5. A man works $6\frac{1}{2}$ hours on Wednesday and $3\frac{3}{4}$ hours on Thursday. How many hours did he work altogether?

Subtracting Fractions

Subtraction of fractions has the same meaning and arises from the same situations as does subtraction of whole numbers. The physical materials used to teach addition of fractions can be used to teach subtraction.

1. Subtracting fractions with the same denominator

From previous use of fraction strips, the child already knows that

$$\frac{2}{3} - \frac{1}{3} = \frac{1}{3} \qquad \frac{4}{5} - \frac{1}{5} = \frac{3}{5} \qquad \frac{5}{8} - \frac{2}{8} = \frac{3}{8}$$

The child now has to see that when he subtracted fractions with the same denominator, in effect he *subtracted the numerators and placed their difference over the common denominator*:

$$\frac{2}{3} - \frac{1}{3} = \frac{2-1}{3} = \frac{1}{3} \qquad \frac{5}{8} - \frac{2}{8} = \frac{5-2}{8} = \frac{3}{8}$$

2. Subtracting fractions with different denominators

Example 1: $\frac{1}{2} - \frac{1}{3}$

(a) Find the lowest common denominator: The LCD for $\frac{1}{2}$ and $\frac{1}{3}$ is 6.

(b) Rename each fraction with an equivalent fraction having a denominator of 6, and subtract:

$$\begin{array}{r} \frac{1}{2} = \frac{3}{6} \\ -\frac{1}{3} = \frac{2}{6} \\ \hline \frac{1}{6} \end{array}$$

Example 2: $\frac{3}{4} - \frac{2}{3}$

(a) The LCD for $\frac{3}{4}$ and $\frac{2}{3}$ is 12.

(b)
$$\begin{array}{r} \frac{3}{4} = \frac{9}{12} \\ -\frac{2}{3} = \frac{8}{12} \\ \hline \frac{1}{12} \end{array}$$

Example 3: $\frac{4}{5} - \frac{3}{10}$

(a) The LCD for $\frac{4}{5}$ and $\frac{3}{10}$ is 10.

(b)
$$\begin{array}{r} \frac{4}{5} = \frac{8}{10} \\ -\frac{3}{10} = \frac{3}{10} \\ \hline \frac{5}{10} = \frac{1}{2} \end{array}$$

3. Subtracting mixed numbers

Example 1: $3\frac{1}{2} - 1\frac{1}{4}$
After renaming the fractions, subtract them and then subtract the whole numbers:

$$\begin{array}{r} 3\frac{1}{2} = 3\frac{2}{4} \\ -1\frac{1}{4} = 1\frac{1}{4} \\ \hline 2\frac{1}{4} \end{array}$$

Example 2: $5 - 1\frac{1}{2}$
This problem needs careful explanation. A good way to proceed is to pose the problem in terms of money: Suppose you have 5 dollars and spend $1\frac{1}{2}$ dollars. How much money do you have left? The child will see that if he changes one of the 5 dollars to 2 half-dollars, he can solve the problem this way:

$$\begin{array}{r} 5 \ = 4 + \frac{2}{2} = 4\frac{2}{2} \\ -1\frac{1}{2} = 1 + \frac{1}{2} = 1\frac{1}{2} \\ \hline 3\frac{1}{2} \end{array}$$

Example 3: $3\frac{1}{4} - 2\frac{1}{2}$
This problem too, needs careful explanation and can be given with the money analogy: Suppose you have $3\frac{1}{4}$ dollars and spend $2\frac{1}{2}$ dollars. How much money do you have left? The child will see that if he changes one of the 3 dollars to 4 quarters, he can solve the problem this way:

$$\begin{array}{r} 3\frac{1}{4} = 3\frac{1}{4} = 2 + \frac{4}{4} + \frac{1}{4} = 2\frac{5}{4} \\ -\ 2\frac{1}{2} = 2\frac{2}{4} = 2 + \frac{2}{4} \qquad = 2\frac{2}{4} \\ \hline \frac{3}{4} \end{array}$$

EXERCISES

1. (a) $\frac{2}{3} - \frac{1}{3}$ (b) $\frac{4}{5} - \frac{1}{5}$ (c) $\frac{2}{3} - \frac{4}{9}$ (d) $\frac{7}{8} - \frac{3}{16}$ (e) $\frac{3}{4} - \frac{5}{8}$

2. (a) $4\frac{3}{4} - 2\frac{1}{4}$ (b) $2\frac{3}{4} - 1\frac{2}{3}$ (c) $1\frac{1}{2} - \frac{2}{3}$ (d) $6\frac{1}{5} - \frac{4}{5}$

3. If you cut off a piece of wood measuring $3\frac{1}{4}$ inches from a piece measuring $5\frac{1}{8}$ inches, how large a piece do you have left?

4. The distance between two cities is $12\frac{1}{5}$ miles. After riding $7\frac{3}{10}$ miles, how many more miles remain in the ride?

Multiplying Fractions

Multiplication is usually taught in three stages: (a) multiplication of a whole number and a fraction (b) multiplication of two fractions (c) multiplication with mixed numbers

1. Multiplying a whole number and a fraction can be interpreted as repeated addition. For example,

$$3 \times \frac{1}{4} \text{ means } \frac{1}{4} + \frac{1}{4} + \frac{1}{4} = \frac{3}{4}.$$

$$\text{That is, } 3 \times \frac{1}{4} = \frac{3}{4}.$$

Since $3 \times \frac{1}{4} = \frac{1}{4} \times 3$ (commutative property, remember?), $\frac{1}{4} \times 3$ also equals $\frac{3}{4}$.

$$\text{That is, } \frac{1}{4} \times 3 = \frac{3}{4}.$$

Since the whole number 3 can be expressed as $\frac{3}{1}$, we have

$$3 \times \frac{1}{4} = \frac{3}{1} \times \frac{1}{4} = \frac{3}{4},$$

and

$$\frac{1}{4} \times 3 = \frac{1}{4} \times \frac{3}{1} = \frac{3}{4}.$$

Similarly,

$$\frac{2}{5} \times 2 = \frac{2}{5} \times \frac{2}{1} = \frac{4}{5},$$

and

$$5 \times \frac{3}{16} = \frac{5}{1} \times \frac{3}{16} = \frac{15}{16}.$$

After working through a number of such exercises, the child will observe that the product of a whole number and a fraction can be found by (1) rewriting the whole number as a fraction with a denominator of 1, then (2) multiplying the numerators to obtain the numerator of the answer, and multiplying the denominators to obtain the denominator of the answer:

$$2 \times \frac{4}{9} = \frac{2}{1} \times \frac{4}{9} = \frac{2 \times 4}{1 \times 9} = \frac{8}{9}$$

$$\frac{3}{16} \times 3 = \frac{3}{16} \times \frac{3}{1} = \frac{3 \times 3}{16 \times 1} = \frac{9}{16}$$

2. Multiplication of two fractions is accomplished by the same procedure:

$$\frac{1}{2} \times \frac{1}{3} = \frac{1 \times 1}{2 \times 3} = \frac{1}{6}$$

$$\frac{2}{3} \times \frac{4}{5} = \frac{2 \times 4}{3 \times 5} = \frac{8}{15}$$

NOTE:
- The word *of* is often used to denote multiplication. For instance, $\frac{1}{4} \times 3$ is sometimes stated as "$\frac{1}{4}$ of 3," and $\frac{1}{2} \times \frac{1}{3}$ as "$\frac{1}{2}$ of $\frac{1}{3}$."
- The child should understand that when a number is multiplied by a fraction smaller than 1, the product is always smaller than the number because a fractional part of the given number is being found.

"Seeing" $\frac{1}{3} \times \frac{1}{2}$

The product of two fractions can be visualized through a geometric shape like the rectangle in Fig. 14-18. Let's see what $\frac{1}{3} \times \frac{1}{2}$ means in this rectangle.

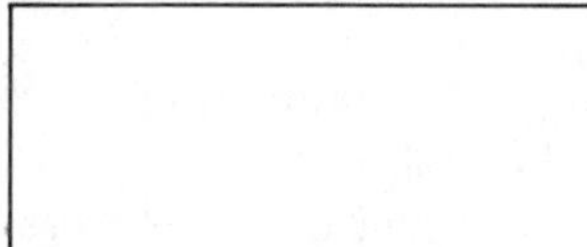

Figure 14-18

$\frac{1}{3} \times \frac{1}{2}$, or $\frac{1}{3}$ of $\frac{1}{2}$ of the rectangle means:

1. **Shade (horizontally) $\frac{1}{2}$ of the rectangle (Fig. 14-19).**

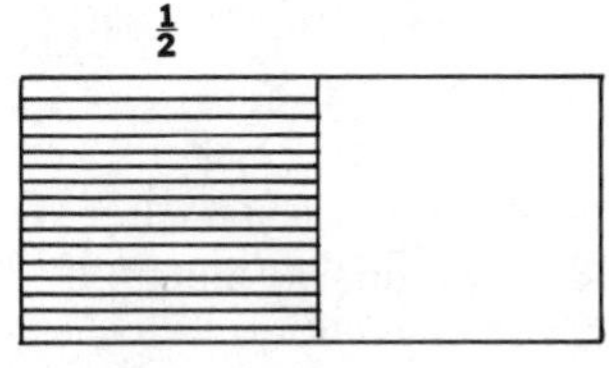

Figure 14-19

2. **Shade (vertically) $\frac{1}{3}$ of the $\frac{1}{2}$ (Fig. 14-20).**

The result is that $\frac{1}{3}$ of $\frac{1}{2}$ of the rectangle (the cross-hatched piece) is $\frac{1}{6}$ of the entire rectangle.

That is, $\frac{1}{3} \times \frac{1}{2} = \frac{1}{6}$.

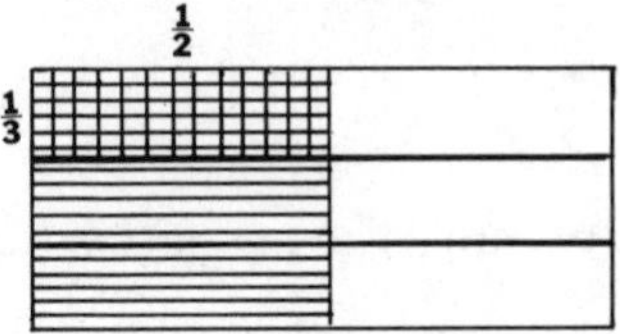

Figure 14-20

3. Multiplication with mixed numbers is accomplished by renaming the mixed numbers as improper fractions and then proceeding as before. For example,

$$2\frac{1}{2} \times \frac{3}{8} = \frac{5}{2} \times \frac{3}{8} = \frac{15}{16},$$

and

$$1\frac{2}{3} \times 2\frac{1}{8} = \frac{5}{3} \times \frac{17}{8} = \frac{85}{24} = 3\frac{13}{24}.$$

4. Simplifying the multiplication of fractions

It is sometimes possible to simplify multiplication of fractions by dividing any numerator and denominator by the greatest common factor. For example, the multiplication $\frac{2}{3} \times \frac{6}{7}$ can be simplified by dividing out the common factor 3 *before* carrying out the multiplication. That is,

$$\frac{2}{\cancel{3}_{1}} \times \frac{\cancel{6}^{2}}{7} = \frac{4}{7}$$

Similarly,

$$\frac{9}{16} \times \frac{8}{15} = \frac{\cancel{9}^{3}}{\cancel{16}_{2}} \times \frac{\cancel{8}^{1}}{\cancel{15}_{5}} = \frac{3}{10}$$

You divide out the common factor *3* from 9 and 15 and the common factor *8* from 8 and 16.

The process of dividing out a common factor is sometimes called *cancellation.* While cancellation can be helpful, it may lead to errors when the child does not really understand the process. Children often cross out numerators and denominators when adding or subtracting. For instance,

$$\frac{\cancel{3}}{4} + \frac{1}{\cancel{3}} \quad \text{or} \quad \frac{4}{\cancel{5}} - \frac{\cancel{5}}{9}$$

It is, therefore, suggested that you delay introducing cancellation until the child understands all the operations on fractions.

EXERCISES

1. (a) $\frac{1}{2} \times 8$ (b) $24 \times \frac{1}{3}$ (c) $\frac{3}{4} \times 12$ (d) $35 \times \frac{2}{5}$ (e) $\frac{3}{4} \times 48$

2. (a) $\frac{1}{2} \times \frac{1}{4}$ (b) $\frac{1}{4} \times \frac{3}{8}$ (c) $\frac{2}{3} \times \frac{3}{4}$ (d) $\frac{3}{5} \times \frac{5}{6}$ (e) $\frac{4}{15} \times \frac{5}{16}$

3. (a) $5\frac{1}{2} \times \frac{2}{5}$ (b) $\frac{4}{9} \times 3\frac{1}{4}$ (c) $24 \times 7\frac{3}{4}$ (d) $2\frac{1}{3} \times \frac{3}{10}$ (e) $2\frac{1}{4} \times 1\frac{3}{8}$
4. If $1\frac{5}{8}$ yards of material is needed to upholster a chair, how many yards are needed to upholster 4 chairs?
5. A man drove his car at an average speed of 40 miles an hour for $2\frac{3}{4}$ hours. How many miles did he drive?

Dividing Fractions

Division involves dividing (a) whole numbers by fractions, (b) fractions by whole numbers, and (c) fractions by fractions. Dividing mixed numbers is the same as dividing fractions by fractions since mixed numbers can be converted to fractions.

Division of fractions means the same as division of whole numbers:

$6 \div 2$ asks: How many twos are contained in 6?

$3 \div \frac{1}{2}$ asks: How many halves are contained in three whole units?

A child will understand this question better if you relate it to money or to fraction strips:

Money: $3 \div \frac{1}{2}$ asks: How many half-dollars are contained in 3 dollars?

The child should have no problem figuring out that 3 dollars contains 6 half-dollars. That is,

$$3 \div \frac{1}{2} = 6.$$

Fraction Strips: The child lays out three whole units and counts the number of half-strips that add up to the same length. He will again have little difficulty determining that there are 6 halves in 3 whole units. That is, $3 \div \frac{1}{2} = 6$.

By the same procedures, the child determines that

$$\frac{1}{2} \div \frac{1}{4} = 2,$$

$$\frac{3}{4} \div \frac{1}{8} = 6,$$

$$6 \div \frac{3}{4} = 8,$$

and

$$2\frac{1}{2} \div \frac{1}{2} = 5.$$

After carefully examining these results, worked out with fraction strips, the child might be asked: Can you think of a way to get these answers *without* using fraction strips?

Eventually, your child should be guided to see that if the *second* fraction is *inverted*, the answer to the division turns out to be the *product* of the two fractions:

By Fraction Strips	*By Inverting and Multiplying*
1. $3 \div \frac{1}{2} = 6$	*1.* $3 \div \frac{1}{2} = \frac{3}{1} \times \frac{2}{1} = \frac{6}{1} = 6$
2. $\frac{1}{2} \div \frac{1}{4} = 2$	*2.* $\frac{1}{2} \div \frac{1}{4} = \frac{1}{2} \times \frac{4}{1} = \frac{4}{2} = 2$
3. $\frac{3}{4} \div \frac{1}{8} = 6$	*3.* $\frac{3}{4} \div \frac{1}{8} = \frac{3}{4} \times \frac{8}{1} = \frac{24}{4} = 6$
4. $6 \div \frac{3}{4} = 8$	*4.* $6 \div \frac{3}{4} = \frac{6}{1} \times \frac{4}{3} = \frac{24}{3} = 8$
5. $2\frac{1}{2} \div \frac{1}{2} = 5$	*5.* $2\frac{1}{2} \div \frac{1}{2} = \frac{5}{2} \times \frac{2}{1} = \frac{10}{2} = 5$

To "invert and multiply" is the quickest and most efficient way to divide fractions. But if it is given to the child as a blind rule to follow, he is likely to make many errors, such as inverting the wrong fraction, because he doesn't understand the reason for the rule.

Unfortunately, the reason for this rule—developed below under **For the Curious**—can be confusing to many children, though it is based on ideas with which they are already familiar.

So, perhaps, the best procedure might be for the child to be convinced that the rule "works" by comparing the answers he gets using money or fraction strips with those he gets using the rule.

Then, when you think the child is ready to understand the reason for the rule, present it very slowly and carefully.

EXERCISES

1. (a) $\frac{3}{4} \div \frac{3}{8}$ (b) $\frac{7}{8} \div \frac{5}{16}$ (c) $\frac{3}{4} \div 5$ (d) $15 \div \frac{5}{6}$ (e) $\frac{3}{5} \div \frac{3}{5}$ (f) $3\frac{1}{4} \div \frac{1}{4}$
   (g) $1\frac{2}{3} \div 2\frac{1}{3}$ (h) $2\frac{1}{4} \div 1\frac{2}{5}$
2. How many $1\frac{1}{4}$-inch strips can you cut from a 15-inch piece of metal?
3. A car travels 117 miles in $2\frac{1}{4}$ hours. What is the average speed of the car?

Why "Invert and Multiply"?

We are going to show that $\frac{1}{3} \div \frac{2}{5} = \frac{1}{3} \times \frac{5}{2}$. That is, that a division can be changed to a multiplication *if you invert the second fraction*.

(continued)

1. **A division can be written as a fraction:**

$$\frac{1}{3} \div \frac{2}{5} \text{ may be written as } \frac{\frac{1}{3}}{\frac{2}{5}}$$

In this fraction, $\frac{1}{3}$ is the numerator, and $\frac{2}{5}$ is the denominator.

2. **When you invert a fraction, you get its *reciprocal*. For instance, the reciprocal of $\frac{3}{7}$ is $\frac{7}{3}$, and the reciprocal of $\frac{2}{5}$ is $\frac{5}{2}$.**

3. **The product of a fraction (nonzero) and its reciprocal is 1. For instance,**

$$\frac{3}{7} \times \frac{7}{3} = 1, \text{ and } \frac{2}{5} \times \frac{5}{2} = 1.$$

4. **If the numerator and denominator of a fraction are multiplied by the same number, the value of the fraction is unchanged.**

We are now going to multiply the numerator and denominator of the fraction $\frac{\frac{1}{3}}{\frac{2}{5}}$ by the same number, $\frac{5}{2}$:

$$\frac{\frac{1}{3}}{\frac{2}{5}} = \frac{\frac{1}{3} \times \frac{5}{2}}{\frac{2}{5} \times \frac{5}{2}} = \frac{\frac{1}{3} \times \frac{5}{2}}{1}$$

5. **Any number divided by 1 results in the same number:**

$$\frac{\frac{1}{3} \times \frac{5}{2}}{1} = \frac{1}{3} \times \frac{5}{2}$$

6. **Therefore,** $\frac{1}{3} \div \frac{2}{5} = \frac{1}{3} \times \frac{5}{2}$

Step 6 says that dividing one fraction by another is the same as multiplying the first fraction by the reciprocal of the second. Thus,

$$\frac{1}{2} \div \frac{2}{3} = \frac{1}{2} \times \frac{3}{2} = \frac{3}{4}$$

and

$$\frac{2}{5} \div \frac{7}{9} = \frac{2}{5} \times \frac{9}{7} = \frac{18}{35}$$

Activities Involving Fractions

1. To reinforce your child's understanding of what a division like $2 \div \frac{1}{4}$ means, cut a strip of any stiff material to measure two units. Then cut about a dozen strips, each measuring $\frac{1}{4}$ of the unit. Ask child to lay out, end to end, enough $\frac{1}{4}$ units to cover the length of the two-unit strip, and then count the number of $\frac{1}{4}$ units needed (Fig. 14-21).

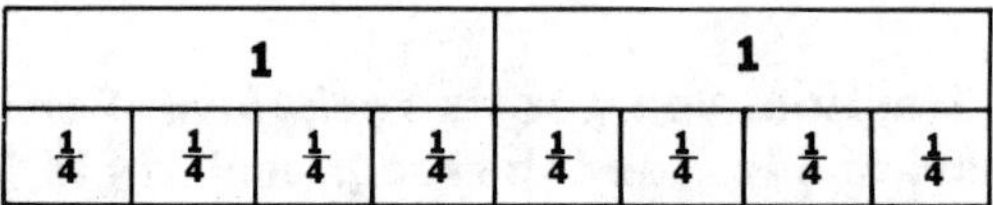

Figure 14-21

The child will observe that when she divided the two units by $\frac{1}{4}$, she was really using the fraction as a *unit of measure.*

2. Prepare a set of cards containing basic fractions (halves, thirds, fourths, fifths, and eighths). Lay out several of these cards in random order (Fig. 14-22), and ask the child to:
   (a) read each fraction aloud
   (b) arrange them in ascending order (from smallest to largest); for instance,

Figure 14-22

3. The cards in #2 can also be used for the following activity: Lay out three or more cards on the table (Fig. 14-23). The object is for the child to add, subtract, multiply, or divide the fractions on the left of the = sign to make the result equal to the fraction on the right of the = sign. For instance,

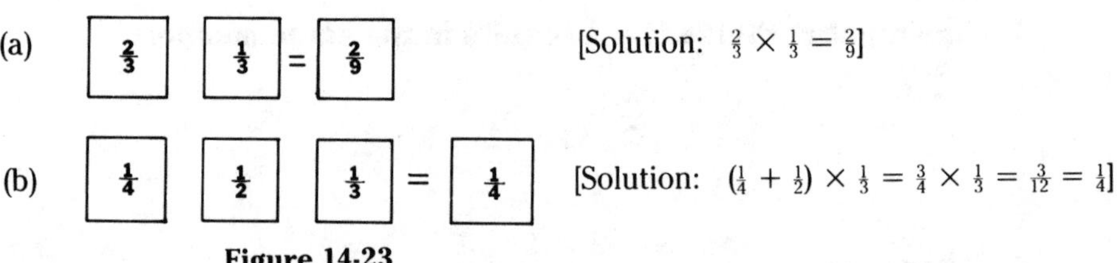

(a) [Solution: $\frac{2}{3} \times \frac{1}{3} = \frac{2}{9}$]

(b) [Solution: $(\frac{1}{4} + \frac{1}{2}) \times \frac{1}{3} = \frac{3}{4} \times \frac{1}{3} = \frac{3}{12} = \frac{1}{4}$]

Figure 14-23

4. Game
   Materials Needed:
   Two spinners marked with fractional numerals
   One spinner marked with operation signs

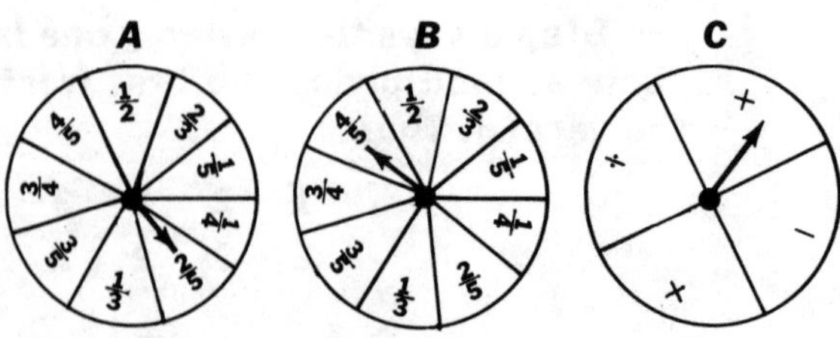

Figure 14-24

Game Action:
(1) The player obtains two fractions from spinners A and B (Fig. 14-24), and an operation sign from spinner C.

(2) The player is required to perform the indicated operation on the two fractions. A correct answer earns 5 points. An incorrect answer loses 2 points.
(3) The play alternates between players; the highest score after a set amount of time wins.

Common Errors with Fractions

Below are common errors children make with fractions. Next to each example is the wrong procedure, shown in brackets, that led to the incorrect answer.

1. Change $\frac{2}{3}$ to an equivalent fraction having a denominator of 9:

$$\frac{2}{3} = \frac{4}{9} \quad \left[\frac{2 \times 2}{3 \times 3} = \frac{4}{9} \text{ instead of } \frac{2 \times 3}{3 \times 3} = \frac{6}{9}\right]$$

2. $\frac{3}{4} = 1\frac{1}{3}$ $\left[\text{Treated } \frac{3}{4} \text{ as if it were } \frac{4}{3}.\right]$

3. $\frac{1}{5} > \frac{1}{3}$ $\left[\text{Thought } \frac{1}{5} > \frac{1}{3} \text{ because } 5 > 3.\right]$

4. $\frac{3}{5} = \frac{4}{6}$ $\left[\textit{Added} \text{ 1 to numerator and denominator: } \frac{3 + 1}{5 + 1}\right]$

5. $\frac{2}{3} + \frac{3}{4} = \frac{5}{7}$ $\left[\text{Added numerators and denominators: } \frac{2 + 3}{3 + 4}\right]$

6. $\frac{3}{5} - \frac{1}{2} = \frac{2}{3}$ $\left[\text{Subtracted numerators and denominators: } \frac{3 - 1}{5 - 2}\right]$

7. $2\frac{1}{5} + 2\frac{3}{5} = 2\frac{4}{5}$ [Didn't add the whole numbers.]

8. $3\frac{1}{2} - 1\frac{3}{4} = 2\frac{1}{4}$ $\left[\text{Subtracted } \frac{1}{2} \text{ from } \frac{3}{4}.\right]$

9. $\frac{2}{3} \times \frac{2}{3} = \frac{4}{3}$ [Didn't multiply the denominators.]

10. $\frac{2}{3} \div \frac{5}{8} = \frac{15}{16}$ $\left[\text{Inverted first fraction instead of second: } \frac{3}{2} \times \frac{5}{8} = \frac{15}{16}.\right]$

11. $\frac{3}{4} \div \frac{7}{10} = \frac{40}{21}$ $\left[\text{Inverted } \textit{both} \text{ fractions: } \frac{4}{3} \times \frac{10}{7}.\right]$

12. $1\frac{1}{2} \times 2\frac{3}{5} = 2\frac{3}{10}$ [Multiplied the whole numbers, then the fractions, separately.]

Comparing Fractions

Which is larger: $\frac{2}{5}$ or $\frac{3}{8}$?

A good way to find out is to rename both fractions with equivalent fractions having the same denominator. Then compare the numerators:

$$\frac{2}{5} = \frac{16}{40}$$

$$\frac{3}{8} = \frac{15}{40}$$

Since $\frac{16}{40}$ is greater than $\frac{15}{40}$, $\frac{2}{5} > \frac{3}{8}$.

An easier way to get the same result is to *cross-multiply.* Then compare the products:

$$\frac{2}{5} \times ? \times \frac{3}{8}$$

$$2 \times 8 \quad ? \quad 3 \times 5$$

$$16 > 15$$

Therefore, $\frac{2}{5} > \frac{3}{8}$

Similarly,

$$\frac{2}{3} \times ? \times \frac{3}{4}$$

$$2 \times 4 \quad ? \quad 3 \times 3$$

$$8 < 9$$

Therefore, $\frac{2}{3} < \frac{3}{4}$

The cross-multiplication method is basically the same as the first method, except that in cross-multiplication you compare numerators without bothering to write down the denominators.

HIGHLIGHTS OF FRACTIONS

The study of fractions should leave your child with the following understandings:

- A fraction is *part* of something. It can be part of a unit or part of a set of objects.
- A fraction can also be used to indicate division; $\frac{2}{3}$ means $2 \div 3$ or $3\overline{)2}$; $19\overline{)15}$ can be written $\frac{15}{19}$.
- When the numerator *and* denominator of any fraction are multiplied or divided by the *same* number (except 0), the value of the fraction is not changed.

Also, at the completion of this chapter, your child should be able to

- Tell which of two fractions is the greater
- Rename fractions with equivalent fractions having any desired denominator
- Find the least common multiple of a set of numbers; also, find the lowest common denominator of two or more fractions
- Add, subtract, multiply, and divide fractions as well as mixed numbers
- Explain and use the following vocabulary:

fraction	mixed number
numerator	equivalent fractions
denominator	least common denominator
improper fraction	

QUESTIONS ON THE CHAPTER

1. Write as a fraction:
   (a) $5\overline{)3}$ (b) $1 \div 6$ (c) two thirds
2. Change:
   (a) $\frac{1}{8}$ to fortieths (b) $\frac{3}{4}$ to eighths (c) $\frac{2}{5}$ to twentieths
3. Reduce each fraction to its lowest terms.
   (a) $\frac{3}{9}$ (b) $\frac{4}{8}$ (c) $\frac{6}{21}$ (d) $\frac{12}{32}$

4. Change to improper fractions.
   (a) $2\frac{1}{4}$ (b) $1\frac{3}{5}$ (c) $3\frac{1}{8}$ (d) $2\frac{4}{9}$

5. Change to mixed numbers.
   (a) $\frac{8}{3}$ (b) $\frac{11}{2}$ (c) $\frac{13}{4}$ (d) $\frac{7}{5}$ (e) $\frac{19}{8}$

6. Add:
   (a) $\frac{1}{2} + \frac{3}{4}$ (b) $\frac{1}{2} + \frac{3}{8}$ (c) $\frac{7}{8} + \frac{5}{6}$ (d) $3\frac{2}{3} + 5$ (e) $2\frac{1}{5} + 3\frac{3}{5}$ (f) $\frac{3}{4} + \frac{7}{8} + \frac{1}{6}$

7. Subtract:
   (a) $\frac{3}{5} - \frac{1}{5}$ (b) $\frac{5}{6} - \frac{1}{2}$ (c) $\frac{7}{8} - \frac{1}{5}$ (d) $\frac{3}{4} - \frac{7}{10}$ (e) $\frac{15}{16} - \frac{3}{4}$ (f) $3\frac{4}{5} - 1\frac{2}{5}$
   (g) $5\frac{1}{3} - 2\frac{2}{3}$ (h) $4\frac{3}{8} - 1\frac{7}{8}$ (i) $5 - 1\frac{3}{4}$ (j) $3 - \frac{4}{5}$ (k) $4\frac{3}{4} - 1\frac{1}{3}$ (l) $2\frac{5}{8} - 1\frac{11}{12}$

8. Multiply:
   (a) $\frac{1}{5} \times \frac{1}{3}$ (b) $\frac{5}{8} \times \frac{3}{4}$ (c) $\frac{3}{5} \times 5$ (d) $\frac{3}{4} \times 2$ (e) $4 \times \frac{7}{8}$ (f) $2\frac{1}{2} \times \frac{4}{5}$
   (g) $1\frac{5}{8} \times 3\frac{2}{5}$

9. Divide:
   (a) $\frac{1}{3} \div \frac{3}{4}$ (b) $\frac{2}{3} \div \frac{5}{16}$ (c) $\frac{3}{4} \div \frac{3}{8}$ (d) $\frac{7}{8} \div 2$ (e) $10 \div \frac{1}{2}$ (f) $4 \div \frac{3}{5}$
   (g) $1\frac{2}{3} \div 2$ (h) $2\frac{1}{2} \div \frac{5}{6}$ (i) $1\frac{1}{6} \div 9\frac{1}{3}$ (j) $5\frac{2}{3} \div 5\frac{2}{3}$

10. Which is greater:
    (a) $\frac{1}{2}$ or $\frac{1}{3}$? (b) $\frac{3}{4}$ or $\frac{4}{5}$? (c) $\frac{3}{2}$ or 2?

    (d) $\frac{3}{8}$ of an inch or $\frac{1}{2}$ of an inch?

    (e) $\frac{7}{5}$ or $\frac{9}{7}$? (f) $\frac{8}{5}$ or $\frac{5}{3}$? (g) $\frac{7}{8}$ or $\frac{8}{7}$?

11. Find the missing numbers. (a) $3 = 2\frac{?}{4}$ (b) $1\frac{2}{3} = \frac{?}{3}$

12. Arrange in ascending order: $\frac{7}{12}$, $\frac{3}{8}$, $\frac{3}{5}$, 2

13. Find the average of $\frac{1}{2}$ and $\frac{1}{3}$.

14. A plumber needed $3\frac{7}{8}$ feet of pipe for one job and $1\frac{3}{4}$ feet for a second job. How many feet of pipe did he need for both jobs?

15. What is the distance around a triangle if its 3 sides measure $1\frac{3}{8}$ inches, $2\frac{11}{16}$ inches, and $4\frac{3}{4}$ inches?

16. Josh weighs $72\frac{1}{2}$ pounds. A year ago he weighed $59\frac{3}{4}$ pounds. How many pounds did he gain this year?

17. Lindsay bought $5\frac{3}{8}$ yards of cloth but used only $4\frac{3}{4}$ yards. How much cloth did she have left?

18. If you need $3\frac{3}{4}$ feet of wood to make a shelf, how many feet will you need to make 5 shelves?

19. A recipe calls for $2\frac{1}{2}$ cups of flour and $1\frac{1}{4}$ cups of sugar. How many cups of flour and sugar are needed for 6 times the recipe?

20. How many pieces of paper measuring $2\frac{3}{8}$ inches in length can be cut out of a sheet 14 inches long?

21. If $1\frac{3}{4}$ pounds of meat cost $3, how much does 1 pound cost?

22. A woman drives 110 miles in $2\frac{1}{4}$ hours. What is her average speed?

23. Find the fraction midway between:
    (a) $\frac{1}{2}$ and $\frac{3}{8}$ (b) $\frac{2}{3}$ and $\frac{4}{5}$

24. What number added to $\frac{2}{3}$ gives $\frac{5}{6}$?

25. What number subtracted from $\frac{3}{4}$ gives $\frac{2}{5}$?

26. What number multiplied by $\frac{2}{5}$ equals $\frac{3}{4}$?

27. Is the value of a fraction changed:
    (a) If you subtract the same number from both the numerator and the denominator? Why?
    (b) If you add the same number to the numerator and the denominator? Why?

28. If the numerators of two fractions are the same but their denominators are different, how can you tell which fraction is greater?

29. Give the fraction of the day you spend on: sleeping; school; watching TV; eating; everything else.

REVIEW

1. What's the place value of the *3* in each of the following numbers:
   (a) 35 (b) 293 (c) 350

2. What number is 14 more than 21?

3. A man bought a car for $8750. If he paid $800 in cash and arranged to pay off the balance in 8 equal monthly payments, what is the amount of each payment?

4. Find *mentally* the average of 9, 6, 7, 10, 8.

5. Find the missing numbers:
   (a) ? × 3 = 21 (b) 5 × ? = 35 (c) 12 ÷ ? = 2

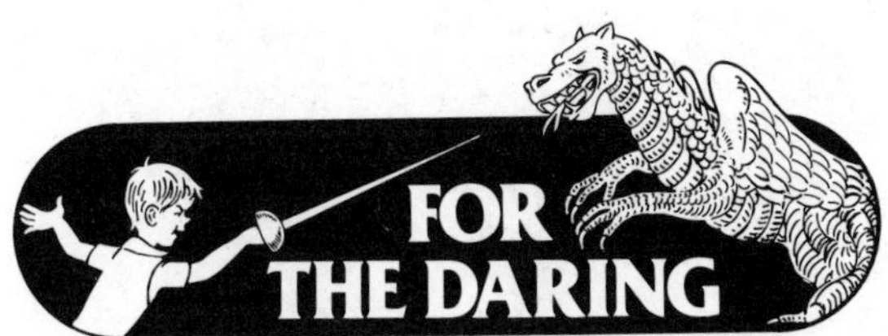

1. A boy spends one third of his money and loses two thirds of the remainder. He then has $12 left. How much did he have at the beginning?

2. A worker is to be paid $25 for each acceptable tape recorder he completes, but is penalized $10 for each defective one. After completing 25 recorders, he was paid $380. How many acceptable recorders did he make?

(continued)

3. How many fractions are there *between* $\frac{1}{4}$ and $\frac{1}{2}$? How can you prove your answer?

4. In a certain town there reside families called Jones, Smith, Carter, and Baker. Determine the number of members in each family from the following clues:
(a) There are more Bakers than Carters.
(b) The number of Bakers is less than 10.
(c) The number of Carters is twice the number of Joneses.
(d) There is an even number of Joneses.
(e) The number of Smiths plus the number of Joneses equals 5.
(f) The number of Smiths is less than 3.

5. Three neighborhood dogs barked continuously last night. Rex, Muff, and Punch began with a simultaneous bark at 10 P.M. Then Rex barked every 4 minutes; Muff, every 2 minutes; and Punch, every 5 minutes.

When Tommy Reese awakened at midnight, did he do so because there was no barking, some dogs were barking, or all the dogs were barking at that hour?

(The answers appear on page 259.)

Chapter 15
Decimals

INTRODUCTION

The widespread acceptance of decimals about 400 years ago was a major breakthrough in computation. Before decimals, computing with fractions like 53/357 could become a bit sticky. But being able to convert them to decimals changed all this. Decimals, an extension of the place value concept, makes computation with fractions very similar to computation with whole numbers.

Though something was known about decimals in ancient China, medieval Arabia, and Renaissance Europe, they were not fully understood or widely used until the latter part of the sixteenth century, when Simon Stevin (1548–1620), a Dutch mathematician, explained the system in detail. "I wish," he declared, "to teach everyone how to perform with an ease, unheard of, all computations necessary between men by integers without fractions."

Stevin's *concept* of decimals is the one we use today, but his way of *writing* them is not. The modern way of writing decimals—using a decimal point to separate the whole and the fractional parts of a number—first appeared in 1616 in a book by the Scotch mathematician John Napier (1550–1617), the inventor of logarithms. But even today, the symbol for a decimal is not universal. In England, the decimal point is positioned higher on the line than in the United States. For instance, in England, 3.14 is written 3·14, while in some European countries a comma is used instead of a decimal point: 3,14.

Decimals are the second leg of a triad consisting of *fractions*, *decimals*, and *percents*. In this chapter you'll see how two of them—fractions and decimals—are related. In the next chapter you'll see how the three are related and how all say the same thing in different ways.

Materials used in this chapter include:

Several sheets of squared grid paper (see page 176)

Several sheets of bird arrays (see page 72)

A centimeter ruler

Play money (dollars, dimes, and pennies)

Two spinners (see page 186)

Decimal strips (see page 186)

THE CHAPTER IN A NUTSHELL

In this chapter you will see how to

- Prepare your child for the study of decimals
- Relate decimals to fractions
- Explain decimals as an extension of the concept of place value
- Teach addition, subtraction, multiplication, and division with decimals
- Recognize and correct common errors made with decimals
- Use a variety of activities to motivate the study of decimals

You will also be given

- A list of understandings and skills your child should be left with at the completion of the chapter
- Questions to test and reinforce your child's comprehension of decimals; also, review questions on earlier material
- Further insight into "terminating" and "repeating" decimals—under **For the Curious**
- Several challenging problems and puzzles—under **For the Daring**

WHAT'S A DECIMAL?

A **decimal** is a fraction with a special denominator. The denominator is always 10, 100, 1000, or some other power of 10. We use place value to indicate which power of 10. For instance, the fraction $\frac{7}{10}$ is written .7; that is,

$$\frac{7}{10} = .7$$

We write the numerator 7, and indicate that the denominator is 10 by placing the **decimal point** in such a position that there is exactly *one* digit to its right.

To indicate a denominator of 100, we place the decimal point so that there are *two* digits to its right. For instance,

$$\frac{7}{100} = .07$$

Note that 0 counts as a digit.

To indicate a denominator of 1000, place the decimal point so that there are *three* digits to its right. For instance,

$$\frac{7}{1000} = .007$$

Four decimal places indicate a denominator of 10,000, and so on.

The child should understand that in any decimal

- The *number* to the right of the decimal point represents the *numerator* of the corresponding fraction.
- The *number of digits* to the right of the decimal point indicates the *denominator*.

Thus,

$$.3 = \frac{3}{10}$$

$$.25 = \frac{25}{100}$$

$$.013 = \frac{13}{1000}$$

$$.0005 = \frac{5}{10000}$$

Since $\frac{3}{10} = \frac{30}{100} = \frac{300}{1000} = \frac{3000}{10000}$, it follows that $.3 = .30 = .300 = .3000$

This shows that any number of zeros may be added at the *end* of a decimal without changing its value.

The child should also note that the value of any decimal is always less than 1, no matter to how many places it is carried. When the value is greater than 1, as in 3.2, the number is called a *mixed decimal number*. Here, the decimal point serves as a "road sign" saying that the whole number has ended and the fractional part of the number is beginning.

Decimals and Place Value

Decimals are an extension of the place value concept to the *right* of the ones place. In this extended system, the value of each place continues to be *one tenth* the value of the place to its immediate *left* (Fig. 15-1):

thousand	hundred	ten	one	tenth	hundredth	thousandth
1000	100	10	1	$\frac{1}{10}$	$\frac{1}{100}$	$\frac{1}{1000}$
1000	100	10	1	.1	.01	.001

Figure 15-1

Note the symmetrical relationships around the *ones* place as the reference point (Fig. 15-2):

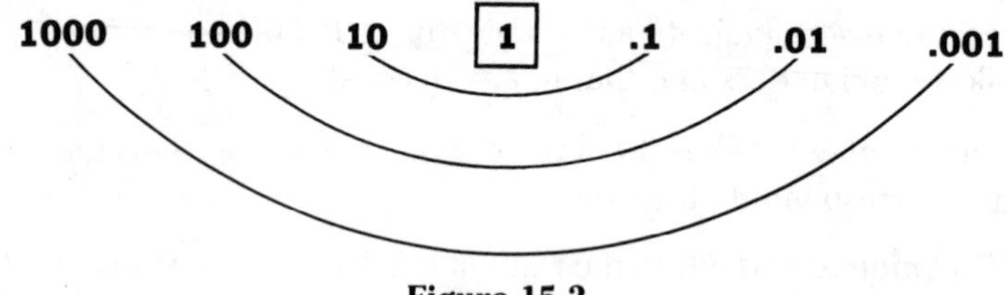

Figure 15-2

Also note that it is the suffix *-ths* that distinguishes the names of the whole number positions from the names of the decimal positions:

ten—ten*ths*; hundred—hundred*ths*; thousand—thousand*ths*

Common Fractions and Decimal Fractions

Fractions like $\frac{1}{4}$ and decimals like .7 are both fractions. But $\frac{1}{4}$ is often called a *common fraction*, while .7 is called a *decimal fraction*. In this book we'll refer to a common fraction as a *fraction*, and to a decimal fraction as a *decimal*.

READINESS FOR DECIMALS

Before your child is ready for the study of decimals, he must have an understanding of the

- Basic operations with whole numbers
- Concept of place value as used with whole numbers
- Concepts related to fractions
- Basic operations with fractions

Then follow the development of decimals as presented on pages 172 and 173.

Reading Decimals

To read a decimal, read it initially as if it were a whole number; then use the name that applies to the position of the last digit. For instance,

.2	is read as	"two tenths."
.06	is read as	"six hundredths."
.375	is read as	"three hundred seventy-five thousandths."

To read a mixed decimal, first read the whole number part, then the fractional part. For instance,

4.9	is read as	"four *and* nine tenths."
35.02	is read as	"thirty-five *and* two hundredths."

EXERCISES

1. Read the following decimals:

   (a) .3 (b) .08 (c) .005 (d) 6.38 (e) .348
   (f) 2.7 (g) 4.0001

2. Write each as a decimal or mixed decimal:

   (a) four tenths (b) two and three tenths
   (c) five hundredths (d) fifteen thousandths
   (e) one and twenty-five hundredths

TEACHING DECIMALS

Tenths

Introduce the concept of *tenths* by giving your child a strip of paper subdivided into 10 equal parts. From earlier experience with fractions the child knows that the entire strip is the whole unit, and each part is $\frac{1}{10}$ of the unit. Since .1 is another way of writing $\frac{1}{10}$, each part is .1 of the entire strip (Fig. 15-3):

1									
$\frac{1}{10}$	$\frac{1}{10}$	$\frac{1}{10}$	$\frac{1}{10}$	$\frac{1}{10}$	$\frac{1}{10}$	$\frac{1}{10}$	$\frac{1}{10}$	$\frac{1}{10}$	$\frac{1}{10}$
.1	.1	.1	.1	.1	.1	.1	.1	.1	.1

Figure 15-3

1. Ask your child to shade .3 of the strip. Repeat with other decimal parts.
2. Write a number like 1.8 and ask the child to represent it with shaded strips. Repeat with other mixed decimals.

Hundredths

Give your child a 10 × 10 sheet of squared grid paper (Fig. 15-4), bird arrays, or ordinary graph paper. He will see that each small square in the sheet represents $\frac{1}{100}$ of the entire sheet. Since .01 is another way of writing $\frac{1}{100}$, each square is .01 of the whole sheet.

1. Ask the child to shade various hundredths parts of the sheet.
2. Ask him to represent 1.50, 2.85, and other mixed decimals on squared grid paper or with bird arrays.

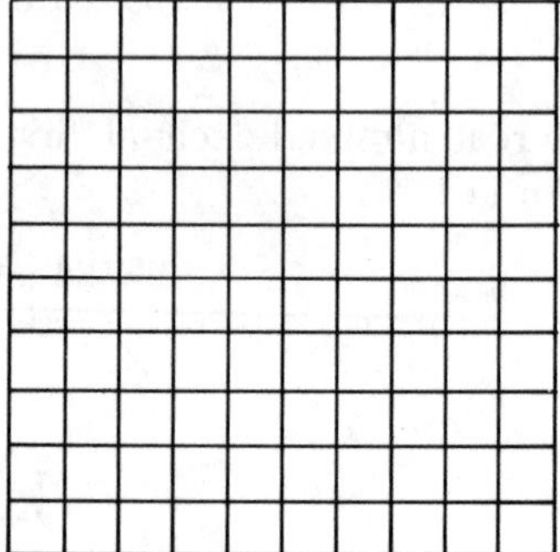

Figure 15-4

As examples of the use of *hundredths*, mention money: cents are hundredths of a dollar; or explain that the ERA (earned run average) for pitchers is given in hundredths of a run; the speed of seagoing ships is given in hundredths of a knot (1 knot = 1 nautical mile per hour).

3. Display different combinations of dollars and cents. Ask the child to write each combination as a decimal.
4. Have your child measure the lengths of small objects with a centimeter ruler to the nearest tenth of a centimeter and then record the lengths as decimals.
5. Write down any five decimals and ask the child to arrange them in ascending order.
6. Ask the child to express decimals as fractions—for instance, 2.07 as $2\frac{7}{100}$.
7. Ask the child to complete sequences like
   (a) .01, .03, .05, ___, ___, ___.
   (b) .2, .5, .8, ___, ___, ___.
   (c) 1.02, 1.07, 1.12, ___, ___, ___.
   (d) 3.9, 3.7, 3.5, ___, ___, ___.

Adding and Subtracting Decimals

1. Adding and subtracting decimals is the same as adding and subtracting whole numbers, except for the presence of a decimal point. Point out to the child that since, with whole numbers, ones are added to ones, tens to tens, hundreds to hundreds (and so on), we can expect to do the same with decimals because both share the same place value system. That's exactly what we do: We add tenths to tenths, hundredths to hundredths, thousandths to thousandths .... In subtraction we follow the same idea.

Introduce these operations through simple problems like the following:

(1) I made one purchase for \$.50 and another for \$.35. How much did I spend altogether?

(2) On Monday, .4 inches of rain fell; on Tuesday, .3 inches more fell. How many inches of rain fell on both days?

(3) If I spend \$3.75 in a supermarket, how much change will I get from a \$5 bill?

(4) Anne lives 1.3 miles from school and Eileen lives .7 miles from school. Who lives farther from school? By how much?

(*NOTE:* Where possible, let your child verify answers with physical materials like play money.)

2. Relate addition and subtraction of decimals to addition and subtraction of fractions:

(a) $\begin{array}{r} .2 = 2/10 \\ +.5 = 5/10 \\ \hline .7 = 7/10 \end{array}$ (b) $\begin{array}{r} .8 = 8/10 \\ -.2 = 2/10 \\ \hline .6 = 6/10 \end{array}$ (c) $\begin{array}{r} .03 = 3/100 \\ +.06 = 6/100 \\ \hline .09 = 9/100 \end{array}$

(d) $\begin{array}{r} .25 = 25/100 \\ -.09 = 9/100 \\ \hline .16 = 16/100 \end{array}$ (e) $\begin{array}{r} .7 = 7/10 \\ +.5 = 5/10 \\ \hline 1.2 = 12/10 \\ = 1\,2/10 \end{array}$ (d) $\begin{array}{r} 2.8 = 2\,8/10 \\ -.5 = 5/10 \\ \hline 2.3 = 2\,3/10 \end{array}$

3. Let your child recall that in adding $\frac{1}{4} + \frac{3}{10}$, we find the lowest common denominator and then write the sum of the numerators over the common denominator:

$$\frac{1}{4} + \frac{3}{10} = \frac{5}{20} + \frac{6}{20} = \frac{5 + 6}{20} = \frac{11}{20}$$

Similarly,

$$\frac{7}{10} + \frac{11}{100} = \frac{70}{100} + \frac{11}{100} = \frac{70 + 11}{100} = \frac{81}{100}$$

In decimal form, this becomes

$$.7 + .11 = .70 + .11 = .81 \quad \text{or} \quad \begin{array}{r} .70 \\ +.11 \\ \hline .81 \end{array}$$

To add several decimals like .3 + .15 + .039, we note that the lowest common denominator is 1000. So we write:

.3 + .15 + .039 = .300 + .150 + .039
and then add vertically:

$$\begin{array}{r} .300 \\ .150 \\ +.039 \\ \hline .489 \end{array}$$

Let's now summarize the procedure for adding and subtracting decimals:

- Write the decimals in a vertical column with all decimal points directly under each other.

- Add enough zeros so that *all* decimals in the example are carried out to the same number of places (giving them the same common denominator).
- Add or subtract as you do whole numbers.
- Place the decimal point in the answer directly beneath the decimal points in the example.

EXERCISES

1. Add:

   (a) .3 + .6 (b) .9 + .5 (c) .08 + 2.3 (d) .5 + .6 + .4

2. Subtract:

   (a) .7 − .2 (b) .35 − .16 (c) .015 − .007 (d) 2 − .35

Multiplying Decimals

Multiplication with decimals is the same as multiplication with whole numbers, but you need to know where to place the decimal point in the product. Relating decimals to their fraction equivalents explains how to do this. For instance,

$$(1)\ \frac{5}{10} \times \frac{7}{10} = \frac{35}{100}$$

$$.5 \times .7 = .35$$

$$(2)\ \frac{3}{100} \times \frac{2}{10} = \frac{6}{1000}$$

$$.03 \times .2 = .006$$

$$(3)\ \frac{5}{100} \times \frac{7}{100} = \frac{35}{10{,}000}$$

$$.05 \times .07 = .0035$$

With many such examples, the child will observe that

> When *tenths* are multiplied by *tenths*, the product is *hundredths.*
>
> When *tenths* are multiplied by *hundredths*, the product is *thousandths.*
>
> When *hundredths* are multiplied by *hundredths*, the product is *ten-thousandths.*

And so on. Eventually, the child will arrive at the generalization that *the number of decimal places in a product equals the sum of the decimal places in the factors.*

This generalization becomes more apparent from the following sequence of examples:

(1)
```
  .4 ← 1 decimal place
 ×2  ← 0 decimal places
 ---
  .8 ← 1 decimal place
```

(2)
```
  .3  ← 1 decimal place
 ×.4  ← 1 decimal place
 ---
  .12 ← 2 decimal places
```

(3) $\begin{array}{r} 2.15 \\ \times \quad .3 \\ \hline .645 \end{array}$ $\begin{array}{l} \leftarrow \text{2 decimal places} \\ \leftarrow \text{1 decimal place} \\ \leftarrow \text{3 decimal places} \end{array}$

(4) $\begin{array}{r} 1.003 \\ \times \quad .0004 \\ \hline .0004012 \end{array}$ $\begin{array}{l} \leftarrow \text{3 decimal places} \\ \leftarrow \text{4 decimal places} \\ \leftarrow \text{7 decimal places} \end{array}$

NOTE:
- Remind your child that zeros are digits, and are to be included when counting decimal places.
- To visualize multiplication with decimals, translate them to fractions. Then illustrate the example with a geometric figure as is done on page 159.

Let's now summarize the procedure for multiplying decimals:

1. Multiply as you do whole numbers.
2. There will be as many decimal places in the answer as there are decimal places in the two numbers being multiplied.
3. Starting at the right and moving toward the left, count off as many decimal places in the answer as required, and put down the decimal point.
4. If you count more places than there are digits in the answer, add as many zeros *to the left of the answer* as are needed to place the decimal point correctly.

SHORT WAY TO MULTIPLY A DECIMAL BY 10, 100, or 1000

Multiply the decimal .003 first by 10, then by 100, and then by 1000 and see what happens:

$$.003 \times 10 = .03$$
$$.003 \times 100 = .3$$
$$.003 \times 1000 = 3$$

Notice that the effect of multiplying by 10 is to move the decimal point 1 place to the right; multiplying by 100 has the effect of moving the decimal point 2 places to the right; multiplying by 1000 has the effect of moving the decimal point 3 places to the right; and so on.

This idea permits us to multiply any number by 10, 100, 1000, . . . quickly. All you do is write down the number to be multiplied and move the decimal point to the *right* as many places as required:

$$3.78 \times 10 = 37.8$$
$$3.78 \times 100 = 378$$
$$.52 \times 10 = 5.2$$
$$2.5 \times 1000 = 2500$$

EXERCISES

1. Multiply:

(a) $.2 \times 9$ (b) $.05 \times .3$ (c) $.02 \times .01$ (d) $5 \times .6$ (e) $\$.59 \times .06$

2. Using the short method, multiply each of the following numbers by 10, then by 100, then by 1000:

   (a) .275 (b) \$1.28 (c) \$.10 (d) .003 (e) 3.9

Dividing Decimals

As with whole numbers, division with decimals is not easily understood by children. Though the same method is used to divide decimals as to divide whole numbers, there is an added difficulty with decimals: what to do with the decimal points in the problem.

Before a child can learn to divide with decimals, she must first be able to

Divide by whole numbers

Express decimals as fractions

Rename fractions with equivalent fractions

Multiply decimals

The child must understand that the effect of moving a decimal point to the right is the same as multiplying by powers of ten.

She should also know that a whole number like 25 may be written as "25."—with the decimal point following the last digit.

1. Dividing by a whole number
The simplest case of division with decimals is one in which the divisor is a whole number. So begin with problems like these:

Example 1: If you pay \$1.50 for 2 cups of popcorn, how much does 1 cup cost?

The child will have no difficulty figuring out that the cost is \$.75. That is,

$$\frac{\$1.50}{2} = 2\overline{)1.50}\ \ (\$\ .75)$$

The answer can be checked by multiplication: \$.75 × 2 = \$1.50
Example 2: If 3 gallons of gas cost \$3.75, how much does 1 gallon cost?

$$\frac{\$3.75}{3} = 3\overline{)3.75}\ \ (\$1.25)$$

CHECK: \$1.25 × 3 = \$3.75

The child should note that when you divide decimals by whole numbers:

- The decimal point in the quotient (answer) is placed *directly above* the decimal point in the dividend.
- There are the same number of decimal places in the quotient as there are in the dividend.
- The answer can be checked by multiplication: The quotient times the divisor should give you the dividend.

2. Dividing by a decimal
Since division by a whole number is easiest, convert a problem with a decimal divisor to an equivalent problem with a whole number divisor. Then proceed as before.

Example 1: In the division .62/.2, change the denominator (divisor), .2, to a whole number by multiplying it by 10. Since we don't want to change the value of the original fraction, we must also multiply the numerator (dividend), .68, by 10:

$$.2\overline{).68} = \frac{.68}{.2} = \frac{.68 \times 10}{.2 \times 10} = \frac{6.8}{2} = \begin{array}{r} 3.4 \\ 2\overline{)6.8} \end{array}$$

CHECK: (*Reminder:* Check the answer against the *original* problem.)

$$3.4 \times .2 = .68$$

Example 2: $1.15 \div 2.3$

$$2.3\overline{)1.15} = \frac{1.15}{2.3} = \frac{1.15 \times 10}{2.3 \times 10} = \frac{11.5}{23} = \begin{array}{r} .5 \\ 23\overline{)11.5} \end{array}$$

CHECK: $.5 \times 2.3 = 1.15$

Note in Examples 1 and 2 that by multiplying the divisor *and* dividend by 10, in effect we moved the decimal points one place to the right in *both* numbers:

$.2\overline{).68}$ became $2\overline{)6.8}$

and

$2.3\overline{)1.15}$ became $23\overline{)11.5}$

We can "see" these movements of the decimal points by the use of arrows:

$.2\overline{).6\,8} = .2\overline{).6_{\wedge}8}$

and

$2.3\overline{)1.1\,5} = 2.3\overline{)1.1_{\wedge}5}$

The arrow heads show the new positions of the decimal points. We then place the decimal point in the quotient directly over the *new* position of the decimal point in the dividend, and complete the division.

Example 3: $.25\overline{)7.5}$

Here we must multiply by 100 to change the .25 to a whole number:

$$.25\overline{)7.5} = \frac{7.5}{.25} = \frac{7.5 \times 100}{.25 \times 100} = \frac{750}{25} = \begin{array}{r} 30 \\ 25\overline{)750} \end{array}$$

We can simplify this operation by writing

$$.25\overline{)7.5} = \begin{array}{r} 30. \\ .25\overline{)7.50} \end{array}$$

CHECK: $30 \times .25 = 7.5$

Note that before we could move the decimal point two places in the dividend, we had to add a 0 at the end of the 7.5.

Example 4: $.95\overline{)1.9095}$

$$.95\overline{)1.9095} = \begin{array}{r} 2.01 \\ .95\overline{)1.9095} \end{array}$$

CHECK: $2.01 \times .95 = 1.9095$

Example 5: $.003\overline{)6}$

$$.003\overline{)6} = .003\overline{)6.000} \quad \text{quotient: } 2\,000.$$

CHECK: 2000 × .003 = 6

SHORT WAY TO DIVIDE A DECIMAL BY 10, 100, OR 1000

To multiply a decimal by a power of 10, we moved the decimal point to the right. To *divide* a decimal by a power of 10, we do the opposite—we move the decimal point to the *left*. Therefore, to divide a number by 10, move the decimal point one place to the left; to divide by 100, move the decimal point two places to the left; to divide by 1000, move the decimal point three places to the left:

$$13.5 \div 10 = 1.35$$
$$13.5 \div 100 = .135$$
$$13.5 \div 1000 = .0135$$

EXERCISES

1. Divide:

   (a) 5 ÷ .25 (b) .03 ÷ 6 (c) .25 ÷ .4 (d) 2.4 ÷ .012 (e) 12 ÷ .4

2. Using the short method, divide each of the following numbers by 10, then by 100, then by 1000:

   (a) 25 (b) 4.2 (c) 62.5 (d) .03 (e) .008

Rounding Decimals

When an exact answer is not possible or not needed, we *round* the answer. For instance,

$$1 \div 3 = \quad 3\overline{)1.000\ldots} \quad \text{quotient: } .333\ldots$$

Here, division does not lead to an exact decimal answer because we keep getting a remainder of 1 no matter how far we carry the division. So we may wish to approximate the answer at some cutoff point. This is called **rounding** the answer.

Example 1: To round .333 to, say, the nearest *tenth*, follow this procedure:

(1) Drop all the digits to the right of the place to which you are rounding. In the case of .333, we drop all the digits to the right of the *tenths* place:

$$.333 \rightarrow .3$$

(2) If the first digit to be dropped is 5, 6, 7, 8, or 9, *increase* the preceding digit by 1.

(3) If the first digit to be dropped is 0, 1, 2, 3, or 4, leave the preceding digit *as it is*. In the case of .333, the first digit dropped was 3; so we leave the tenths digit as is:

$$.3\,③ \rightarrow .3$$

Therefore, .333 rounded to the nearest tenth is .3.

Example 2: Round 3.476 to the nearest tenth.

$$3.476 \rightarrow 3.4\,⑦ \rightarrow 3.5$$

Since the first digit dropped after the tenths place was 7, we increased the 4 to 5.

To round to the nearest hundredth, follow the same procedure after dropping all the digits to the right of the hundredths place.

Example 3: Round 5.213 to the nearest hundredth.

$$5.213 \rightarrow 5.21\,③ \rightarrow 5.21$$

EXERCISES

1. Round to the nearest tenth: (a) .48 (b) 3.519 (c) 62.075
2. Round to the nearest hundredth: (a) .649 (b) .2147 (c) 6.008 (d) .098

Changing Fractions to Decimals

Since one of the meanings of a fraction is that it indicates division, a fraction can be changed to its equivalent decimal by dividing the numerator by the denominator:

Example 1: $\frac{1}{5} = 5\overline{)1.0}$ with quotient .2

```
   .2
5)1.0
  1 0
```

Example 2: $\frac{3}{4} = 4\overline{)3.00}$ with quotient .75

```
   .75
4)3.00
  2 8
    20
    20
```

Example 3: $\frac{1}{3} = 3\overline{)1.000\ldots}$ with quotient .333 . . .

```
   .333 . . .
3)1.000 . . .
   9
   10
    9
    10
     9
     1
```

Example 4: $\frac{2}{11} = 11\overline{)2.0000\ldots}$ with quotient .1818 . . .

```
    .1818 . . .
11)2.0000 . . .
   1 1
     90
     88
      20
      11
       90
       88
        2
```

Note that the resulting decimals either *terminate*, as in Examples 1 and 2, or *repeat*, as in Examples 3 and 4. In a terminating decimal, the division always ends with a remainder of 0. In a repeating decimal, the division cannot be completed no matter to how many decimal places the division is carried out because there continues to be a remainder. In Example 3, the remainder will always be 1, resulting in an endless supply of 3's; in Example 4, the remainder will continue to alternate between 9 and 2, resulting in an endless supply of 18's.

Will It Terminate or Repeat?

We said that when a fraction is converted to a decimal, the result is either a *terminating* or a *repeating* decimal. Is there a way to predict which it will be? For instance, can you tell, without actually dividing, whether $\frac{7}{30}$ will result in a terminating or repeating decimal?

Yes, there is a way. Reduce the fraction to its lowest terms, and then look at the denominator. If its factors consist only of twos, fives, or a combination of twos and fives, then the fraction will result in a *terminating* decimal. For instance, in the fraction $\frac{1}{4}$, the denominator 4 has only twos as factors ($4 = 2 \times 2$). Therefore, $\frac{1}{4}$ will result in a terminating decimal: $\frac{1}{4} = .25$. In the fraction $\frac{7}{20}$, the factors of the denominator 20 are a combination of twos and a five ($20 = 2 \times 2 \times 5$). Therefore, $\frac{7}{20}$ will result in a terminating decimal: $\frac{7}{20} = .35$.

But if the denominator contains factors *other* than twos and fives, the fraction will result in a *repeating* decimal. For instance, in the fraction $\frac{7}{30}$, the denominator 30 contains the factors 2, 5, and 3 ($30 = 2 \times 3 \times 5$). It will therefore result in a repeating decimal:

```
      .233 . . .
  30)7.000 . . .
     6 0
     1 00
       90
       100
        90
        10
```

(continued)

Why does this way of predicting work? It works because the only way a fraction can become a terminating decimal is if its denominator is, or can be expressed as, 10, 100, 1000, or some other power of 10. And the only factors that can produce 10, 100, 1000, etc., are 2, 5, or a combination of twos and fives.

Common Errors with Decimals

Apart from computational errors, the most common errors with decimals involve wrong placement of the decimal point. The examples below illustrate some of these errors, with the correct answers given alongside.

(1)	$.03 + 2.1 = 2.4$	[2.13]
(2)	$15 - 1.2 = .3$	[13.8]
(3)	$.2 \times .7 = 1.4$	[.14]
(4)	$.05 \times .7 = .35$	[.035]
(5)	$.8 \div .02 = .4$	[40]
(6)	$15 \div .5 = 3$	[30]
(7)	$.30 \times .2 = .6$	[.060]
(8)	$.800 \div .4 = 200$	[2.00]

Activities Involving Decimals

1. Ask your child to construct an addition table using tenths instead of whole numbers (Fig. 15-5).

2. Ask the child to construct a multiplication table using tenths instead of whole numbers.

+	.1	.2	. . .	.9
.1				
.2				
⋮				
.9				

Figure 15-5

3. Prepare two spinners, one marked with the decimals .1 to .9, the other with the whole numbers 1 to 9 (Fig. 15-6). Have the child dial both spinners and then add, subtract, divide, or multiply the two numbers.

Figure 15-6

4. Prepare a set of decimal strips (Fig. 15-7) consisting of several *units*, and batches of .1 to .9 strips. Ask the child to perform decimal divisions with these strips in the following way:

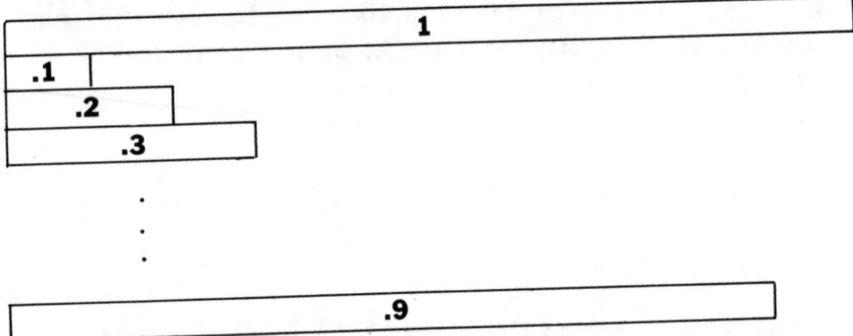

Figure 15-7

To perform the division 1.2 ÷ .4, the child lays out 1.2 units on the table. Then, using a .4 strip as a unit of measure, the child counts the number of .4 strips needed to cover the 1.2 units (Fig. 15-8).

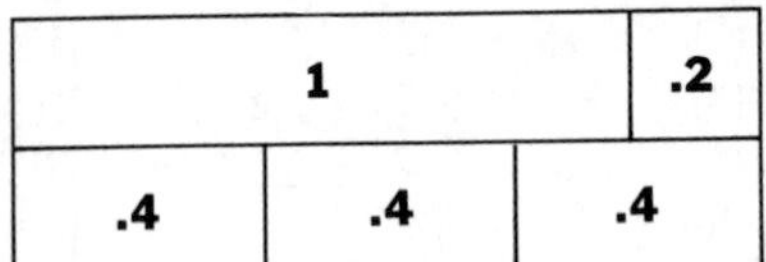

Figure 15-8

Repeat with other decimal divisions.

5. Ask the child to complete the table in Fig. 15-9 showing the decimal equivalents of the fractions listed.

Fraction	$\frac{1}{2}$	$\frac{1}{3}$	$\frac{2}{3}$	$\frac{1}{4}$	$\frac{3}{4}$	$\frac{1}{5}$	$\frac{2}{5}$	$\frac{4}{5}$	$\frac{1}{8}$	$\frac{3}{8}$	$\frac{5}{8}$	$\frac{7}{8}$
Decimal												

Figure 15-9

6. Prepare two spinners as shown in Fig. 15-10. The child obtains a number and an operation from the two spinners. The object is for him to make up a problem, using this operation, that will result in this number as the answer.

   For instance, if the number is .6 and the operation is +, a problem that uses addition and has an answer of .6 might be .2 + .4 = .6.

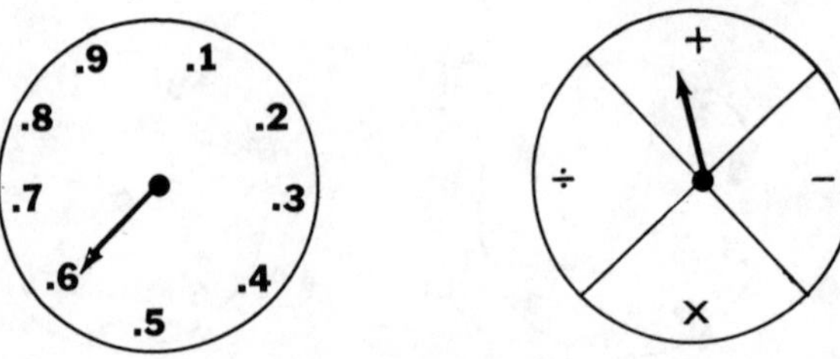

Figure 15-10

HIGHLIGHTS OF DECIMALS

The study of decimals should leave your child with the following understandings:

- A decimal is a fraction with a special denominator—10, 100, 1000, or some other power of 10.
- The decimal point shows where the whole number ends and the fractional part begins.
- Decimals are an extension of the place value concept used with whole numbers.
- The *number* after the decimal point is the *numerator* of the decimal fraction; the *number of digits* (including zeros) after the decimal point indicates the *denominator.*
- The decimal part of any number is always less than 1, no matter how many digits follow the decimal point.
- A mixed decimal is a number containing a whole number and a decimal fraction. It is always equal to or greater than 1.
- Zeros may be added at the *end* of a decimal without changing its value.

Also, at the completion of this chapter, your child should be able to

- Read decimals
- Identify one decimal place as tenths, two places as hundredths, three as thousandths, and so on
- Add, subtract, multiply, and divide decimals
- Convert decimals to fractions and fractions to decimals
- Multiply and divide, mentally, any decimal by a power of 10
- Round a decimal to the nearest tenth, hundredth, thousandth
- Explain and use the following vocabulary:

decimal	to round a decimal
decimal point	terminating decimal
mixed decimal	repeating decimal

QUESTIONS ON THE CHAPTER

1. Add:
   (a) .3 + .4 (b) .6 + .9 (c) .06 + .97
   (d) 9.3 + 7.2 (e) .25 and .386 (f) 5 and .08

2. Subtract:
   (a) .9 − .2 (b) take .9 from 2.5
   (c) 3 − .75 (d) from $5 subtract $1.72
   (e) take .37 from 1.815 (f) from 17 take .375

3. Multiply:
   (a) .3 × .7 (b) .6 × .8 (c) .72 × .4
   (d) 25 × .6 (e) 4.3 × 7 (f) .67 × .04

4. Divide:
   (a) 7.5 by 3 (b) .6 by .2 (c) $.5\overline{)4.25}$
   (d) $.03\overline{)73.5}$ (e) $.09\overline{)45}$ (f) $.008\overline{)9.84}$
   (g) $.15\overline{)4.5}$ (h) $.003\overline{).47}$ (to nearest tenth)
   (i) $.3\overline{)17}$ (to nearest hundredth)

5. In a 5-week period, a city had these weekly rainfalls in inches: 2.32, .07, 1.9, .6, and .09. What was the total rainfall for the 5 weeks?

6. June spends $3.19 in a store. How much change should she get from a $10 bill?

7. If the normal body temperature is 98.6°F, how many degrees above normal is a child's temperature if it reads 102°?

8. If a man spends $12.95 a week for transportation to his job, how much does he spend on transportation for 26 weeks?

9. At $1.17 per gallon of gasoline, how much will it cost to fill a tank holding 18 gallons?

10. A man earns $21,500 a year and gets paid in 12 equal monthly amounts. How much does he earn each month?

11. A woman walked 8.06 miles in 3.1 hours. How many miles did she walk per hour?

12. Change the following fractions to decimals:
    (a) $\frac{5}{8}$ (b) $\frac{13}{40}$ (c) $\frac{15}{16}$ (d) $\frac{19}{8}$ (e) $5\frac{1}{3}$

13. Change the following decimals to fractions:
    (a) .24 (b) .065 (c) .001

14. Which of the following represents 24 thousandths?
    (a) .024 (b) .240 (c) .00024 (d) 24000

15. Which is greater?
    (a) 2.8 or .97 (b) 3.25 or 3.250 (c) .2 or .195

16. Which is equal to one fourth?
    (a) .14 (b) 1.4 (c) .25 (d) .41

17. Round each of the following to the nearest tenth then to the nearest hundredth:
    (a) 15.068 (b) .125 (c) .00561

REVIEW

1. Locate $\frac{5}{3}$ on the number line.

2. Which is smaller, $\frac{5}{7}$ or $\frac{2}{3}$?

3. Find the sum of $\frac{1}{3}$, $\frac{3}{4}$, and $\frac{5}{8}$.

4. $\frac{1}{2} \div \frac{3}{8}$

5. $2 - \frac{3}{4}$

6. The *product* of two numbers means ______.

7. The answer to a division is called the ______.

8. Find $\frac{3}{4}$ of 20.

9. 9 is what part of 12?

10. $\frac{1}{6}$ of what number is 18?

1. **Two joggers are approaching each other on a straight road, the first running at 12 miles an hour, and the other at 8 miles an hour. When they are 20 miles apart, an energetic horsefly alights on the nose of one jogger, then immediately dashes off to the nose of the other.**

The horsefly shuttles back and forth between the two noses at a steady rate of 35 miles an hour. What distance does the horsefly cover by the time the joggers meet?

2. **If a grasshopper halves the distance to a wall on every jump, how many jumps will it need to reach a wall 15 feet away?**

3. **I count the lines on a page in a book. Counting by threes, I get a remainder of 2; by fives, I get a remainder of 2; by sevens I get a remainder of 5. How many lines are on the page?**

(The answers appear on page 260.)

Chapter 16
Percent

INTRODUCTION

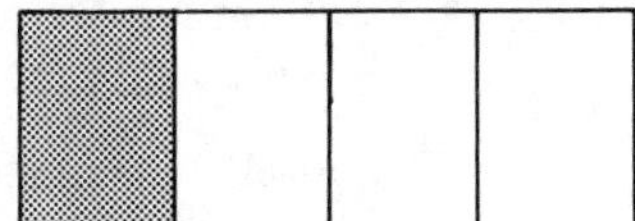
Figure 16-1

Percent is the third leg of the triad mentioned in Chapter 15. Look at the rectangle in Fig. 16-1. So far your child has learned two ways to express what part the shaded piece is of the entire rectangle: $\frac{1}{4}$ and .25. In this chapter he will learn a third way: 25%. That is, the shaded piece is 25 percent of the entire rectangle.

It's important for your child to understand that $\frac{1}{4}$, .25, and 25% are three different ways of saying the same thing, but that each way is more appropriate in certain situations. It is therefore useful to be able to change from any one form to either of the other two.

The words *percent* and *percentage*, though often used interchangeably, do not mean the same thing. For instance, in the sentence 25% of 32 = 8, 25 is the *percent*, 8 is the *percentage*, and 32 is called the *base*. A *percentage* is the answer you get from finding a percent of a number:

$$\underset{\substack{\uparrow \\ \text{percent}}}{25\%} \quad \text{of} \quad \underset{\substack{\uparrow \\ \text{base}}}{32} \quad = \quad \underset{\substack{\uparrow \\ \text{percentage}}}{8}$$

We again stress the importance of using physical materials to make a mathematical concept meaningful to the child. To help explain the meaning of percent, the following **materials** will be used in this chapter:

About a dozen sheets of 10 × 10 squared graph paper

Two pegboards, pegs, and rubber bands

THE CHAPTER IN A NUTSHELL

In this chapter you will see how to

- Explain the meaning of percent
- Teach your child to express a fraction, decimal, and percent in terms of each other
- Teach your child to solve problems involving percent
- Use a variety of activities to reinforce your child's competence with percent

You will also be given

- A list of understandings and skills your child should be left with at the completion of the chapter
- Questions to test and reinforce your child's comprehension of percent; also, review questions on earlier material
- Several challenging problems and puzzles under **For the Daring**

THE MEANING OF PERCENT

The word *percent* means "hundredths," or "for each hundred" (from the Latin words *per centum*). So,

25 percent means "25 hundredths," or $\frac{25}{100}$, or .25, and
5 percent means "5 hundredths," or $\frac{5}{100}$, or .05.

The symbol for "percent" is %. So,

25 percent may be written 25%
5 percent may be written 5%
and

$$25\% = \frac{25}{100} = .25$$
$$5\% = \frac{5}{100} = .05$$

To say that 35 percent of the people in a stadium are women means that 35 out of every 100 people there are women. To say that 2 percent of all TV sets are defective means that 2 out of every 100 sets are defective.

Fractions, decimals, and percents are all fractions, differing only in their denominators:

A **fraction** can have **any counting number** for a denominator.

A **decimal** is a fraction with **any power of 10** for a denominator.

A **percent** is a fraction with **only 100** for a denominator.

READINESS FOR PERCENT

Before beginning the study of percent, your child should be able to

- Multiply and divide fractions
- Multiply and divide decimals
- Multiply any number by a power of 10 by moving the decimal point to the right
- Divide any number by a power of 10 by moving the decimal point to the left
- Change a decimal to its equivalent fraction
- Change a fraction to its equivalent decimal

Your child should also know that when no decimal point appears in a number, it is understood to be to the right of the number. For instance, 5 is the same as 5., and 125 is the same as 125.

TEACHING PERCENT

Getting Started

1. To motivate the study of percent:

   (a) Turn to the sports page of a newspaper and ask your child whether he knows what the percents given for teams or individual players mean.
   Then turn to other sections of the paper reporting information involving percents, like discount sales, and ask whether the child understands the information.
   (b) Ask what the child thinks the following statements mean:
   (1) 50% of the children at a party are girls.
   (2) The cost of pizza has gone up 25%.
   (3) The chances of rain are 100%.
   (c) Ask him to give other examples of percents.

2. Explain the meaning of *percent* as given on page 192.

   To help with the explanation, give your child three 10 × 10 sheets of graph paper, one at a time, marked as shown in Figs. 16-2, 16-3, and 16-4.

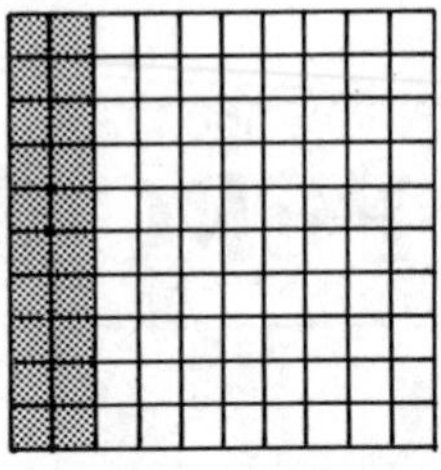

Figure 16-2

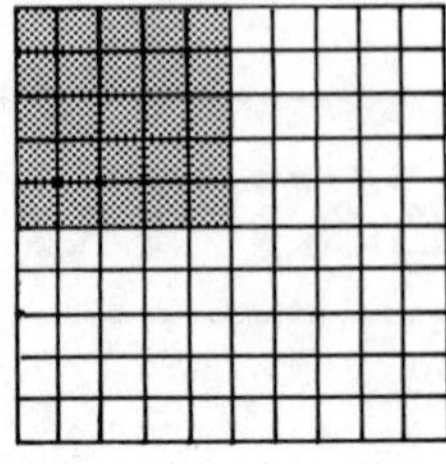

Figure 16-3

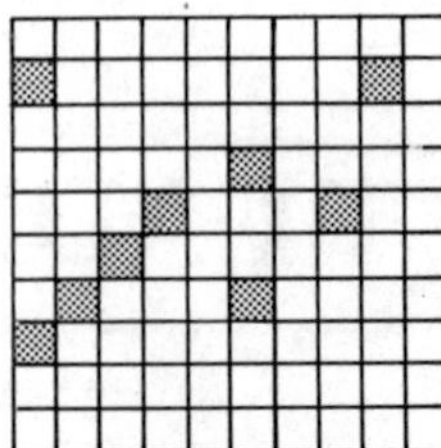

Figure 16-4

For each figure, ask the child to tell:

(a) What percent of the squares is shaded. Have him write the answer as a percent, fraction, and decimal.
(b) What percent is *not* shaded. Have him write the answer as a percent, fraction, and decimal.
(c) What percent is shaded *and* unshaded. Let the child conclude that

$$\frac{100}{100} = 100\% = 1$$

that is, 100% corresponds to the whole unit, 1.

3. Repeat the procedures in #2 with a 10 × 10 pegboard, pegs, and rubber bands.

(a) Insert all 100 pegs into the pegboard. Then put rubber bands around different groups of pegs, and ask the child to tell and then write the percent of pegs *enclosed* by the rubber band and the percent *outside* the rubber band.
(b) Ask the child to section off with a rubber band a certain percent of the pegs in the board.
(c) Remove all pegs from the board. Then ask the child to insert different percents of pegs into the board.

4. Try the following exercise when you think your child is ready for it.

   Place two pegboards on the table: pegboard A, with 100 pegs, and pegboard B, with 36 pegs.

(a) Have your child section off 50% of the pegs in pegboard A.
   Then ask: "How would you section off 50 percent of the pegs in pegboard B?"
   Let him see that since 50% is $\frac{1}{2}$ of the pegs, he would section off $\frac{1}{2}$ of 36, or 18 pegs.
   Let the child also see that 18 pegs in a board with 36 pegs *corresponds* to 50 pegs in a board with 100 pegs:

$$\frac{18}{36} = \frac{50}{100}$$

   Each is one half the total number of pegs in the board.
   Repeat this exercise with different percents.
(b) Keeping pegboard A with 100 pegs as the standard of comparison, change the number of pegs in pegboard B. Then repeat activity (a).

Fraction-Decimal-Percent Conversions

To solve percent problems, we often need to convert fractions, decimals, and percents to each other. For instance, we need to know that

$$\tfrac{1}{4} = .25 = 25\%$$

We'll now see how such equivalences can be found for *any* fraction, decimal, or percent.

REMINDER:
- A short way to *multiply* by 100 is to move the decimal point 2 places to the *right*.
- A short way to *divide* by 100 is to move the decimal point 2 places to the *left*.

1. Changing a percent to a decimal

Example 1: Change 15% to a decimal.

Since *percent* means "hundredths,"

$$15\% = 15 \times .01 = .15$$

and

$$\begin{aligned} 32\% &= 32 \times .01 = .32 \\ 4\% &= 4 \times .01 = .04 \\ 2.5\% &= 2.5 \times .01 = .025 \\ 100\% &= 100 \times .01 = 1.00 \\ 175\% &= 175 \times .01 = 1.75 \end{aligned}$$

So to change a percent to a decimal, move the decimal point two places to the left and remove the % symbol. In effect we exchange the % symbol for movement of the decimal point two places to the left.

Omitting the middle step, we can write directly

$$\begin{aligned} 15\% &= .15 \\ 1\% &= .01 \\ 125\% &= 1.25 \end{aligned}$$

2. Changing a decimal to a percent

Example 2: Change .75 to a percent.

If to change a percent to a decimal we move the decimal point two places to the left and remove the % symbol, then to change a decimal back to a percent we do the opposite: Move the decimal point two places to the *right* and *add* the % symbol. For instance,

$$\begin{aligned} .75 &= 75\% \\ .01 &= 1\% \\ 2.8 &= 280\% \end{aligned}$$

3. Changing a percent to a fraction

Example 3: Change 35% to a fraction.

Express the 35% as a fraction with a denominator of 100, and reduce to lowest terms:

$$35\% = \frac{35}{100} = \frac{7}{20}$$

Similarly, $$20\% = \frac{20}{100} = \frac{1}{5}$$

$$5\% = \frac{5}{100} = \frac{1}{20}$$

$$3\% = \frac{3}{100}$$

$$125\% = \frac{125}{100} = \frac{5}{4} \text{ or } 1\frac{1}{4}$$

NOTE: $$12\tfrac{1}{2}\% = \frac{12\frac{1}{2}}{100} = \frac{\frac{25}{2}}{100} = \frac{25}{2} \times \frac{1}{100} = \frac{25}{200} = \frac{1}{8}$$

4. Changing a fraction to a percent

Example 4: Change $\frac{1}{5}$ to a percent.

First change the $\frac{1}{5}$ to a decimal, then change the decimal to a percent, as in #2 above:

$\frac{1}{5} = .20 = 20\%$

Similarly, $\frac{3}{4} = .75 = 75\%$

$\frac{1}{8} = .125 = 12.5\%$ or $12\frac{1}{2}\%$

EXERCISES

1. Write as a percent:

   (a) thirty-eight hundredths (b) nine hundredths (c) 85 out of 100

2. Write each of the following percents as a decimal.

   (a) 29% (b) 98% (c) 9% (d) 130% (e) 3.25%

3. Write each of the following decimals as a percent.

   (a) .45 (b) .09 (c) .375 (d) 2.5

4. Write each of the following percents as a fraction.

   (a) 25% (b) 80% (c) 4% (d) 62.5% (e) 125%

5. Write each of the following fractions as a percent.

   (a) $\frac{3}{10}$ (b) $\frac{2}{5}$ (c) $\frac{3}{8}$ (d) $\frac{1}{3}$ (e) $\frac{1}{50}$

6. Have your child develop the conversions below. After she fully understands their meaning, ask her to *memorize* them since they are the most commonly used equivalents:

Frac.	Dec.	%	Frac.	Dec.	%	Frac.	Dec.	%
$\frac{1}{2}$	.50	50%	$\frac{2}{5}$	.40	40%	$\frac{7}{8}$	$.87\frac{1}{2}$	$87\frac{1}{2}\%$
$\frac{1}{3}$	$.33\frac{1}{3}$	$33\frac{1}{3}\%$	$\frac{3}{5}$	.60	60%	$\frac{1}{10}$	.10	10%
$\frac{2}{3}$	$.66\frac{2}{3}$	$66\frac{2}{3}\%$	$\frac{4}{5}$	.80	80%	$\frac{3}{10}$	.30	30%
$\frac{1}{4}$	.25	25%	$\frac{1}{8}$	$.12\frac{1}{2}$	$12\frac{1}{2}\%$	$\frac{7}{10}$	.70	70%
$\frac{3}{4}$	.75	75%	$\frac{3}{8}$	$.37\frac{1}{2}$	$37\frac{1}{2}\%$	$\frac{9}{10}$	.90	90%
$\frac{1}{5}$	.20	20%	$\frac{5}{8}$	$.62\frac{1}{2}$	$62\frac{1}{2}\%$			

Solving Problems Involving Percents

There are basically three types of percent problems:

1. finding a percent of a number
2. finding what percent one number is of another
3. finding a number when a percent of it is known

Before your child can learn to solve such problems, he must be able to

- Express fractions, decimals, and percents in terms of each other
- Find the value of the unknown (*n*) in each of the following types of equations (sentences):

(1) $2n = 10$, $n = \frac{10}{2}$, $n = 5$

(2) $n \times 3 = 12$, $n = \frac{12}{3}$, $n = 4$

(3) $3 = .06n$, $n = \frac{3}{.06}$, $n = 50$

Your child should also know that in solving percent problems:

The word *is* means "equals" (=).

The word *of* means "times" (×).

The percent is changed to a decimal or fraction, whichever is easier.

A letter like *n* is used to represent the unknown.

1. Finding a percent of a number

Example 1: Find 15% of 140.
Translation: 15% times 140
Solution: Change 15% to .15 and multiply: $140 \times .15 = 21$
Answer: 21 is 15% of 140.

Example 2: Find 25% of 56.
Solution: Change 25% to $\frac{1}{4}$ and multiply: $\frac{1}{4} \times 56 = 14$.
Answer: 14 is 25% of 56.

EXERCISES

1. What is 25% of 32?
2. Find 2% of 150.
3. What is $33\frac{1}{3}\%$ of 21?
4. Find 6% of \$85.
5. What is $12\frac{1}{2}\%$ of 160?

Before teaching your child the next type of percent problem, it would be well to let him "warm up" with questions like the following:

(1) What *fractional part* of 21 is 7?

Though the answer, $\frac{7}{21}$, can be gotten by inspection, the child can also obtain it by translating the question into an equation and solving for n:

Question:	What fractional part	of	21	is	7?
	↓	↓	↓	↓	↓
Translation:	n	$\times$	21	$=$	7
Solving for n:			$21n$	$=$	7
			n	$=$	$\frac{7}{21}$ or $\frac{1}{3}$

(2) What fractional part of 30 is 6?
(3) 8 is what fractional part of 32?
(4) What fractional part of 42 is 7?
(5) 11 is what part of 88?

2. Finding what percent one number is of another

Example 3: 3 is what percent of 12?
Form a fraction showing what fractional part 3 is of 12. Then change the fraction to a percent:

$$\frac{3}{12} = \frac{1}{4} = 25\%$$

Answer: 3 is 25% of 12.
CHECK: 25% of 12 $= \frac{1}{4} \times 12 = 3$.

Example 4: 7 is what percent of 20?

$$\frac{7}{20} = .35 = 35\%$$

Answer: 7 is 35% of 20.
CHECK: 35% of 20 $= .35 \times 20 = 7$.

EXERCISES

1. 2 is what percent of 6?
2. 30 is what percent of 40?
3. What percent of 7 is 2?
4. What percent of 80 is 15?
5. 5 is what percent of 4?

3. Finding a number when a percent of it is known

Example 5: 3 is 5% of what number?

A good way to solve this type of problem is to translate it into an equation, letting n stand for the unknown number:

	3	is	5%	of	what number?
	↓	↓	↓	↓	↓
Translation:	3	=	.05	×	n
Solution:	3	=	$.05n$		

$$n = \frac{3}{.05} = .05\overline{)3.00}\quad 60.$$

$$n = 60$$

Answer: 3 is 5% of 60.

CHECK: 5% of 60 = 60 × .05 = 3.

Example 6: 6 is 25% of what number?

Translation: $6 = .25 \times n$ or $6 = \frac{1}{4} \times n$

$$n = \frac{6}{.25} \qquad n = 6 \div \frac{1}{4}$$

$$n = 24 \qquad n = 24$$

Answer: 6 is 25% of 24.

CHECK: 25% of 24 = $\frac{1}{4} \times 24 = 6$.

EXERCISES

1. 6 is 3% of what number?
2. 15 is 25% of what number?
3. 12% of what number is 96?
4. $33\frac{1}{3}$% of what number is 7?
5. 36 is 75% of what number?

HIGHLIGHTS OF PERCENT

The study of percent should leave your child with the following understandings:

- *Percent* means "hundredths" or "for each hundred." The symbol for percent is **%**.

- Percent is another way of expressing a fraction or a decimal.
- To multiply or divide with a percent, we first rename it with its equivalent fraction or decimal, whichever is easier.
- Problems involving percent are basically one of the three following types:
  (1) Find 25% of 40.
  (2) 3 is what percent of 15?
  (3) 5 is 50% of what number?

Also, at the completion of this chapter, your child should be able to

- Express fractions and decimals as percents
- Express percents as fractions and decimals
- Find a percent of a given number
- Find what percent one number is of another
- Find a number from a given percent of it
- Solve verbal problems involving percents

QUESTIONS ON THE CHAPTER

1. Express 6 hundredths as a fraction, decimal, and percent.
2. Express the following percents as decimals and fractions:
   (a) 10% (b) 30% (c) 25% (d) $12\frac{1}{2}\%$ (e) $33\frac{1}{3}\%$
3. Express the following fractions as percents:
   (a) $\frac{1}{5}$ (b) $\frac{1}{6}$ (c) $\frac{3}{5}$ (d) $\frac{2}{3}$ (e) $\frac{7}{8}$
4. Express the following decimals as percents:
   (a) .125 (b) $.62\frac{1}{2}$ (c) $.66\frac{2}{3}$ (d) 1.0
5. What is meant by the following statements?
   (a) A shirt is made of 55% cotton and 45% polyester.
   (b) There were 65% men and 35% women at a ball game.
   (c) A girl spends 25% of a day in school and 30% on sleep. What percent does she spend on other activities?
6. A player has a batting average of .275. What percent of the times she was at bat did she get a hit?
7. If you spend 20% of your allowance on a movie, what *part* of your allowance did you spend on the movie?
8. If $12\frac{1}{2}\%$ of the children in your class wear eyeglasses, what *part* of the class wears glasses?

9. A store has a sale giving a $\frac{2}{5}$ reduction on all shirts. What is the percent of the price reduction?

10. (a) What does 100% class attendance mean? (b) Can you have 150% class attendance? (c) Can the cost of a sweater increase by 150%? Explain what this means.

11. Find: (a) 15% of 60 (b) 5% of \$120 (c) 120% of 84

12. (a) 3 is what percent of 12? (b) What percent of 5 is 4?
(c) 17 is what percent of 17? (d) 25 is what percent of 20?

13. (a) 8% of what number is 16? (b) 25% of what number is 13?
(c) 15 is 100% of what number? (d) 2.5% of what number is 175?

14. If you receive 75% on a test having 20 questions, how many questions did you answer correctly?

15. How much money can you save if you get a 25% discount on a purchase of \$50?

16. How many questions out of 15 can you get wrong and still receive a mark of 80%?

17. There were 20 women and 30 men in a bus. What percent of the passengers were men? Women?

18. A team wins nine games and loses six. What percent of the games did the team win? Lose?

19. Bob bought a shirt at a 25% discount. If he paid \$12, what was the regular price of the shirt?

20. If 6 girls make up 20% of the members of a club, how many members are there in the club?

21. Break down your family's average weekly budget for food, recreation, utilities, clothing, and miscellaneous. Ask your child to compute the percent that each expense item is of the entire budget.

22. Is 15% of 42 the same as 42% of 15? Prove your answer.

REVIEW

1. In the number 98,063, which digit represents
(a) tens? (b) thousands?

2. Subtract .07 from 2.

3. Find the product of $\frac{3}{8}$ and $\frac{2}{9}$.

4. Which is smaller: $\frac{3}{7}$ or $\frac{4}{9}$?

5. Round \$18.3865 to the nearest cent.

6. In each of the following equations (sentences) find the value of n:
(a) $.35 \times 4 = n$ (b) $.15n = 60$ (c) $24 = n \times 4$

7. Find the cost of $2\frac{1}{4}$ pounds of beef at \$2.75 a pound.

8. (a) What is $\frac{3}{4}$ of 20? (b) $\frac{2}{3}$ of what number is 12?
   (c) What part of 21 is 7?

1. Is a discount of 10% followed by a discount of 5% the same as a discount of 5% followed by a discount of 10%? Is it the same as a discount of 15%?

2. (a) Use 3 sixes to make 7.
(b) Use 3 fours to make 11.
(c) Use 3 fives to make 5.

3. Three men, traveling with their wives, came to a river they wished to cross. The one available boat could accommodate only two people at a time. Since the husbands were very jealous, no woman could be with a man unless her own husband was present. How did they cross the river?

(The answers appear on page 260.)

Chapter 17
Geometry

INTRODUCTION

There is a story about Ptolemy Soter, the first king of Egypt, who studied geometry under Euclid. Finding the subject difficult, the king asked Euclid if there weren't some easier way to learn geometry. To this Euclid replied, "Oh, King, in the real world there are two kinds of roads—roads for the common people to travel upon and those reserved for the king. In geometry there is no royal road."

The geometry in this chapter is not difficult, but not all of it—especially its pace—is intended for every child. The definitions and theorems, and some of their ramifications ensconced in **For the Curious** sections, will, however, provide *you* with important background for teaching your child geometry.

In previous chapters we were concerned with **numbers**; in this chapter we are concerned with **points**, since all geometric figures are sets of points. Both numbers and points are abstract concepts which, to be understood by children, need to be related to physical objects. Though formal proofs for the many conclusions stated in this chapter are beyond the scope of this book, physical verification through measurement is suggested in many instances.

Because of the nature of the material, this chapter is crowded with definitions and theorems. So take lots of time with it, use only what your child is ready for, and let your child arrive at as many conclusions and insights as possible—on his or her own—through measurement and experimentation.

The **materials** used in this chapter include:

Several sets of geometric shapes cut from cardboard, or any other stiff material, including squares, rectangles, triangles, parallelograms, trapezoids, and circles

A set of geometric solids (available commercially), including a rectangular prism, a cube, a pyramid, a cone, a cylinder, and a sphere (see page 220)

A protractor

Five tops of jars of different diameters (see page 226)

A tape measure

A special chart needed for an activity described on page 226

A transparent grid ruled up in square units (see page 229)

Special cardboard cutouts of squares and triangles as described on page 242

A special set of cutouts as described on page 207

THE CHAPTER IN A NUTSHELL

In this chapter you will see

- What geometry is about and why your child should study it
- How to prepare your child for the study of geometry through informal activities
- How to explain:
  - Certain basic concepts of geometry
  - The definitions and characteristics of familiar geometric figures
  - Angle measurement
  - Finding the perimeter, area, and volume of geometric figures
  - Finding the circumference and area of a circle
  - The meaning of congruent figures
  - The meaning of similar figures
  - The meaning of symmetry

You will also be given

- A list of understandings and skills your child should be left with at the completion of the chapter
- Questions to test and reinforce your child's comprehension of basic concepts in geometry; also, review questions on earlier material
- Several challenging problems and puzzles under **For the Daring**
- Additional facts about familiar geometric figures; the history of a famous number; reasons for certain formulas used in the chapter; an "impossible problem"; and insights into the most famous theorem in mathematics—under **For the Curious**

THE MEANING OF GEOMETRY

Geometry uses "geometric figures" to describe, compare, and relate objects in the environment. These figures are not real objects; they only *represent* real objects. They are classified into several broad categories: (1) points; (2) lines, line segments, and rays; (3) plane figures; (4) space figures. Each category will be discussed later.

Geometry also concerns itself with spatial measurements like the number of square feet in a lawn, the number of cubic feet in a freezer, and the distance between two points in space.

Geometry also reveals interesting, at times beautiful, and sometimes unexpected relationships within and between geometric figures. For instance, the area of *any* circle is always a little more than three times the length of its diameter. Or that if you wish to tile a floor without overlapping and without leaving gaps, you can do so with equilateral triangles or hexagons, but not with equilateral pentagons or octagons.

A famous and vital aspect of geometry is that it serves as a model of deductive reasoning—the kind of reasoning by which we conclude that *if* something is true, then something else *must also* be true. (Remember Sherlock Holmes?) Starting with a few simple ideas *assumed* to be true, and a few rules of logic, geometry builds up a deductive system of increasing complexity. Euclid's geometry was the first example of such a system and became the standard for all such systems.

Not every concept in geometry is formally defined. For instance, a *point* is undefined, but acquires its meaning from a description of it as an "exact location" and as represented by a dot on a piece of paper. We then use such undefined terms to formally define other terms like *segment* and *triangle.*

To understand the need for undefined terms, imagine being asked to explain what a triangle is. You can say that "a triangle is the union of three line segments which join three non-collinear points." But what are "line segments"? You then explain that "a line segment is a piece of a line consisting of two points on the line and all the points between them." But what are "points"?

You can see that sooner or later you are driven against a wall because with each new explanation you are confronted with the same question: What are the definitions of the terms used in the last definition? This is why we have to begin with some *undefined* terms with which to *define* other terms.

Geometry also uses *postulates*, sometimes called *axioms.* A postulate is a statement *assumed* to be true. An example of a postulate is, "Through any two points there is exactly one line." Or, "Through a point not on a line there can be only one parallel to the given line." Definitions and postulates are used to prove that other statements, sometimes called *theorems*, are also true.

Proof in geometry, as in other branches of mathematics, is built upon definitions, postulates, and theorems which serve as the link in the chain of reasoning. Each step in a proof must follow from the previous steps, and a reason must be given for each statement made. Acceptable reasons include (1) facts provided in the statement to be proved, (2) definitions, (3) postulates, and (4) theorems that have already been proved.

Why Should Your Child Study Geometry?

A child's first mathematical experiences in infancy are geometrical rather than arithmetical. From the earliest days children find themselves in an inherently geometric environment, surrounded by objects having shape, size, and location. The day they enter school they already bring with them certain geometric concepts and skills. For instance, a three-year-old can already draw a circle, a four-year-old can already make a square, and a five-year-old can already copy an equilateral triangle.

As children gain self-confidence, their study of geometry becomes more challenging and rigorous. From exploration and experimentation with physical models, they move on to proving a few generalized conclusions. This whole process stimulates imaginative and logical thinking and develops spatial intuition. There is no more natural way to reveal the connection between mathematics and the real world than by capitalizing upon the child's early curiosity about shapes, and the relations between shapes and patterns, through the study of geometry.

READINESS FOR GEOMETRY

To learn to describe, compare, and relate geometric figures, the child needs experience with real objects—not pictures of circles or triangles, but physical ones. A set of Attribute Blocks, available commercially, is ideal for this purpose. You can, if you wish, cut out your own figures from cardboard or some other stiff material. Cut out squares, rectangles, triangles, parallelograms, trapezoids, and circles.

Also helpful would be a set of geometric solids, also available commercially, containing a rectangular prism, a cube, a pyramid, a cone, a cylinder, and a sphere. Your child can engage in many "Readiness" activities with these materials.

1. Let your child engage in unstructured play with all these shapes to get a feel for them.
2. Ask the child to describe the characteristics of each figure. For instance, a square is flat, has four sides—each side the same length—and has four "square corners." Tell the child that a square corner is called a *right angle*. Cite as an example of a right angle the way the hands of a clock look at 3 o'clock. Then ask him to stretch out his arms in such a way as to form a right angle.
3. Let him learn the names of the figures and develop informal definitions of each. (Save the "solids" for later.)
4. Let the child compare figures and see how they are alike and how they differ. For example, a rectangle and a square have four sides and four right angles, but only the square has all four sides the same length.
5. Blindfold the child, place various geometric shapes in his hands, and ask him to identify each shape.

6. Ask the child to trace the various shapes on a piece of paper and then cut out the tracings. Later have her draw, without tracing, various geometric shapes.
7. Cut a square from a stiff piece of paper. Then cut it along the diagonal to form two triangles (Fig. 17-1). Ask the child to form one triangle with the two pieces:

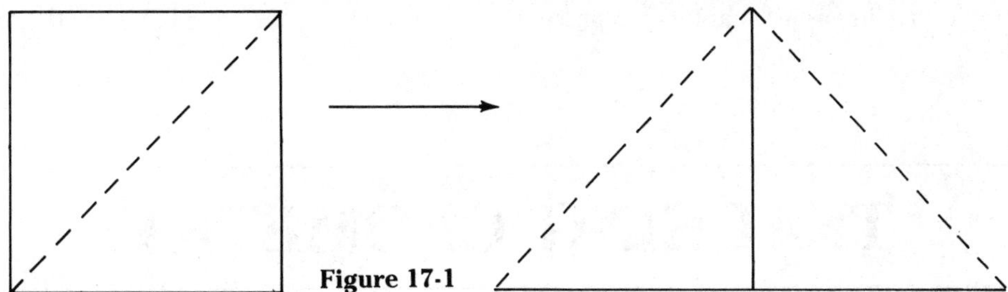

Figure 17-1

8. Prepare the set of cutouts shown in column A. Give them to your child and ask him to form with them the shapes shown in column B.

A *From these*	**B** *Make this*

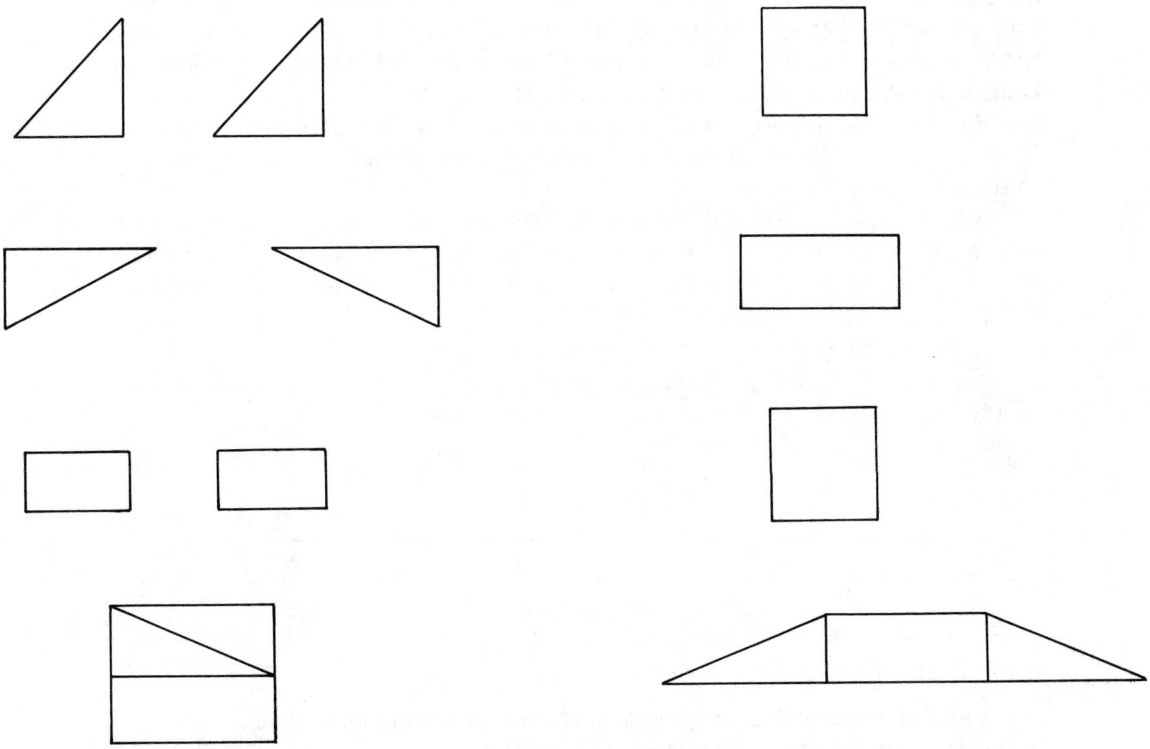

In providing these experiences with objects, you must guard against your child's concluding that a "triangle" (or any other figure) can be held in the hand, is a certain color, is always equilateral, or must always be positioned horizontally. The child should understand that a triangle is an abstraction, has no color, is not made of oak tag, does not have to have sides of equal length, and can appear in any position. What your child is holding in his hand is only a physical representation of a triangle.

The abstractness of a triangle is not an easy concept for a child to grasp. So work at it slowly and carefully.

This background will prepare your child for more formal instruction in geometry, when understanding will become more refined and definitions more precise.

The material in the next section, while presented informally, is more concentrated than in earlier chapters. So present it slowly, and as your child becomes ready for it.

TEACHING GEOMETRY

Building Blocks of Geometry

Let's start with a description of a few basic ideas in geometry: point, line, and plane.

A **point**, the basic building block of all geometric figures, may be thought of as an *exact location* in space. It has no size, can't be seen, and has only position. Physically, a point is suggested by the head of a pin or the tip of a pencil, and is usually represented by a dot. When the dot is erased the point is still there since the location is still there. A point is named by a capital letter like A or C.

We can think of *space* as the set of *all* points, and of the universe as completely filled with points. Geometric figures like triangles and circles can then be seen as subsets of these points in space.

A **line** is a set of points which may be thought of as a string, tightly stretched indefinitely in both directions. When drawing a picture of a line, we can therefore draw only part of it, but show that it extends indefinitely in both directions by attaching arrows to both ends (Fig. 17-2).

A line has neither width nor thickness, only length. It's named either by a single lowercase letter like *c*, or by any two points on the line like A and B. We write *line* c; *line* AB, or *line* BA (Fig. 17-3).

Figure 17-2

Figure 17-3

A part of a line, with definite endpoints like C and D in Fig. 17-4, is called a *line segment* or *segment*. We write *segment* CD or *segment* DC.

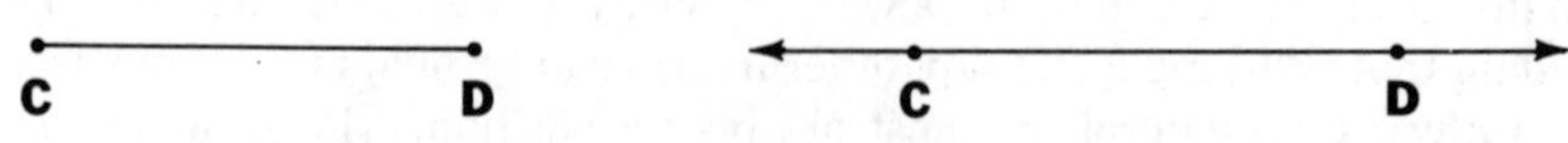

Figure 17-4

A part of a line with only *one* endpoint and extending indefinitely in the other direction is called a **ray** (Fig. 17-5). A physical example of a ray is a beam of light emanating from a pin point source. If, as in Fig. 17-5, the source is at C and the beam shines in the direction of F, we denote the ray by *ray* CF.

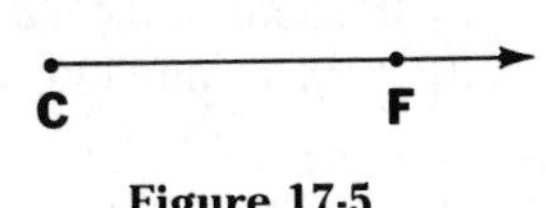

Figure 17-5

A **plane** is a set of points suggested by any flat surface like a sheet of metal, a wall, a floor, or a desk top—if you can imagine the surface stretching indefinitely in every direction. A plane has no thickness, only length and width. It's often named by three of its points which do not lie on the same line, for instance, *plane* CDE in Fig. 17-6.

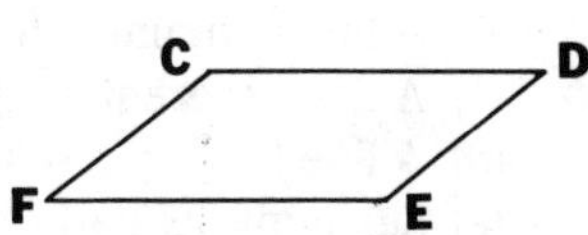

Figure 17-6

Lines that meet are **intersecting lines**. Lines that lie in the same plane and do not intersect are **parallel lines**. Planes that do not intersect are **parallel planes**. Two intersecting lines (or segments, or rays) that form a right angle are **perpendicular** to each other.

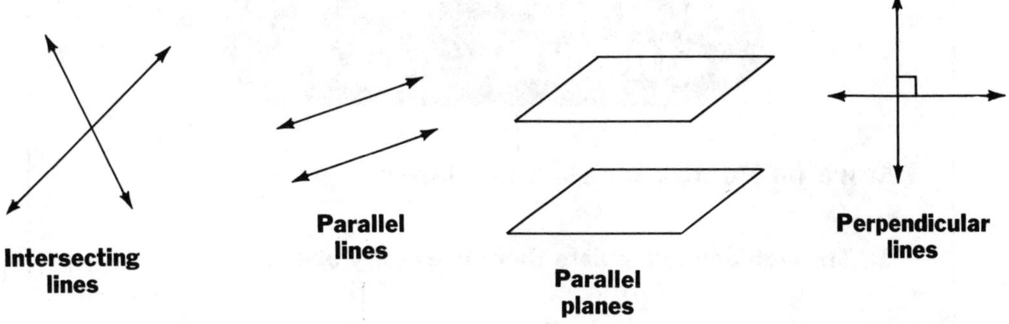

EXERCISES

1. Name each of the following:

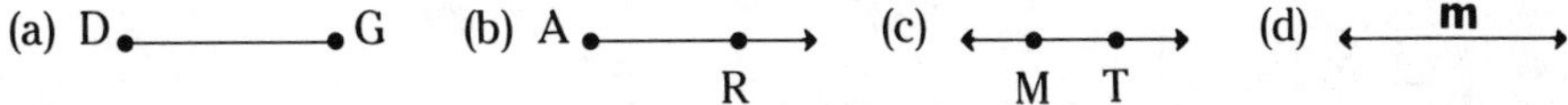

2. Draw:

   (a) segment CD (b) line AB (c) ray PQ
   (d) line segment RF (e) ray XY
   (f) two parallel lines (g) two intersecting lines
   (h) two perpendicular lines (i) two parallel planes

3. In how many points can two lines intersect?

4. If two lines are parallel to a third, are they parallel to each other? Draw these lines before answering the question.

5. If two lines are perpendicular to a third, are they perpendicular to each other? Draw these lines before answering the question.

6. Which geometric ideas are suggested by the following?

   (a) a clothesline stretched between two hooks
   (b) a speck of dust
   (c) a pair of opened scissors

(d) the ceiling in your bedroom
(e) the floor and ceiling in your bedroom
(f) the floor and one wall in your bedroom

7. True or false? Give reasons for your answers.

    (a) A line contains a million line segments.
    (b) A line contains infinitely many line segments.
    (c) A line segment contains infinitely many points.
    (d) A line has two endpoints.
    (e) Segment PQ and segment QP are the same segment.
    (f) If two lines are perpendicular to a third line, then they are parallel to each other.

More on Points, Lines, and Planes

1. Through any two points there is exactly one line.

2. Through any three points not on one line there is exactly one plane.

3. If two lines intersect, they intersect at exactly one point.

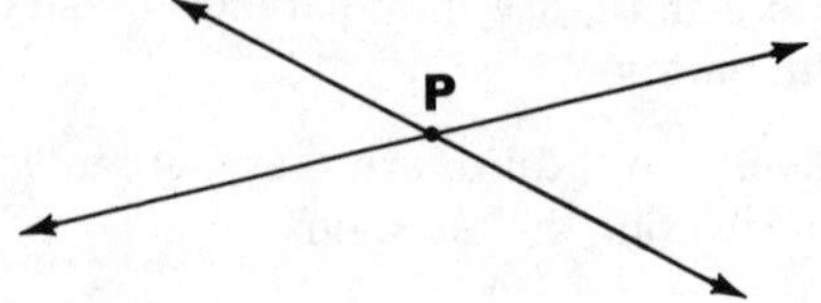

GEOMETRIC FIGURES

ANGLES

A figure formed by two rays having the same endpoint is called an **angle** (Fig. 17-7). The common endpoint is the **vertex** of the angle, and the two rays are its **sides.** The symbol for "angle" is ∠.

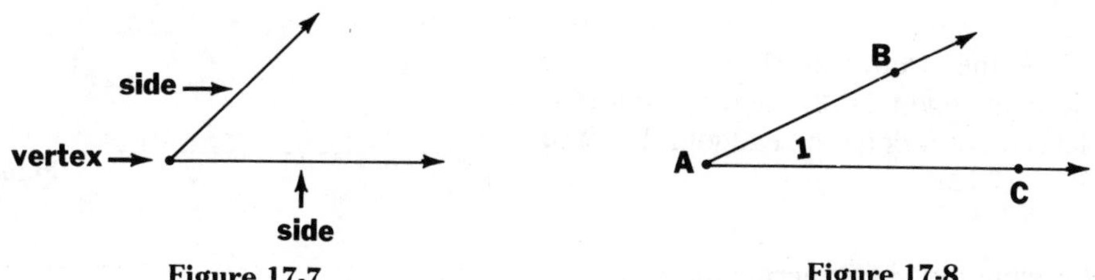

Figure 17-7

Figure 17-8

The angle formed by ray AB and ray AC in Fig. 17-8 is called ∠BAC or ∠CAB, the middle letter always naming the vertex. It can also be called ∠A or ∠1. Angles are measured in units called **degrees,** denoted by °.

In measuring an angle, we measure the "spread" between the rays. The size of an angle does *not* depend on the length of its sides, but on the *amount of rotation* of one of the rays about its endpoint while the other is held fixed (Fig. 17-9). A different angle is being formed as the ray turns from one position to another. One complete rotation forms a 360° angle; half a rotation, 180°; a quarter rotation, 90°.

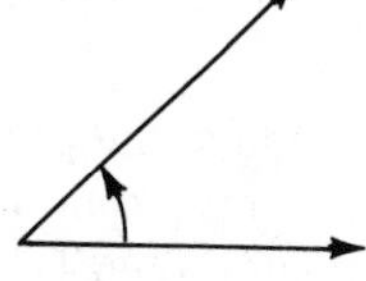

Figure 17-9

To measure or draw an angle, we use an instrument called a **protractor,** marked from 0° to 180° (Fig. 17-10). [Most protractors have two scales—one read in a clockwise direction, the other in a counterclockwise direction.]

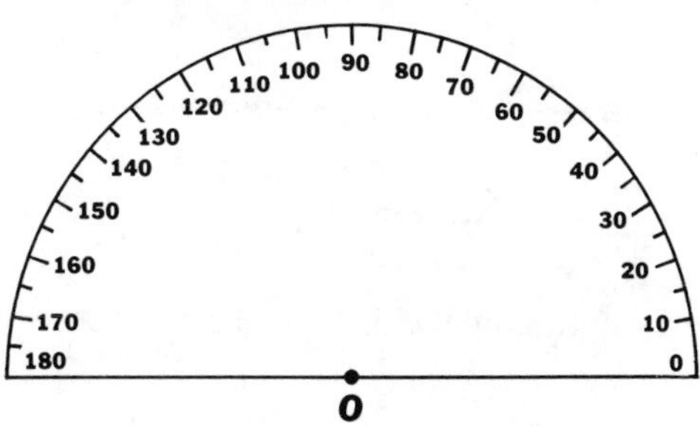

Figure 17-10

To measure an angle:

1. Place the straight edge of the protractor on one side of the angle with its center mark at the vertex of the angle (Fig. 17-11).
2. Read the number of degrees at the point where the other side of the angle cuts the protractor, using the scale with its zero on one side of the angle. ∠BOA in Fig. 17-11 measures 55°; ∠COA measures 120°.

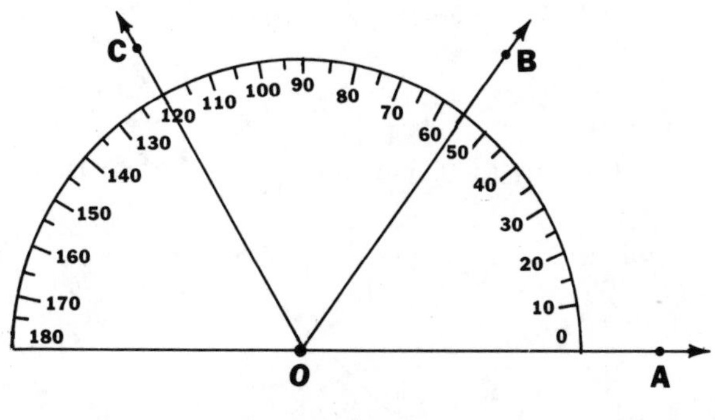

Figure 17-11

3. If the protractor is too large for the sides of the angle, extend them past the protractor.

To draw an angle:

1. Draw a ray (like ray PA in Fig. 17-12) to represent one side of the angle.

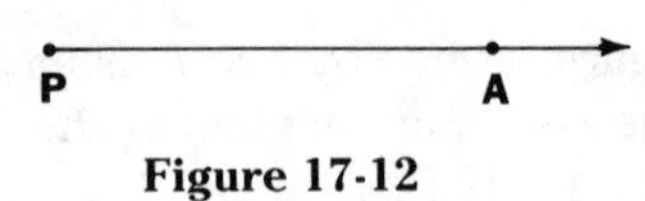

Figure 17-12

2. Place the protractor so that its straight edge falls on the ray and its center mark on the endpoint P (Fig. 17-13).

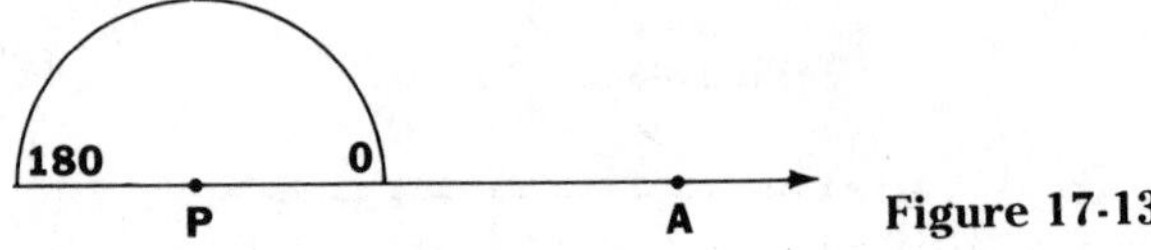

Figure 17-13

3. Counting on the scale having its zero on the ray, locate the required number of degrees, say 70°, and indicate its position with a dot (like point B in Fig. 17-14).

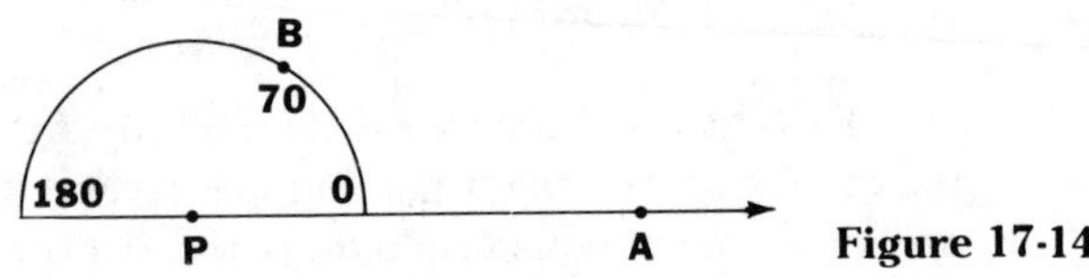

Figure 17-14

4. Remove the protractor and draw the ray from vertex P through point B (Fig. 17-15).

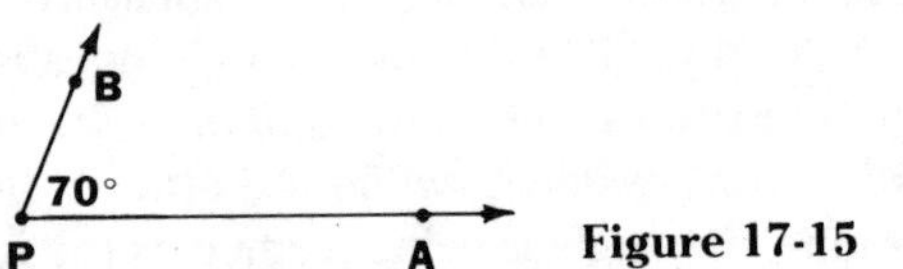

Figure 17-15

A 90° angle is a **right** angle; an angle less than 90° is **acute;** an angle greater than 90° is **obtuse.**

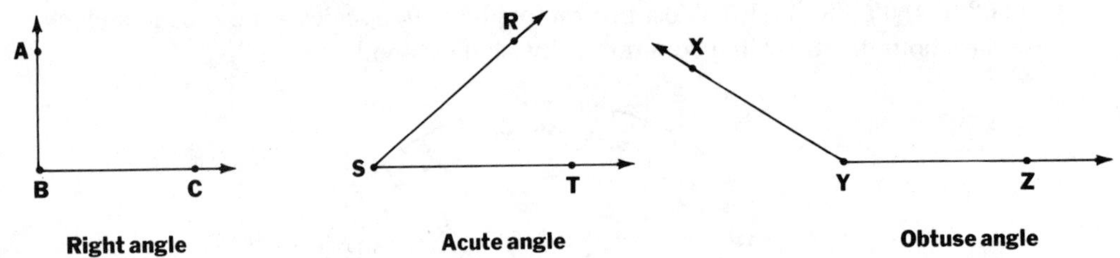

We sometimes indicate a right angle with a small square at its vertex:

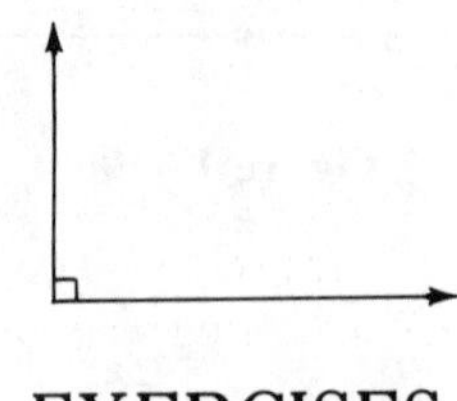

EXERCISES

1. Name the angle and tell whether it's right, acute, or obtuse.

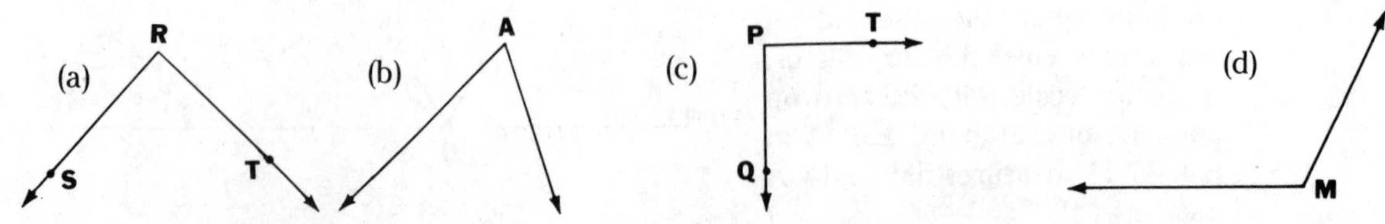

2. In each figure below name all the angles and tell whether each is right, acute, or obtuse.

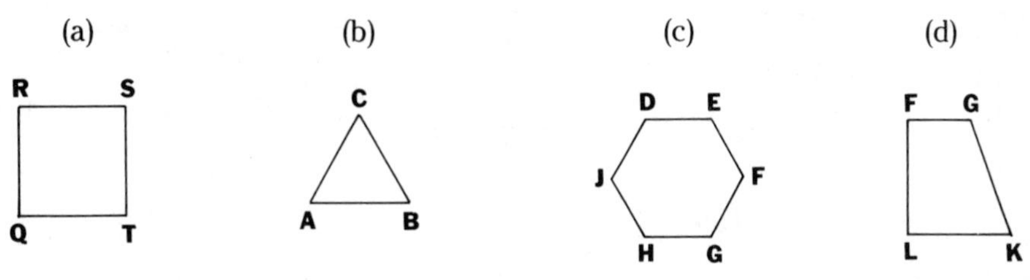
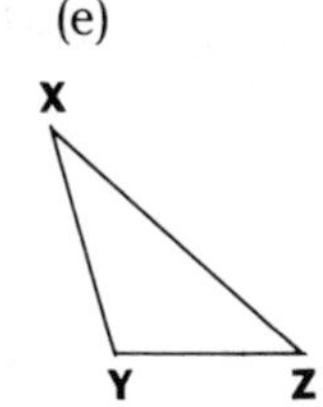

3. Draw a figure with

   (a) three acute angles   (b) four right angles
   (c) one right angle and two acute angles
   (d) two right angles, one obtuse angle, and one acute angle

4. For each angle in question 1

   (a) estimate the number of degrees it measures
   (b) verify your estimate with a protractor

5. How many degrees are in the angle formed by the hands of a clock at:

   (a) 3 o'clock   (b) 2 o'clock   (c) 6 o'clock

6. Use your protractor to draw the following angles:

   (a) 90°   (b) 30°   (c) 45°   (d) 60°   (e) 120°

PLANE FIGURES

To make it easier to classify the endless variety of geometric figures, we separate them into two groups:

1. Those whose points all lie in one plane, called **plane figures,**
2. Those whose points do not lie in one plane, called **space figures.**

We'll start with plane figures and deal with space figures later.

Polygons

An important classification of plane figures is the set of polygons. A **polygon** is a closed figure consisting of three or more line segments called its **sides** (Fig. 17-16). Intersecting sides meet in a point called a **vertex** (plural: **vertices**).

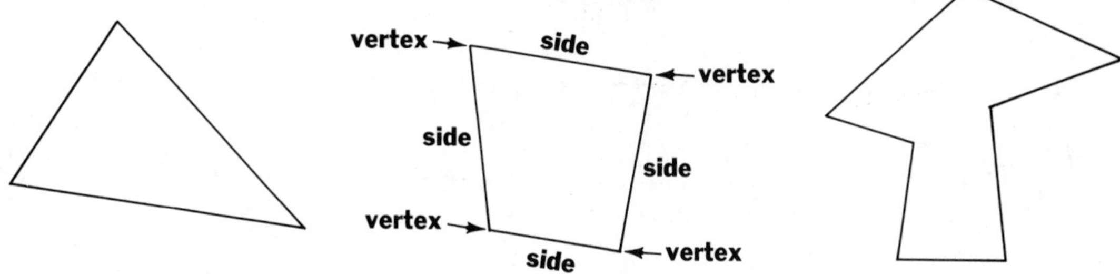

Figure 17-16

The shapes in Fig. 17-17 are *not* polygons because they are either not closed or do not consist entirely of line segments.

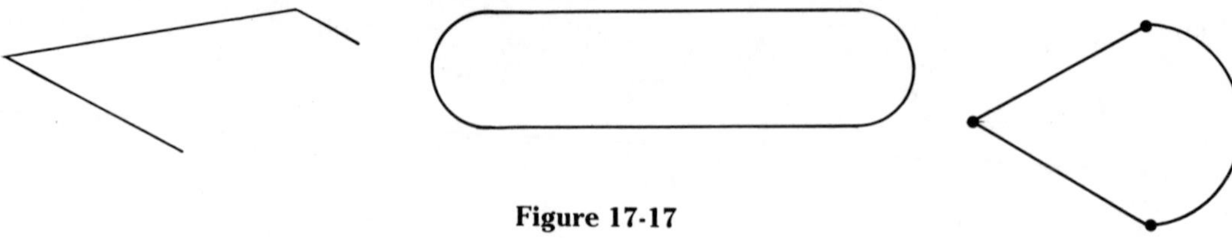

Figure 17-17

A polygon with equal sides and equal angles is called a **regular polygon.** The figures in Fig. 17-18 are examples of regular polygons.

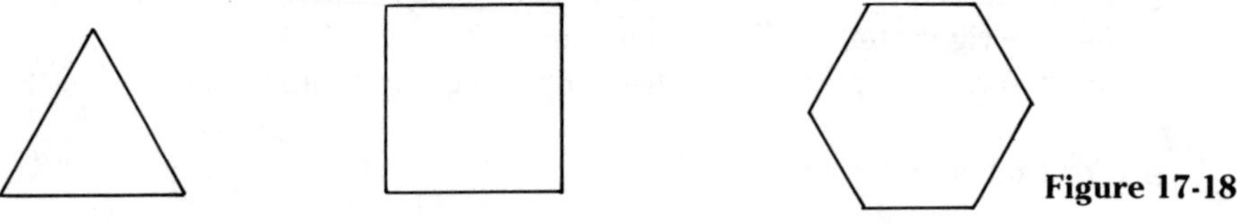

Figure 17-18

A line segment connecting two nonadjacent vertices of a polygon is called a **diagonal.** In Fig. 17-19, segments BD and EC are diagonals.

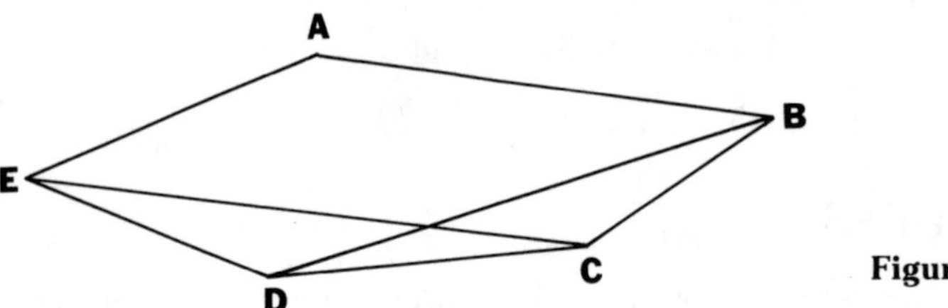

Figure 17-19

A polygon with three sides is a **triangle.**

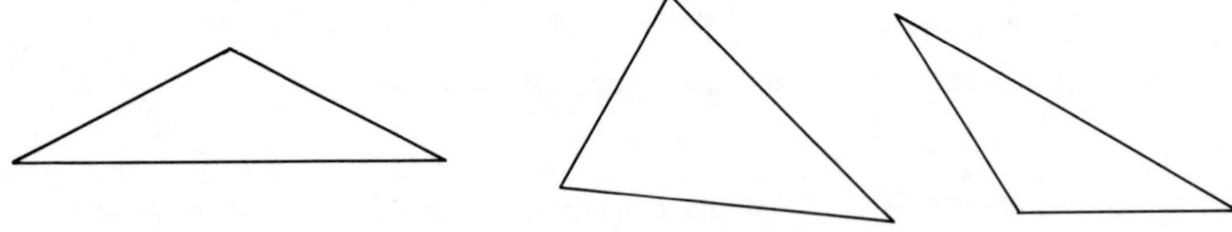

A polygon with four sides is a **quadrilateral.**

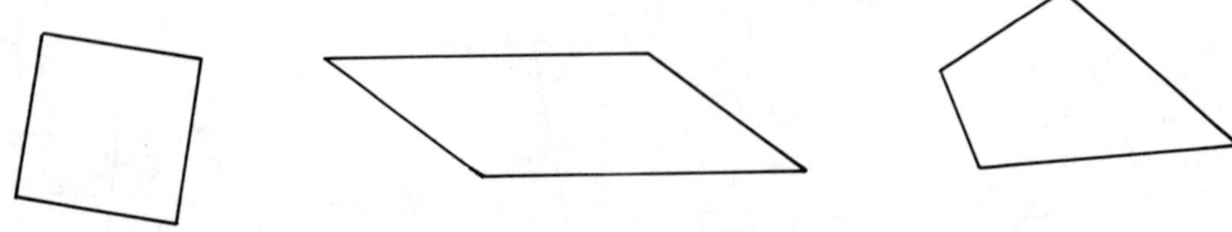

A polygon with five sides is a **pentagon.**

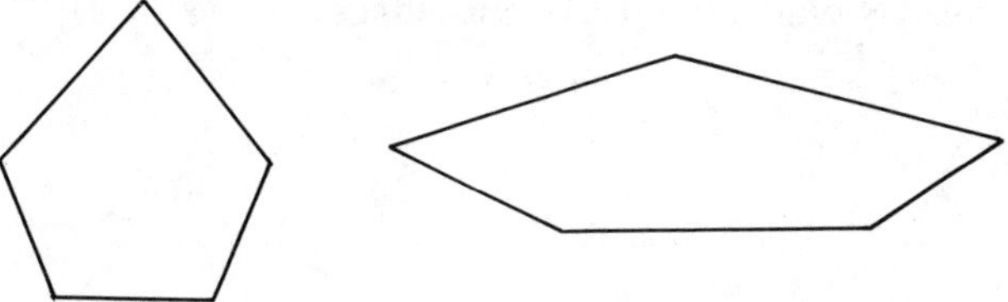

A polygon with six sides is a **hexagon.**

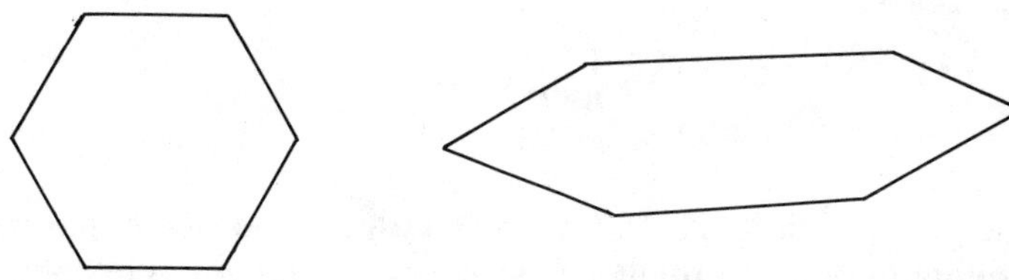

A polygon with eight sides is an **octagon;** with ten sides, a **decagon;** and with twelve sides, a **dodecagon.**

EXERCISES

1. For each polygon listed below, draw two different examples of it. Then tell the number of sides and angles each contains:

   (a) quadrilateral   (b) triangle   (c) pentagon
   (d) hexagon   (e) octagon

2. Is the number of angles in a polygon the same as the number of vertices?

3. How many angles are in a polygon with 10 sides? With 15 sides?

4. Identify by name all the geometric shapes you can find in your room.

> *NOTE:* From here on we will need to refer to lengths of segments and the measures of angles. To keep the language simple and uncluttered, we will use the following notation:
>
> AB = CD   will mean "segment AB and segment CD have equal lengths"
>
> ∠1 = ∠2   will mean "∠1 and ∠2 have equal measures"
>
> ∠A = 35°   will mean "∠A is an angle that measures 35°"

Triangles

A triangle has three sides and three angles. In Fig. 17-20, the points A, B, C are the vertices of the triangle, and the segments AB, BC, and CA are its sides. The sides and angles of a triangle are sometimes referred to as its "parts." The symbol for a triangle is △. The triangle shown in Fig. 17-20 is denoted by △ABC.

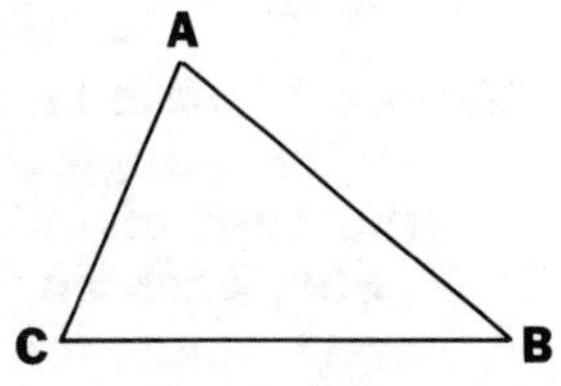

Figure 17-20

If all three sides of a triangle are the same length, the triangle is **equilateral.** If two sides are the same length, the triangle is **isosceles** (Fig. 17-21).

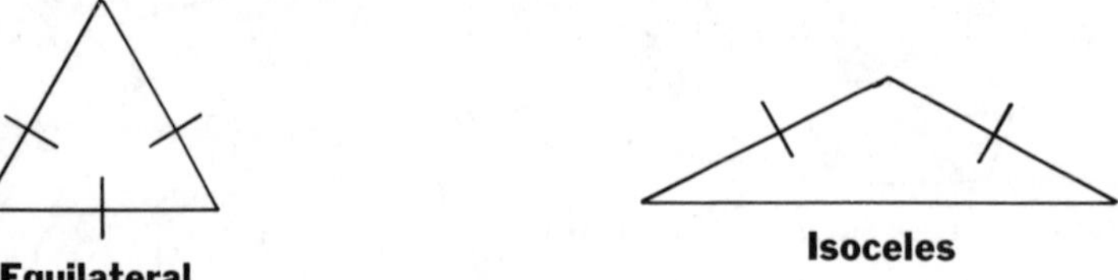

Figure 17-21

A triangle with a right angle is a **right triangle.** The side opposite the right angle is the **hypotenuse.** The other two sides are the **legs**—sometimes called the **altitude** and **base.** The hypotenuse is the longest side of the triangle.

The sum of the angles of any triangle is 180°.

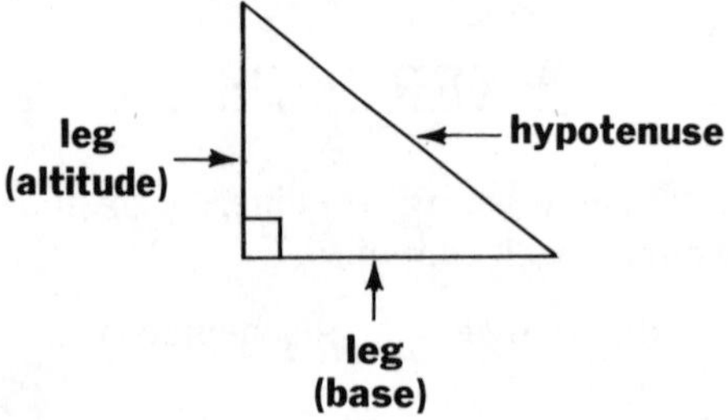

Figure 17-22

More on Triangles

1. **If a triangle is isosceles, the angles opposite the equal sides are equal. That is, if AB = AC then ∠B = ∠C (Fig. 17-23).**

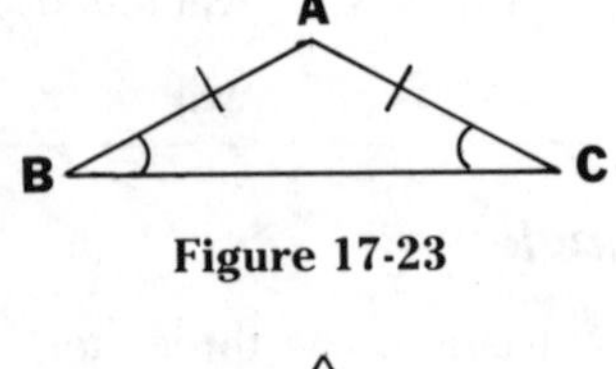

Figure 17-23

2. **If a triangle is equilateral, then all three angles are equal. Since the sum of all three angles is 180°, each angle of an equilateral triangle measures 60° (Fig. 17-24).**

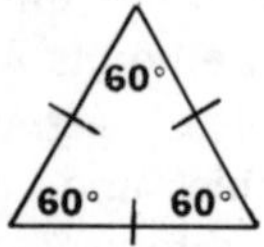

Figure 17-24

(continued)

3. Note that every equilateral triangle is isosceles, but not every isosceles triangle is equilateral.

4. The segment joining the midpoints of two sides of a triangle is parallel to, and half as long as, the third side (Fig. 17-25). That is, if M is the midpoint of segment AB, and P is the midpoint of segment BC, then MP is parallel to AC and MP = $\frac{1}{2}$ AC.

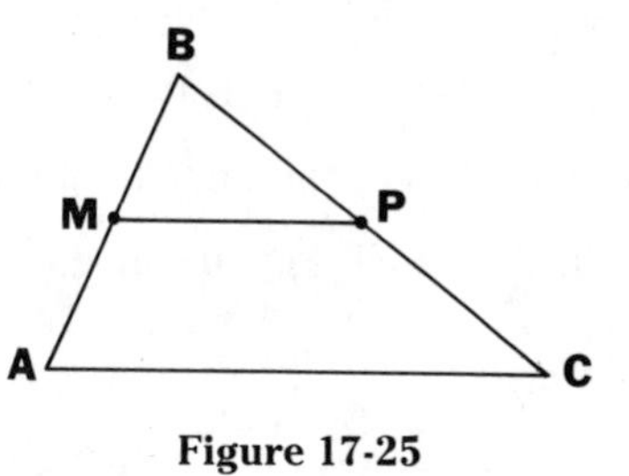

Figure 17-25

Special Quadrilaterals

Certain quadrilaterals are of great importance because of special properties they possess.

A quadrilateral with two pairs of parallel sides is a **parallelogram.** It can be shown that if both pairs of opposite sides are parallel, then the opposite sides also have the same length. In parallelogram ABCD in Fig. 17-26,

AB is parallel to DC, and AB = DC

BC is parallel to AD, and BC = AD

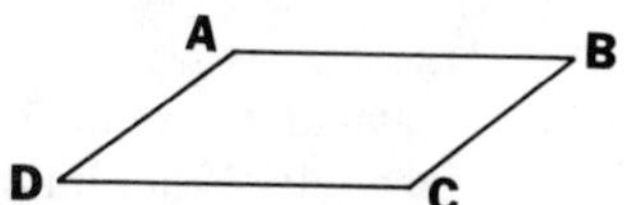

Figure 17-26

A **rectangle** is a parallelogram with four right angles.

A **square** is a rectangle with equal sides.

A **trapezoid** has only one pair of opposite sides parallel.

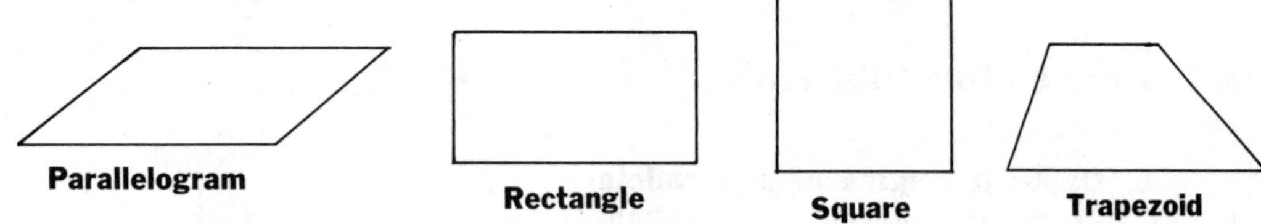

EXERCISES

1. What is true of *all* triangles?
2. How many triangles can you find in the figure at the right? Name them.

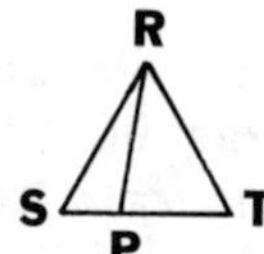

3. How many rectangles can you find in the figure at the right? Name them.

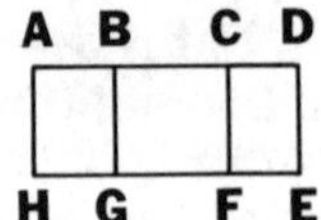

4. Are all squares rectangles?
5. Are all rectangles squares?
6. Are all trapezoids quadrilaterals?
7. Tell as much as you can about each of the following figures:

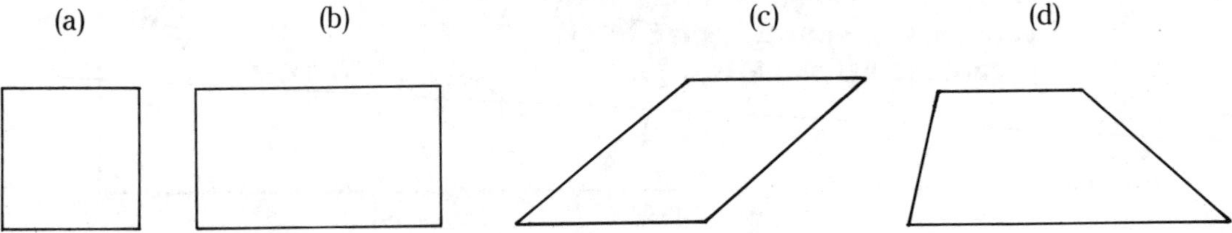

8. Are all rectangles parallelograms?
9. Are all squares parallelograms?
10. What is the greatest number of right angles a triangle can have?
11. If two angles of a triangle measure 70° and 80°, what is the measure of the third angle?
12. If each of the two equal angles of an isosceles triangle measures 45°, what is the measure of the third angle? What kind of triangle is it?

More on Quadrilaterals

1. Opposite angles of a parallelogram are equal. In parallelogram ABCD, ∠A = ∠C, and ∠B = ∠D.

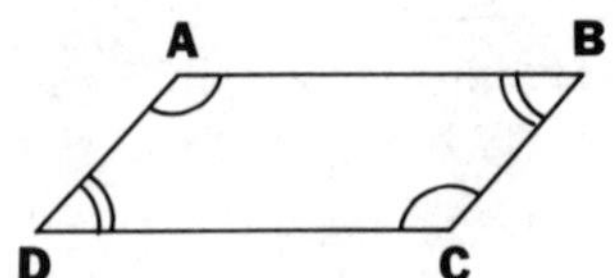

2. The diagonals of a parallelogram bisect each other. In parallelogram ABCD, AM = MC, and BM = MD.

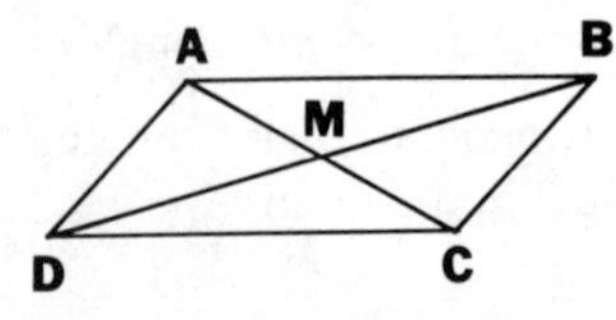

(continued)

3. **If two sides of a quadrilateral are both parallel and equal, then the quadrilateral is a parallelogram. That is, if in quadrilateral EFGH, EF is parallel to HG, and EF = HG, then EFGH is a parallelogram.**

4. **If the opposite sides of a quadrilateral are equal, then the figure is a parallelogram. That is, if in quadrilateral MNPR, MN = RP and MR = NP, then MNPR is a parallelogram.**

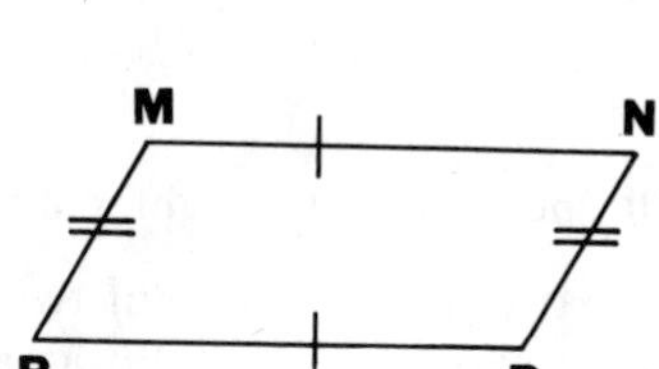

5. **The diagonals of a square are equal. In square LMNO, LN = MO.**

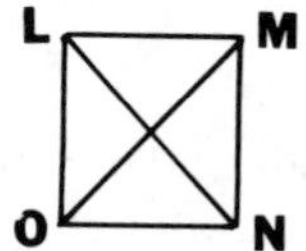

6. **The diagonals of a rectangle are equal. In rectangle ABCD, AC = BD.**

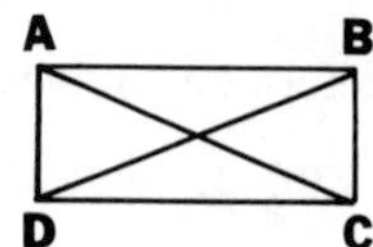

Circles

A **circle** is a set of points that are the same distance from a *center* point. All the points on the circle in Fig. 17-27 are the same distance from the center point O.

The distance from the center of a circle to the circle is the **radius** (plural: **radii**). The distance across a circle through its center is the **diameter**. A diameter is twice the radius of the circle.

A **chord** is a segment with both endpoints on the circle. A **tangent** to a circle is a line that intersects the circle in only one point. In Fig. 17-30, line CD is a tangent intersecting the circle at point P.

A tangent to a circle is perpendicular to the radius drawn to the point of contact. Radius OP is perpendicular to tangent CD at point P.

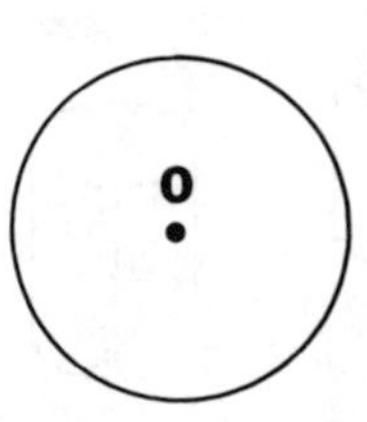

Figure 17-27

Figure 17-28

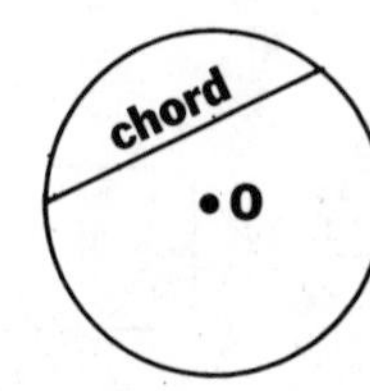

Figure 17-29

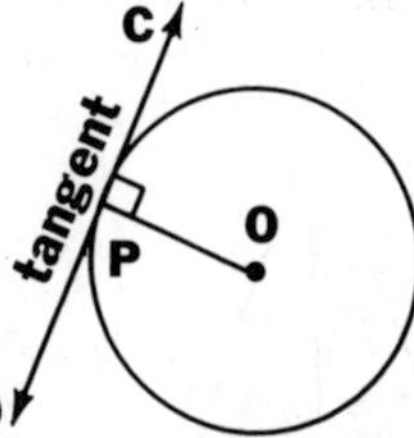

Figure 17-30

An **arc** is a part of a circle. If the endpoints of an arc are the endpoints of a diameter, then the arc is a **semicircle.**

The distance around a circle is its **circumference.**

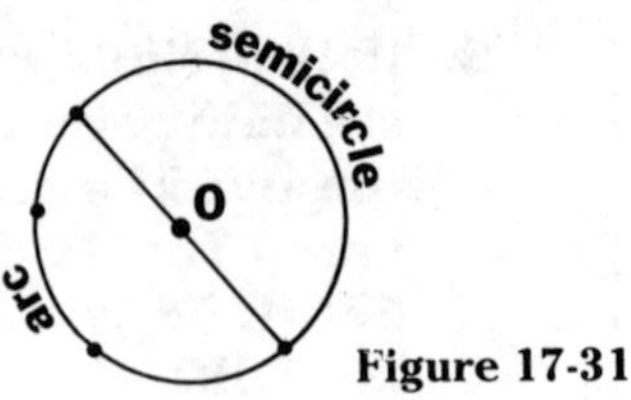

Figure 17-31

EXERCISES

1. In the circle at the right, name

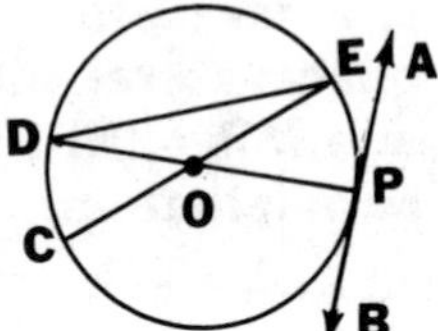

(a) two radii (b) two diameters
(c) a chord (d) an arc
(e) a tangent (f) a semicircle
(g) the center

2. Draw a circle with center P, radius PT, chord XY, and diameter TM.

3. Draw two circles with the same center. (Such circles are called **concentric** circles.)

4. Find the diameter of a circle whose radius measures

(a) 5 inches (b) 21 centimeters (c) 4.5 yards

5. Find the radius of a circle whose diameter measures

(a) 6 centimeters (b) 18 feet (c) 3.8 inches

SPACE FIGURES

The geometry of space figures is sometimes called **solid geometry**. Space figures, like plane figures, are sets of points. Objects such as a sugar cube or a block of wood are examples of solids. Geometrical solids are abstractions of such physical objects.

Figures like rectangles, triangles, and trapezoids have *two* dimensions—usually called length and width. They are all *flat* figures; if laid on a table, every part of the figure would touch the table. We called such figures **plane figures**. But objects like a sugar cube, a block of wood, or a tin can have not only width and length but also height or thickness. These figures have *three* dimensions. If you lay any of them on a flat surface, not every part of the figure would touch the surface. Figures that have three dimensions are called **space figures** or **solids.** Examples of such figures are shown in Fig. 17-32.

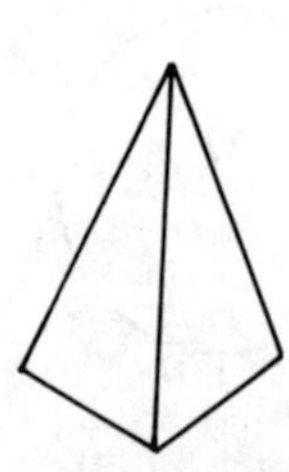

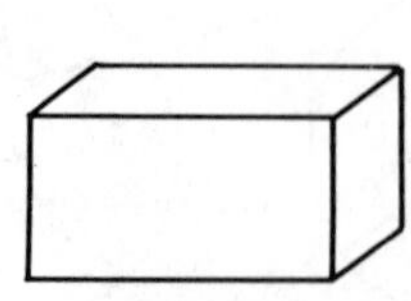
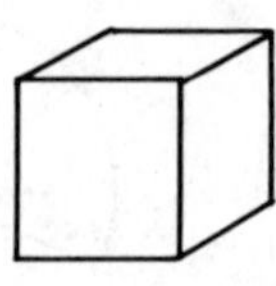
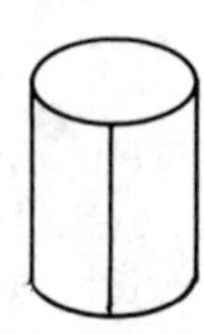

Figure 17-32

Cylinders, Cones, and Spheres

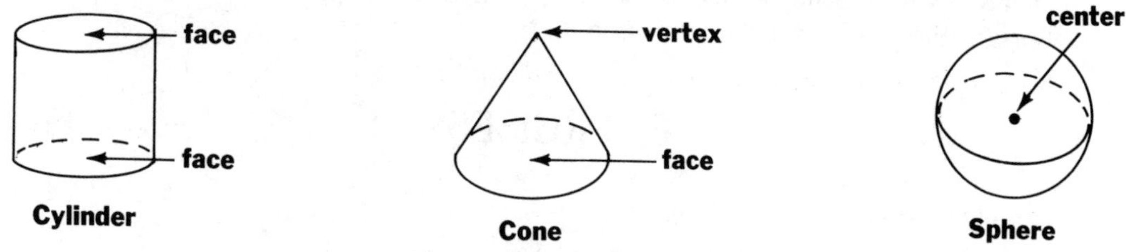

Figure 17-33

A **cylinder** has two circular **faces,** as bases, connected by a curved surface. The two faces are parallel circles of the same size. Examples of cylinders are cans, drums, and pipes.

A **cone** has one circular face as a base, and a **vertex.** Examples of cones are ice cream cones, and paper cups in the shape of cones.

A **sphere** has a curved surface on which every point is the same distance from the center. An example of a sphere is a ball. (See Fig. 17-33.)

EXERCISES

1. Name objects in your room that suggest a

   (a) cylinder (b) cone (c) sphere

2. Which geometric figure is suggested by each of the following?

   (a) a basketball (b) the tip of a pencil (c) a jar (d) a rolled up carpet
   (e) a telephone pole (f) the sun (g) a megaphone

Cubes, Rectangular Prisms, and Pyramids

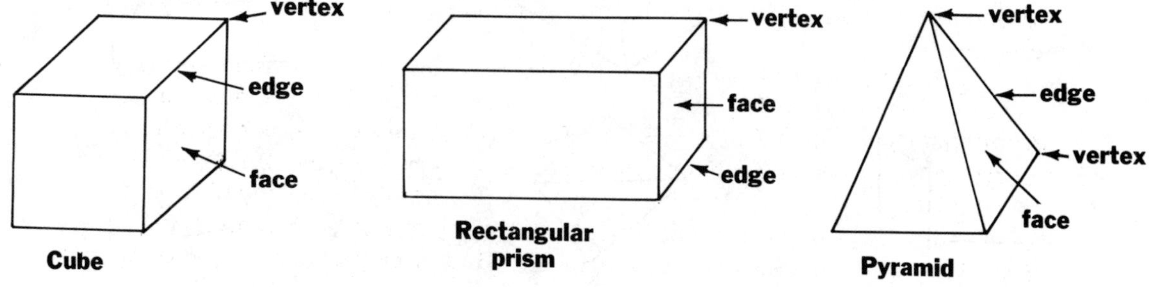

Figure 17-34

The faces of all these figures (Fig. 17-34) are polygons—squares, rectangles, or triangles. The part of the figure where faces meet is an **edge**. The part of the figure where edges meet is a **vertex.**

A **cube** has six squares for its faces. All the edges have the same length. Examples of a cube are sugar cubes and dice. The length, width, and height of a cube are equal.

A **rectangular prism** has six rectangular faces. Examples are bricks and boxes.

A **pyramid** has any polygon for its base, and triangular faces that meet in a common vertex. Examples are the famous pyramids in Egypt and buildings like the Transamerica Pyramid in San Francisco.

EXERCISES

1. Name objects in your room that suggest

   (a) a cube (b) a rectangular prism (c) a pyramid

2. Which geometric figure is suggested by each of the following?

   (a) a chest of drawers (b) a cinder block (c) dice
   (d) a tent that looks like this:

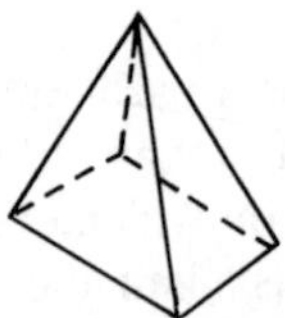

3. How does a cube differ from a rectangular prism?

4. Name each figure below.

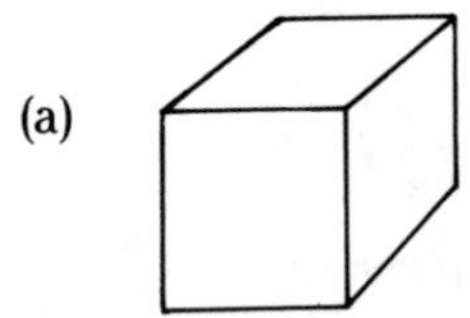

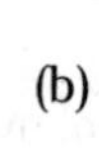

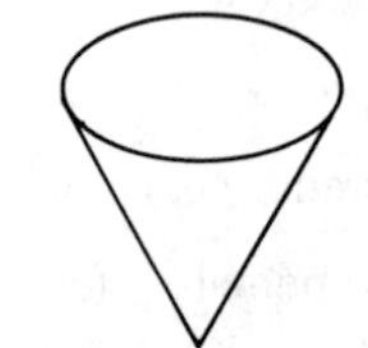

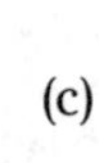

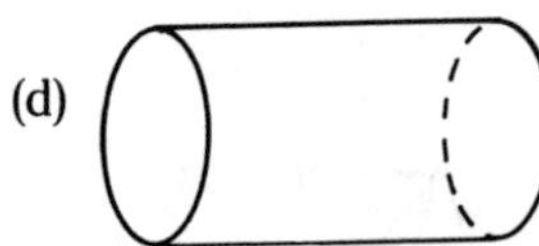

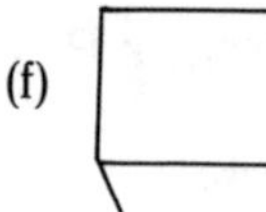

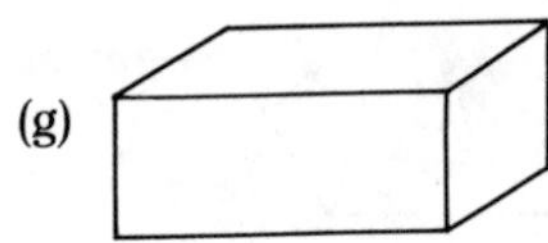

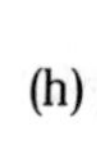

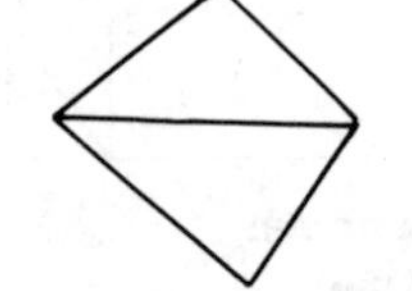

(i)

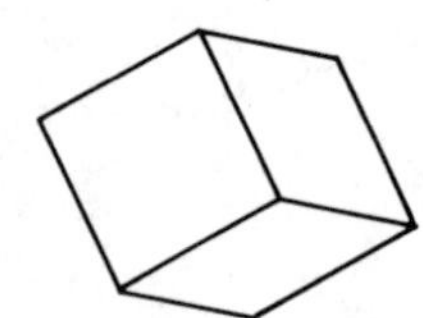

MEASURING LENGTH

PERIMETER

The distance around a figure is its *perimeter*. You find the perimeter of the rectangle in Fig. 17-35 by adding the lengths of the sides: 3 in. + 9 in. + 3 in. + 9 in. = 24 in.

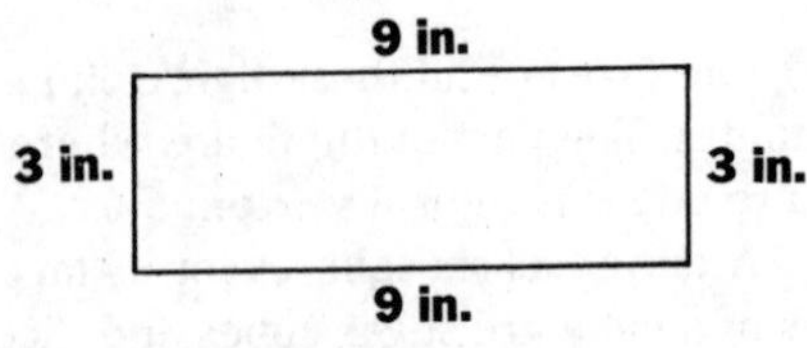

Figure 17-35

Since a rectangle has two equal lengths and two equal widths (Fig. 17-36), its perimeter is twice the length added to twice the width. Letting P stand for the perimeter, l for the length, and w for the width, we have this formula for finding the *perimeter of a rectangle:*

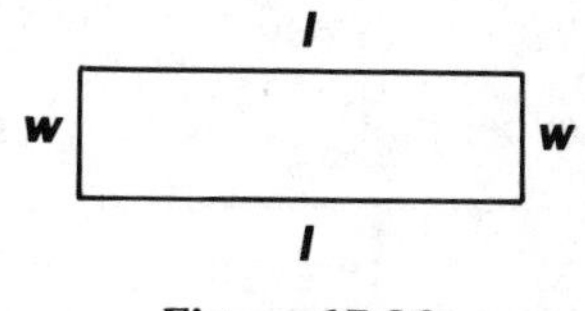

Figure 17-36

$$P = 2l + 2w$$

Example: Find the perimeter of a rectangle 9 inches long and 3 inches wide.

Solution: $P = 2l + 2w$
$P = (2 \times 9) + (2 \times 3)$
$P = 18 + 6$
$P = 24$ inches

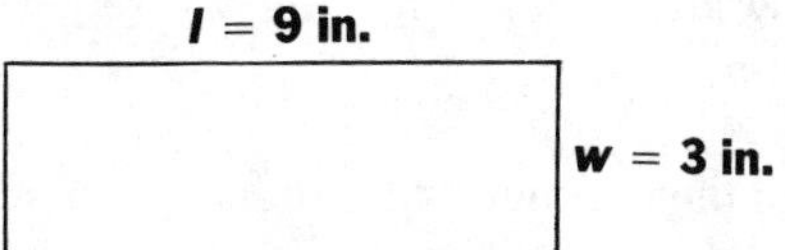

In a square, where the four sides are the same length, the perimeter is four times the length of its side. Letting s stand for the side, we have a formula for finding the *perimeter of a square:*

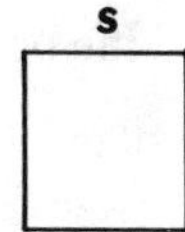

$$P = 4s$$

Example: Find the perimeter of a square whose side is 12 meters.

Solution: $P = 4s$
$P = 4 \times 12$
$P = 48$ meters.

s = 12 m

The perimeter of a triangle is the sum of the lengths of its sides. Letting a, b, c stand for the lengths of the sides of a triangle, the formula for finding the *perimeter of a triangle* is

$$P = a + b + c$$

Example: Find the perimeter of a triangle whose sides measure 3 centimeters, 4 centimeters, and 6 centimeters.

Solution: $P = a + b + c$
$P = 3 + 4 + 6$
$P = 13$ centimeters

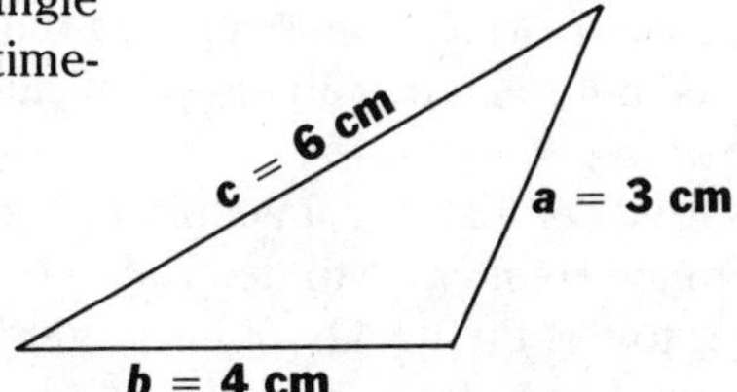

NOTE: To measure the perimeter of an irregularly shaped object, wrap a string around the object and then measure the length of the string.

EXERCISES

1. Find the perimeters of rectangles with these dimensions:

	(a)	(b)	(c)	(d)
Length	**5 cm**	**18 ft**	**$3\frac{1}{2}$ yd**	**$5\frac{1}{4}$ in.**
Width	**3 cm**	**11 ft**	**2 yd**	**$3\frac{1}{2}$ in.**

2. How many feet of fencing are required for a rectangular garden measuring 25 feet by 58 feet?

3. Find the perimeter of a square with one side measuring:

   (a) 7 meters (b) 12 feet (c) $5\frac{3}{4}$ inches

4. Find the perimeter of a triangle measuring:

   (a) 10 in., 14 in., 22 in. (b) 125 ft, 50 ft, 120 ft (c) $7\frac{1}{2}$ in., $8\frac{3}{4}$ in., $5\frac{2}{3}$ in.
   (d) 4 ft, 6 ft, 6 ft (e) 15 in., 15 in., $3\frac{1}{2}$ in.

5. Find the perimeter of an equilateral triangle with one side measuring

   (a) 14 cm (b) 3.25 in. (c) $5\frac{3}{8}$ ft

6. If the perimeter of a square is 12 yards, what is the length of its side?

7. If the perimeter of an equilateral triangle is 57 cm, what is the length of each side?

CIRCUMFERENCE

To find the perimeter of a triangle, or any polygon, we measure its sides and add the lengths. But how can we find the distance around a circle? Can we measure the length of a curve the way we measure the length of a line segment? Can we use a ruler to measure the *circumference* of a circle?

We can, of course, wrap a tape measure around the circle and that way measure its circumference. But this is not always convenient, accurate, or even possible. There is, however, an easy way to find the circumference if we know the length of a particular line segment of the circle—its diameter. There is a fascinating relation between the diameter of every circle and its circumference.

This relation involves a new and amazing number—a number that has intrigued people for thousands of years. Though first referred to in the Old Testament, used by the ancient Greeks, and still used today in connection with circles, this number has turned out to be equally useful in situations that have nothing to do with circles or even with geometry.

It's used, for instance, in public opinion polls to predict the winner of an election, by insurance companies to determine the premium you pay on a policy, by a manufacturer to test the quality of his product, by the television industry to measure the popularity of one of its programs, and by scientists to understand the laws of heredity.

The number is named by the Greek letter π (*pi*) and has an approximate value of $\frac{22}{7}$ or 3.1416. The relation between the circumference (C) of a circle, the length of its diameter (d), and π is that the circumference of a circle equals π times its diameter:

$$C = \pi d$$

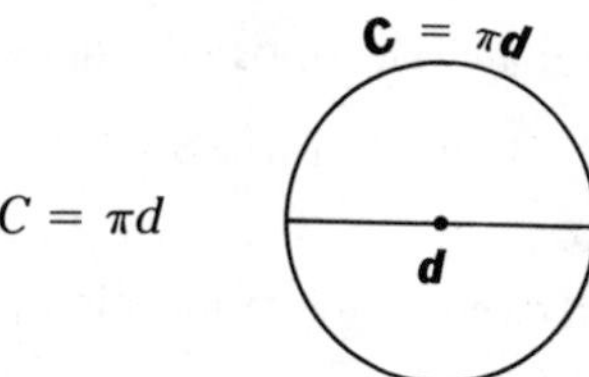

This formula says that the circumference of *any* circle is a little more than three times the length of its diameter. For the value of π we often use $\frac{22}{7}$ or 3.14, whichever is easier.

Since the diameter of a circle is twice its radius (r), we can replace the *d* in the formula with *2r*:

$$C = \pi \times 2r \quad \text{or} \quad C = 2\pi r$$

So to find the *circumference of a circle*, we use the formula

$$C = \pi d \quad \text{or} \quad C = 2\pi r$$

(whichever is easier).

NOTE: From the formula $C = \pi d$, we get $d = C/\pi$. This tells us that if we know the circumference (C) of a circle, we can find its diameter (d) by dividing C by π. (See Example 3 below.)

Example 1: Find the circumference of a circle if its diameter is 9 inches.

Solution: $C = \pi d$
$C = 3.14 \times 9$
$C = 28.26$ in.

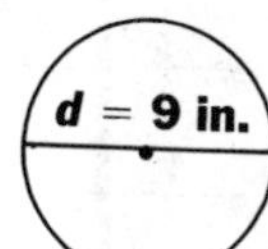

Example 2: Find the circumference of a circle if its radius is 21 feet.

Solution: $C = 2\pi r$
$C = 2 \times 3.14 \times 21$
$C = 131.88$ ft

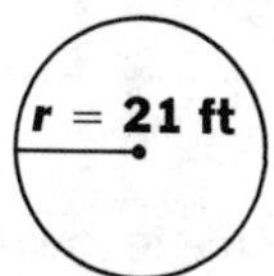

(If you choose to use $\frac{22}{7}$ for the value of π, you get

$$C = 2\pi r$$
$$C = 2 \times \frac{22}{7} \times 21$$
$$C = 132 \text{ ft})$$

Example 3: If the circumference of a circle is 54 feet, what is its diameter?

Solution: $d = \dfrac{C}{\pi}$

$d = \dfrac{54}{3.14}$

$d = 17.20$ ft

C = 54 ft

EXERCISES

1. What is the circumference of a circle whose diameter is:

   (a) 8 cm (b) 23 yd (c) 42 in. (d) 5.8 ft

2. Find the circumference of a circle whose radius is:

   (a) 2 m (b) 9 in. (c) 14 ft

3. What is the circumference of a clock with a 10-inch radius?

4. Find the diameter of a circle whose circumference is 42 cm. What is its radius?

EXPERIMENT: Your child can establish experimentally the formula $C = \pi d$, and the value of π as about 3.14.

Materials Needed:

Five circular objects (like tops of jars) of different diameters
A tape measure
A chart:

Object #	C	d	C/d
1			
2			
3			
4			
5			

Procedure

1. Ask your child to use the tape to measure the circumference (C) and diameter (d) of each object and enter the results in the chart.
2. To obtain the values for the column C/d, have the child divide the value of C by the value of d (to two decimal places) for each object.
3. If the measurements and calculations are accurate, the value of C/d will, in each instance, be about 3.14—the value of π.

   Another way to express the conclusion that

   $$C/d = 3.14$$
   or
   $$C/d = \pi$$

   is to say that "π is the *ratio* of the circumference to the diameter."

4. If $C/d = \pi$, then $C = \pi d$.

A Famous Number

***Pi*, or π, one of the most famous of all numbers, dates back to the Old Testament, where it is given a value of 3. Egyptian mathematicians gave it a more accurate value of 3.16, while the familiar value of 3.1416 was established nearly 2000 years ago.**

Late in the eighteenth century, π was shown to be an *irrational number*—a number whose decimal form consists of an endless string of digits with no repeating cycle of digits (see page 184). Over the centuries many people tried to find a more accurate value for π by calculating it to more and more decimal places. Though, for convenience, we often use 3.14 as its value, a better approximation—accurate to 8 decimal places—is 3.14159265. . . .

The German mathematician Ludolph van Ceulen (1540–1610) computed π to 35 decimal places; this was considered such an extraordinary achievement that the number was engraved on his tombstone.

In 1949, the ENIAC computer calculated the value of π accurate to 2037 decimal places and, in 1958, another computer was used to calculate π to 10,000 places. The fantastic accuracy to 500,000 decimal places was reached in 1967!

The preoccupation with calculating π to so many decimal places is all the more remarkable in light of at least one authoritative opinion that if π, expressed to only 10 decimal places, were used to compute the circumference of the earth, the result could be correct to inches!

MEASURING AREA

Besides length, we often need to know the *amount of surface* enclosed by a geometric figure. For instance, to calculate how much paint is needed to paint a rectangular wall, you need to know how much surface the wall covers. Likewise, the amount of cement needed to make a circular walk depends on how much surface the walk will cover.

Amount of surface is measured in *square units*.

Fig. 17-37 shows a *square inch*, a square measuring 1 inch on each side. The number of such square units needed to cover the surface of a geometric figure is its *area*.

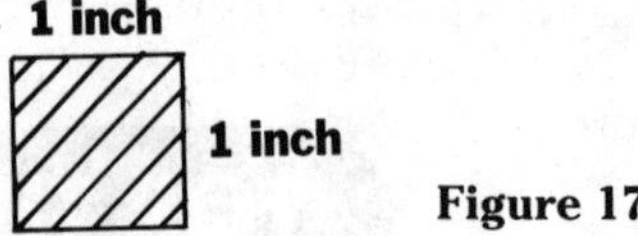

Figure 17-37

Since 6 square inches are needed to cover the surface inside the rectangle in Fig. 17-38, we say that the area of this rectangle is *6 square inches*.

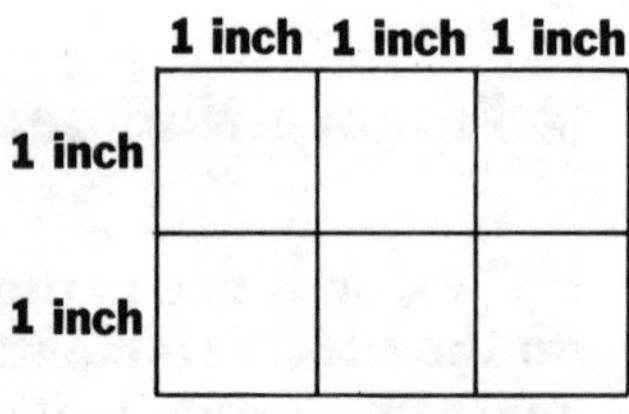

Figure 17-38

Other square units commonly used to measure area are: square centimeters, square feet, square yards, square miles, square meters, and square kilometers.

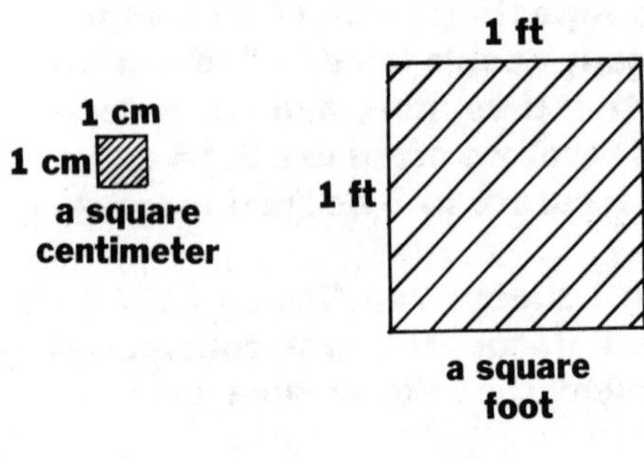

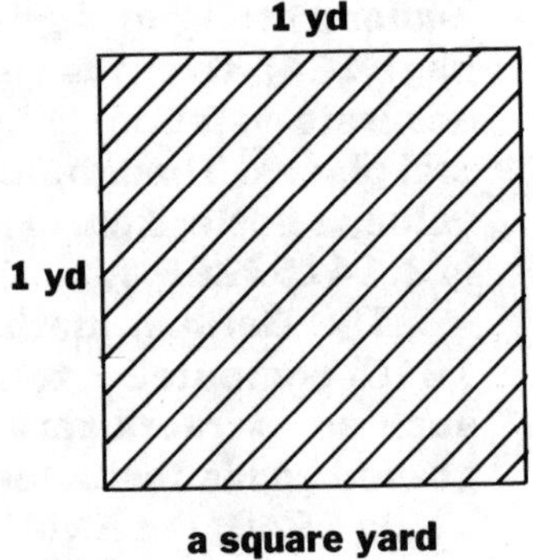

When we say that the area of a window is 8 square inches, we mean that 8 of these [1 inch × 1 inch square] are needed to cover the surface inside the window (Fig. 17-39).

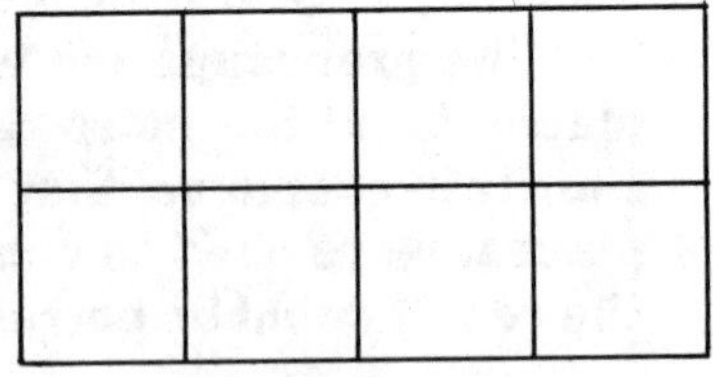

Figure 17-39

Activities to Reinforce Your Child's Understanding of the Concept of *Area*

1. Draw several geometric figures on a sheet of paper:

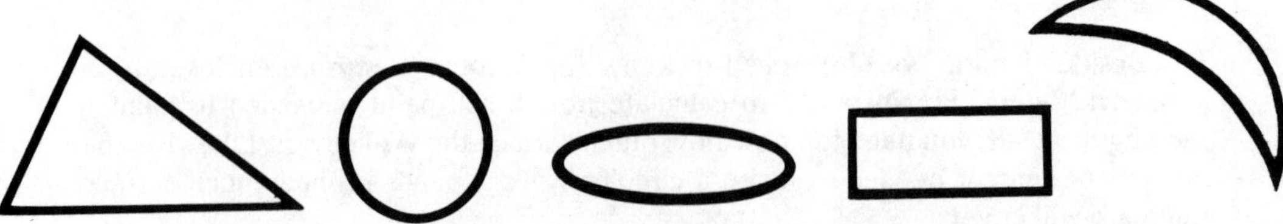

Figure 17-40

Then give your child a transparent grid ruled up in square units. Ask her to place the grid over each figure, one at a time. By counting the number of square units needed to cover each shape (Fig. 17-41), the child obtains an approximation of its area.

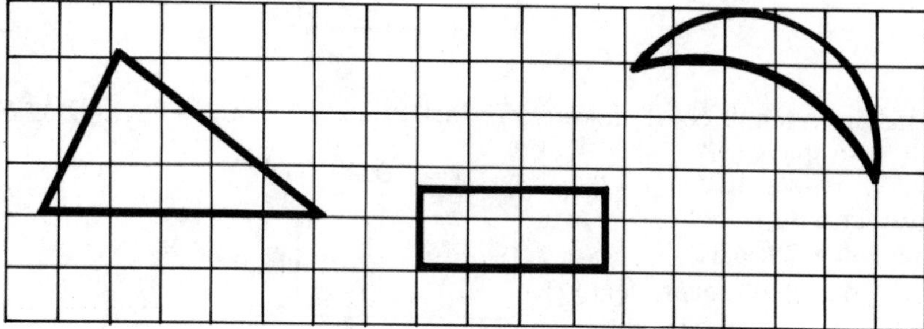

Figure 17-41

2. Have your child draw his own shapes on squared paper and count the number of squares covered by each shape to obtain an approximation of its area.

AREA OF A RECTANGLE

Your child first determines the area of a rectangle by the use of a grid. You then lead him to a shorter and more accurate method: *The number of square units in one row is multiplied by the number of rows.* For instance, in the rectangle in Fig. 17-42 there are three rows, each row having four square units. The area of the rectangle is, therefore,

$$3 \times 4 \text{ square units} = 12 \text{ square units}$$

Let the child verify that this method works by actually counting the number of squares in the rectangle.

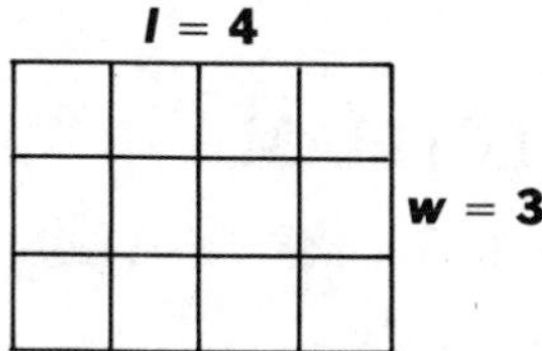

Figure 17-42

This leads to the formula for finding the **area of a rectangle**: The area (A) of a rectangle is equal to *the length (l) times the width (w)*:

$$A = lw$$

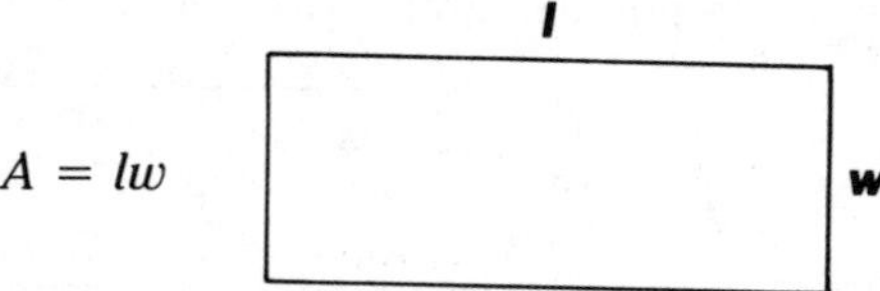

Example 1: Find the area of a rectangle 5 inches long and 3 inches wide.

Solution: $A = lw$
$A = 5 \times 3$
$A = 15$ square inches

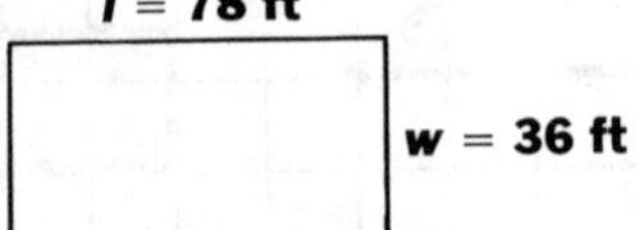

Example 2: A tennis court measures 78 feet by 36 feet. What is the playing area of the court?

Solution: $A = lw$
$A = 78 \times 36$
$A = 2808$ square feet

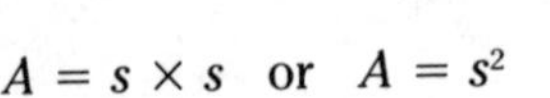

AREA OF A SQUARE

A square is a special kind of rectangle: a rectangle with all sides equal. So if we call the side of the square *s*, its area is

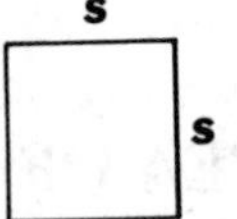

$$A = s \times s \quad \text{or} \quad A = s^2$$

That is, the area of a square is equal to *the square of the length of a side.*

Example: Find the area of a square one side of which is 5 cm.

Solution: $A = s^2$
$A = 5^2$
$A = 5 \times 5$
$A = 25$ square centimeters

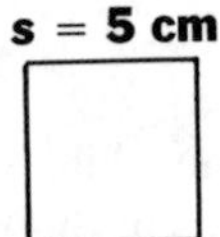

AREA OF A PARALLELOGRAM

The area of a parallelogram is equal to *the base (b) times the height (h)*:

$$A = bh$$

(The reason for this formula is explained below.)

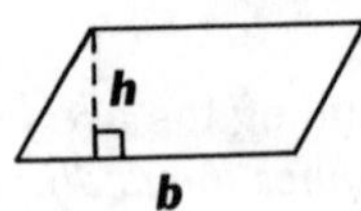

Example: Find the area of a parallelogram with a 15-inch base and a height of 4 inches.

Solution: $A = bh$
$A = 15 \times 4$
$A = 60$ sq. in.

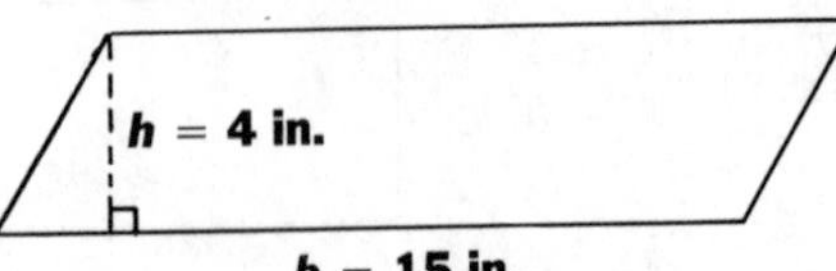

Why Is $A = bh$ the Formula for the Area of a Parallelogram?

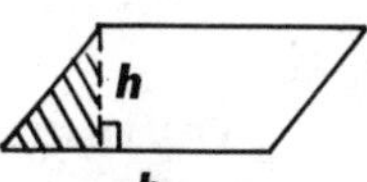

Figure 17-43

The reason is that every parallelogram can be converted into a rectangle with the same area. For instance, if from the parallelogram in Fig. 17-43 we snip off the shaded triangle on the left and attach it to the right as shown in Fig. 17-44, we will have converted the parallelogram in Fig. 17-43 to the rectangle in Fig. 17-44. Both have the same area, since all we did was to *relocate* one piece—the triangle.

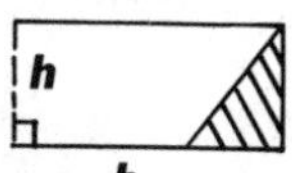

Figure 17-44

Let's now call the length of the rectangle its *base* (b), and its width the *height* (h). Since we already know that the area of this rectangle is $A = bh$, the area of the parallelogram must also be $A = bh$.

AREA OF A TRIANGLE

The area of a triangle is equal to *one half the base (b) times the height (h).* That is,

$$A = \tfrac{1}{2}\,bh$$

h

b

(The reason for this formula is explained below.)

Example: Find the area of a triangle with a height of 8 inches and base of 12 inches.

Solution:
$A = \tfrac{1}{2}\,bh$
$A = \tfrac{1}{2} \times 12 \times 8$
$A = 48$ sq. in.

$h = 8$ in.

$b = 12$ in.

Why Is A = $\frac{1}{2}$ *bh* the Formula for the Area of a Triangle?

The reason is that every triangle can be regarded as *half* of a parallelogram.

Let's start with the triangle in Fig. 17-45. If you take two such triangles and put them together, as in Fig. 17-46, you form a parallelogram.

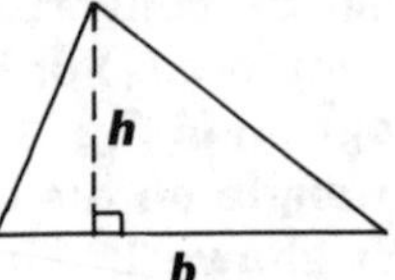

Figure 17-45

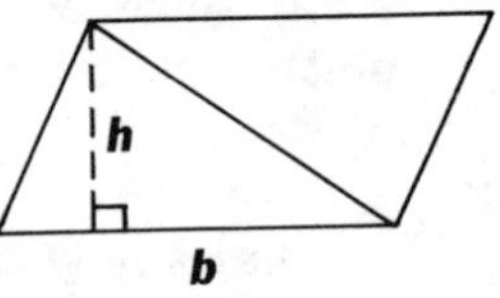

Figure 17-46

[Reason: The opposite sides in Fig. 17-46 are equal. And on page 218 you saw that if the opposite sides of a quadrilateral are equal, it is a parallelogram.]

The area of the triangle in Fig. 17-45 is half the area of the parallelogram in Fig. 17-46, because the parallelogram is made up of two such triangles.

So if the area of the parallelogram is $A = bh$, the area of the triangle is half that, or $A = \frac{1}{2}bh$.

EXERCISES

1. What's the difference between an *inch* and a *square inch*? Between a *centimeter* and a *square centimeter*?

2. In each example below, do you measure *length* or *area*?

   (a) the distance around a block
   (b) the space inside a quadrilateral
   (c) the amount of screening needed to cover a door
   (d) the amount of molding needed for the edge of a table
   (e) the amount of glass needed for a window

For each example in #3 through #6, draw a diagram and explain what your answer means.

3. Find the areas of rectangles with the following dimensions:

	(a)	(b)	(c)	(d)
Length	8 cm	9 miles	16 in.	1.2 km
Width	6 cm	11 miles	$3\frac{1}{8}$ in.	.05 km

4. Find the area of a square one side of which measures:

   (a) 4 in. (b) 6 cm (c) 3.5 in. (d) $2\frac{1}{4}$ ft

5. Find the areas of parallelograms with the following dimensions:

	(a)	(b)	(c)	(d)	(e)	(f)
Base	5 in.	7 yd	16 cm	$3\frac{1}{4}$ ft	$5\frac{1}{2}$ ft	9.3 in.
Height	7 in.	10 yd	5 cm	12 ft	$7\frac{1}{4}$ ft	5.7 in.

6. Find the areas of triangles having these measurements:

	(a)	(b)	(c)	(d)	(e)	(f)
Base	12 in.	8 yd	5 cm	1 ft	$3\frac{1}{2}$ in.	$8\frac{1}{4}$ in.
Height	8 in.	5 yd	3 cm	3 ft	8 in.	$3\frac{1}{2}$ in.

7. How many square meters of wood are needed to make a square tabletop with each side measuring 4 meters?

AREA OF A CIRCLE

Let's try to approximate a formula for finding the area of a circle.

In Fig. 17-47, a circle with radius r is inscribed in a square which is divided into four smaller squares. Since the side of each little square is r, the area of each little square is r^2. The area of three little squares is $3r^2$, and the area of all four little squares is $4r^2$.

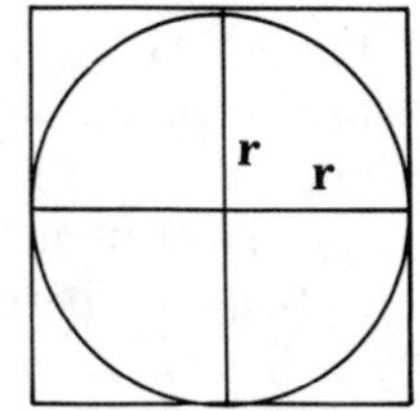

Figure 17-47

If you look at Fig. 17-47, you will see that the area of all four little squares is greater than the area of the circle, while the area of three little squares is not enough to cover the surface of the circle. So the area of the circle would seem to lie somewhere between $3r^2$ and $4r^2$.

The formula for the area of a circle is πr^2, or about $3.14r^2$—which is about what we expected.

The area of a circle is equal to π *times the radius squared*:

$$A = \pi r^2$$

Use $\frac{22}{7}$ or 3.14 as the value for π. For greater accuracy, use 3.1416.

Example: Find the area of a circle with a radius of 6 feet.

Solution:
$A = \pi r^2$
$A = 3.14 \times (6)^2$
$A = 3.14 \times 36$
$A = 113.04$ sq. ft

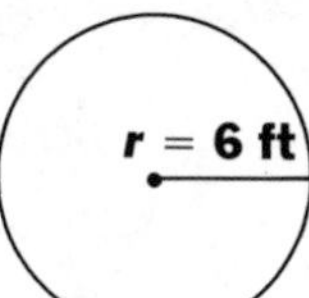

EXERCISES

1. Find the area of a circle with a radius of:

   (a) 3 in. (b) 7 m (c) 15 ft (d) 35 yd (e) 2.5 cm

2. Find the area of a circle with a diameter of:

   (a) 8 m (b) 10 in. (c) 9 ft

3. Find the area of a circular mirror with a diameter measuring 20 inches.

MEASURING VOLUME

Fig. 17-48 shows a **cubic inch:** a cube measuring 1 inch in length, width, and height. Other commonly used cubic units are a cubic centimeter, a cubic foot, and a cubic yard.

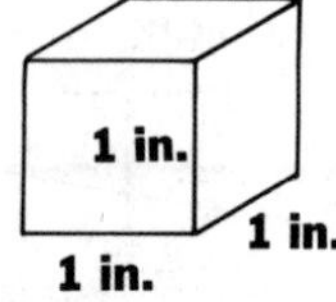

Figure 17-48

The **volume** of a three-dimensional figure is the number of cubic units it contains. The volume of a box is the number of cubic units needed to fill it.

When you want to paint the outside of a box and you need to know how much paint it will take, you are concerned with its *area.* When you want to know how much can be stored in the box, you are concerned with its *volume.*

VOLUME OF A RECTANGULAR PRISM

One way to measure the volume of the box in Fig. 17-49 is to *count* the number of cubic inches it contains: 12 cubic inches.

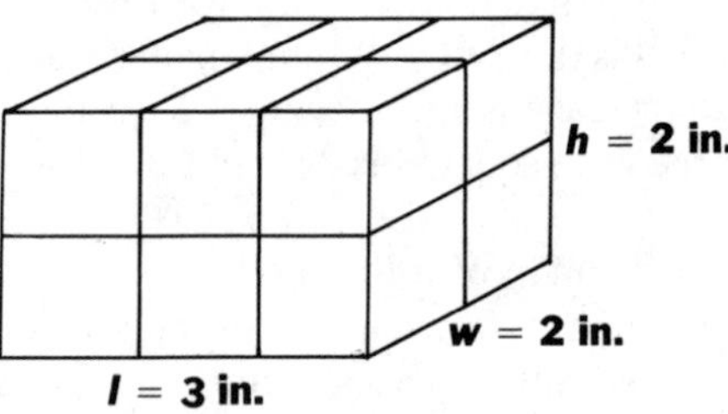

Figure 17-49

You can also find the volume by *multiplication:* There are two layers of cubes. Each layer contains 3×2, or 6 cubic inches. Two such layers of cubes will contain $2 \times (3 \times 2) = 12$ cubic inches. This illustrates that the volume of a rectangular solid is equal to the *length times the width times the height:*

$$V = lwh$$

In finding the volume of a rectangular prism we are, in effect, multiplying the area of the base ($l \times w$) by the height (h).

Example: Find the volume of a freezer 5 feet long, 3 feet wide, and 6 feet high.

Solution: $V = lwh$
$V = 5 \times 3 \times 6$
$V = 90$ cubic feet

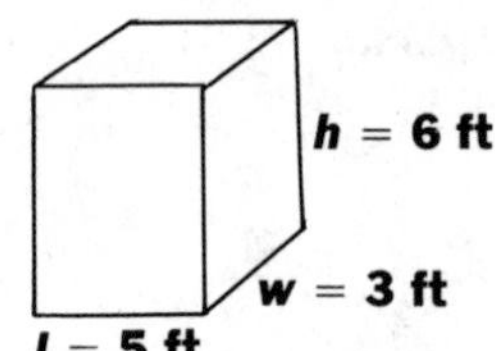

VOLUME OF A CUBE

A cube is a special kind of rectangular prism—one where the length, width, and height are the same. If we call the edge of the cube e, then the formula for finding its volume is

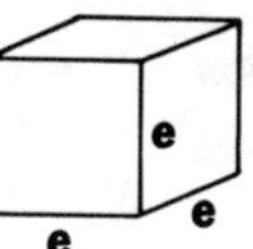

$$V = e \times e \times e \text{ or } V = e^3$$

Example: Find the volume of a cube measuring 4 centimeters on its side (or edge).

Solution: $V = e^3$
$V = 4 \times 4 \times 4$
$V = 64$ cubic cm

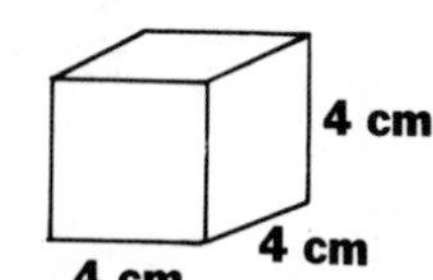

EXERCISES

1. What's the difference between an inch, a square inch, and a cubic inch?

2. Find the volume of a box with the following dimensions:

	(a)	(b)	(c)	(d)	(e)
Length	5 in.	8 cm	5 ft	4 yd	3.4 m
Width	2 in.	5 cm	12 ft	5 yd	2.5 m
Height	4 in.	7 cm	6 ft	$3\frac{1}{2}$ yd	12 m

3. Find the volumes of cubes whose edges measure:
(a) 5 ft (b) 14 in. (c) $2\frac{1}{2}$ ft (d) .3 cm

4. Find the volume of a storage closet 6 feet long, 3 feet deep, and 8 feet high.

5. Find the volume of an ice cube measuring 2.8 centimeters.

VOLUME OF A CYLINDER

When we found the volume of a rectangular prism, we multiplied the area of the base by the height and got the formula

$$V = lwh$$

We follow the same idea in finding the volume of a cylinder: multiply the area of the base by the height:

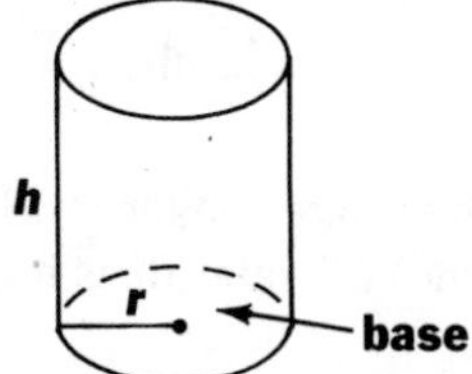

$$V = \text{base} \times \text{height}$$

Since the base of a cylinder is a circle, its area is πr^2. Therefore, the formula for finding the *volume of a cylinder* becomes

$$V = \pi r^2 h$$

Example: Find the volume of a can having a radius of 6 inches and a height of 10 inches.

Solution: $V = \pi r^2 h$
$V = 3.14 \times (6)^2 \times 10$
$V = 3.14 \times 36 \times 10$
$V = 1130.4$ cubic in.

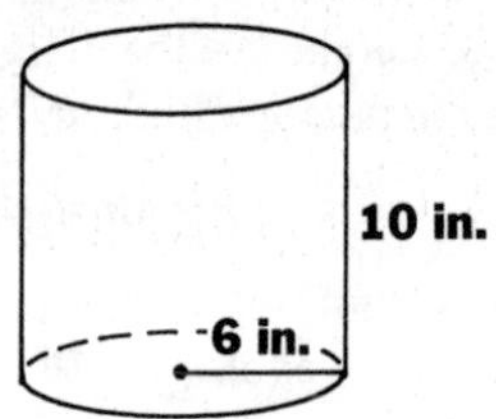

VOLUME OF A SPHERE

In more advanced mathematics books it is shown that the volume of a sphere is equal to $\frac{4}{3}\pi$ *times the cube of the radius.* That is,

$$V = \frac{4}{3}\pi r^3$$

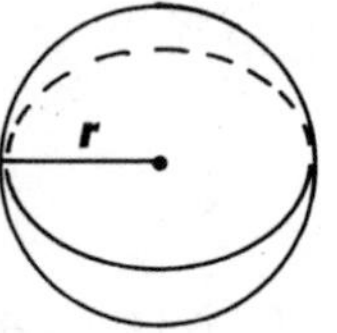

Example: Find the volume of a golf ball with a radius of 2 inches.

Solution: $V = \frac{4}{3}\pi r^3$
$V = \frac{4}{3} \times \frac{22}{7} \times (2)^3$
$V = \frac{4}{3} \times \frac{22}{7} \times 2 \times 2 \times 2$
$V = \frac{4}{3} \times \frac{22}{7} \times 8$
$V = \frac{704}{21}$
$V = 33.52$ cubic in.

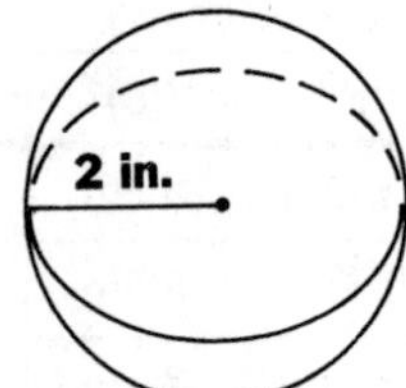

EXERCISES

1. Find the volume of a cylinder with the following dimensions:

	(a)	(b)	(c)	(d)
Radius	3 ft	14 in.	8 cm	7 m
Height	5 ft	10 in.	20 cm	3 m

2. Find the volume of a sphere if:

(a) r = 1 m (b) r = 2 cm (c) r = 3 in. (d) r = 10 yd

3. A hot water boiler is 18 inches in diameter and 50 inches high. How many cubic inches of water does it hold?

4. How many cubic inches are there in a ball with a 6-inch radius?

An Impossible Problem

Here is a cube one edge of which is 2 inches. The volume of this cube is

$$V = e^3$$
$$V = 2^3$$
$$V = 2 \times 2 \times 2$$
$$V = 8 \text{ cubic in.}$$

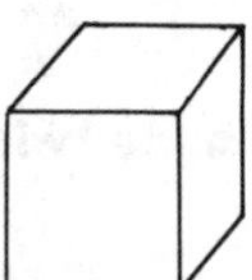

QUESTION: Can you construct, with ruler and compass only, another cube that has *twice* the volume of the first cube?

This sounds like a pretty simple thing to do, yet it can't be done. It's one of the famous "impossible problems" of antiquity—known as the problem of "duplicating the cube." About 150 years ago, a 16-year-old mathematical genius named Evariste Galois (1811–1832) proved that duplication of the cube is impossible with ruler and compass alone. (Galois was killed at the age of 20 in a duel over a girl.)

Let's start with a cube having *a* as an edge. We wish to construct a second cube, with edge *b*, whose volume is twice that of the first cube.

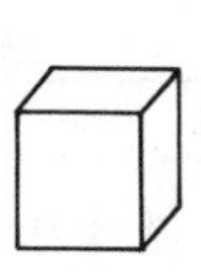

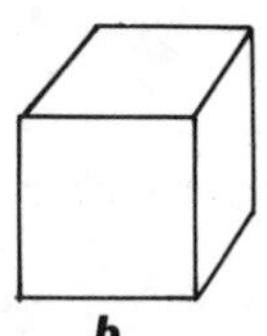

The volume of the first cube is a^3; that of the second is b^3. If the volume of the second cube is to be twice that of the first, then

$$b^3 = 2a^3$$

or

$$b = a\sqrt[3]{2}$$

(Read "*b* is equal to *a* times the cube root of 2.")

This means we have to construct a cube with edge *b* whose length is $\sqrt[3]{2}$ (the cube root of 2 is 1.25992 ...) times the length of *a*. Galois proved that this is impossible to do just with ruler and compass.

(continued)

There is a story that this problem originated in a visit to the Delphic oracle. At a time when a raging epidemic was killing many Athenians, the oracle said that the epidemic would end only if the solid gold cubical altar to Apollo were doubled in size. The Athenians at once enlarged the altar by doubling its sides, but the epidemic did not subside.

The masons and architects did not realize that by *doubling* the side of the cube they made its volume *eight* times as great. Let's see why.

Let a = the edge of the smaller cube

and b = the edge of the larger cube.

If b is made twice the length of a, then

$$b = 2a$$

and

$$b^3 = (2a)^3$$
$$b^3 = (2a) \times (2a) \times (2a)$$
$$b^3 = 8a^3$$

The last line says that the volume of the cube with edge b is eight times the volume of the cube with edge a.

By using instruments other than compass and ruler, the Greek mathematicians finally succeeded, the oracle was appeased, and the epidemic ended.

CONGRUENT FIGURES

If we stack a pile of pennies, one on top of the other, they would match perfectly because all pennies have the same size and shape. A deck of cards, too, can be stacked perfectly without any corner sticking out because all have the same size and shape.

Objects having the same size and shape are *congruent* to each other. Pennies are congruent to each other, and so are cards.

Below are pairs of congruent figures:

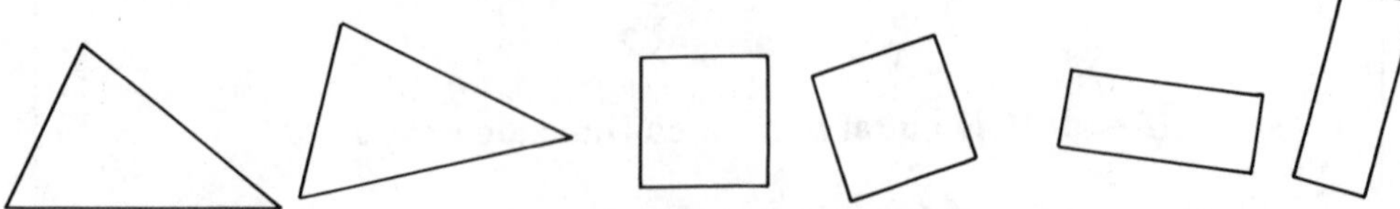

The triangles are the same size and shape. If you cut them out of this page and placed one on top of the other, they would match perfectly. The two triangles are, therefore, congruent to each other. So are the squares, and so are the rectangles.

The concept of congruence is important not only in mathematics. When manufacturers mass-produce a product they are producing congruent objects. The same model cars are congruent to each other; so are the same model radios, pens, screen doors, and basketballs.

Testing whether objects or geometric figures are congruent by superimposing one on top of the other is not always possible or desirable. How would you use this test with basketballs? Obviously, we need better ways to check for congruence. Fortunately, congruence has certain properties that make the test by superposition unnecessary.

CONGRUENT TRIANGLES

The congruent figures that interest us most are triangles. We are often interested in them in order to find out whether particular line segments or angles are equal. For instance, to find out whether, in Fig. 17-50, BC = EF or ∠1 = ∠2, we see whether △ ABC and △ DEF are congruent.

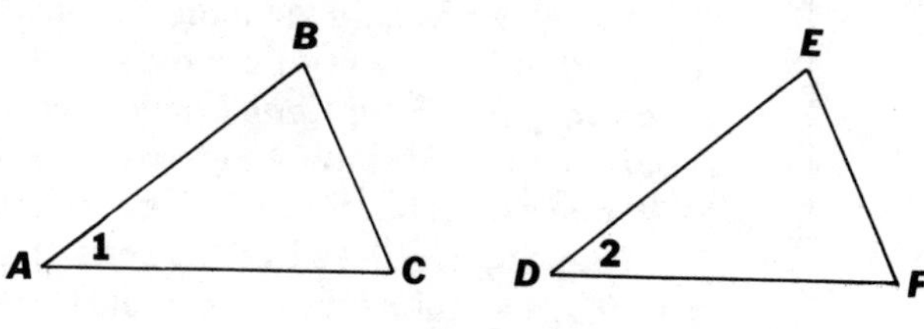

Figure 17-50

If they are, then their matching parts are equal. If segment BC matches segment EF, then they must be equal. If ∠1 matches ∠2, then they, too, are equal.

Here are three standard ways of telling whether two triangles are congruent:

1. If the three sides of one triangle are equal to the corresponding sides of the other triangle:

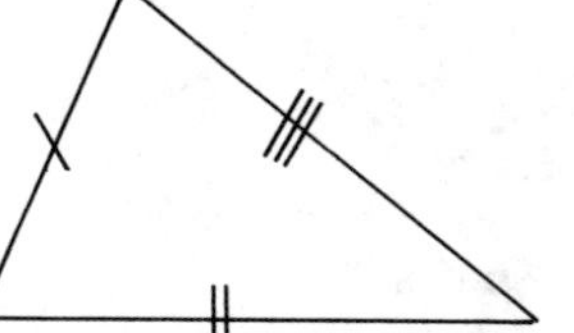
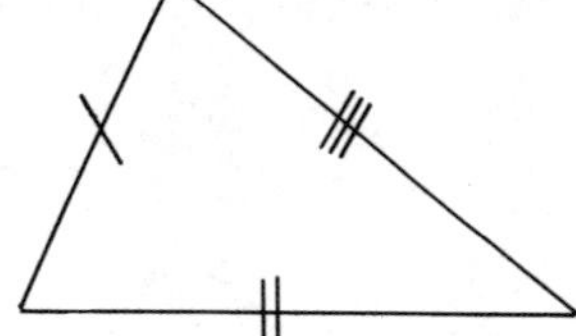

2. If two sides and the included angle of one triangle are equal to the corresponding parts of the other triangle:

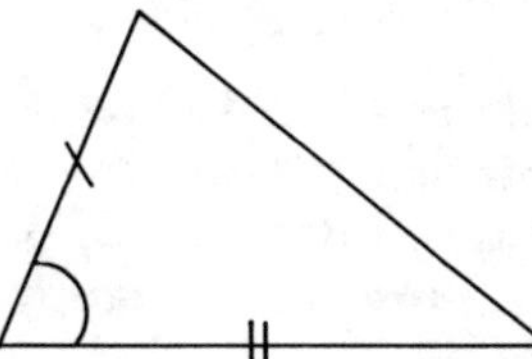
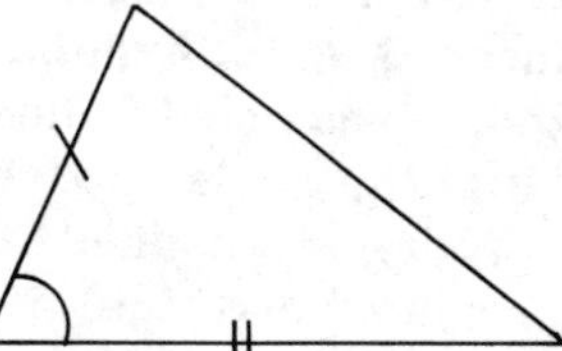

3. If two angles and the included side of one triangle are equal to the corresponding parts of the other triangle:

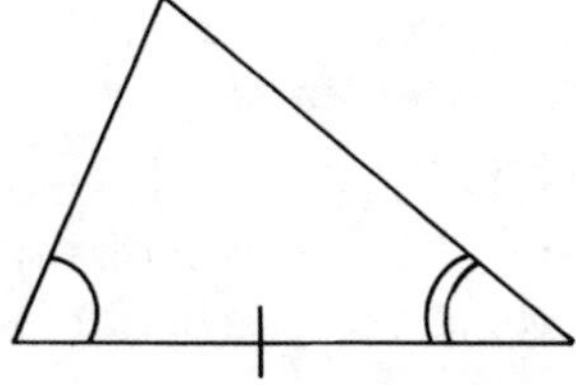
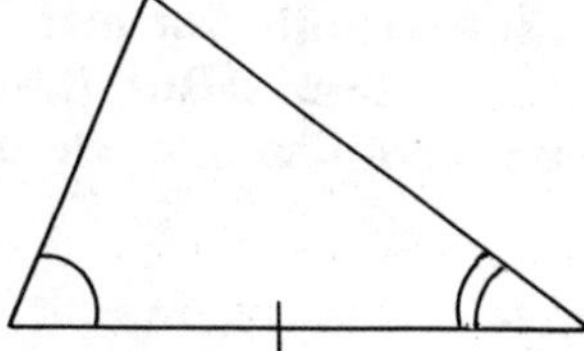

The Most Famous Theorem in Mathematics

Every spring, in ancient Egypt, the Nile River overflowed, washing out property boundaries that later had to be redrawn. Since very few Egyptians knew how to construct the right angles required for reestablishing the boundaries, consultants known as "rope stretchers" had to be called in to do the job.

These specialists tied 13 equally spaced knots in a rope. They then drove a stake into the ground through the 1st and 13th knots; another stake through the 4th knot; and a third stake through the 8th knot (Fig. 17-51).

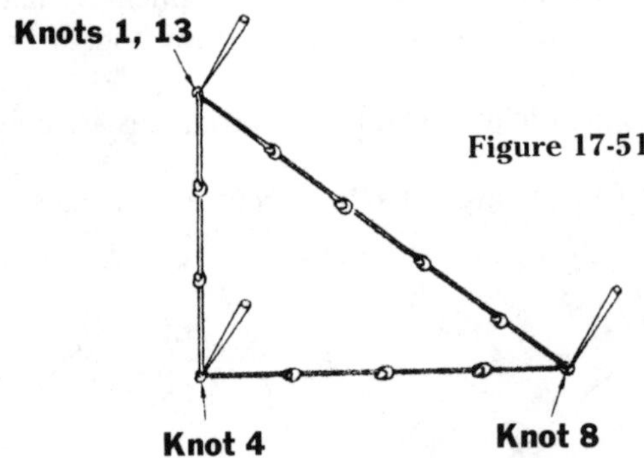

Figure 17-51

The rope was stretched as tightly as possible before each stake was driven in. The knotted rope now formed a triangle with sides measuring 3 units, 4 units, and 5 units.

And, lo and behold, the rope stretchers found to their amazement that the angle opposite the 5-unit side was always a right angle! Though they didn't understand why forming a 3, 4, 5 triangle resulted in a right angle, they were perfectly happy since they obtained the result expected of them.

At about the same time, the Hindus also needed to construct right angles, but they were a step ahead of the Egyptians. They discovered that in addition to the number combination (3, 4, 5) forming a right angle, there were other combinations that achieved the same result—for instance, (5, 12, 13), (12, 35, 37), (8, 15, 17), and others. But they, too, didn't understand why these number combinations yielded a right angle.

(continued)

The answer was finally found about 2500 years ago by the Greek mathematician Pythagoras and became known as the Pythagorean Theorem—the most famous and, perhaps, most useful theorem in mathematics. The theorem is as beautiful today as on the day Pythagoras first discovered it.

The theorem gives us a way to find the length of any side of a right triangle if we know the lengths of the other two sides.

We use the *converse* of the theorem to test whether or not a triangle is, in fact, a right triangle.

***REMINDER:* A RIGHT TRIANGLE is a triangle with a right angle. The side opposite the right angle is the HYPOTENUSE. The other two sides are the LEGS.**

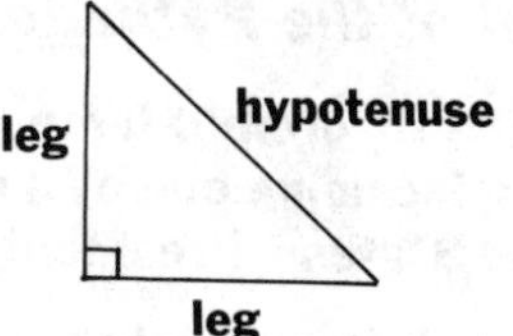

The Pythagorean Theorem

In any right triangle, the square of the length of the hypotenuse is equal to the sum of the squares of the lengths of the legs.

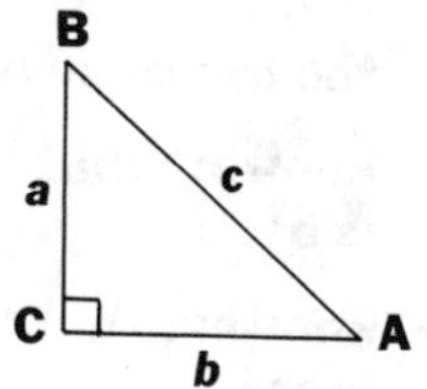

That is, if △ABC is a right triangle with right angle at C, and c is the length of the hypotenuse, and *a* and *b* are the lengths of the legs, then

$$c^2 = a^2 + b^2$$

Converse of the Theorem

If the square of the length of one side of a triangle is equal to the sum of the squares of the lengths of the other two sides, then the triangle is a right triangle whose right angle is opposite the longest side.

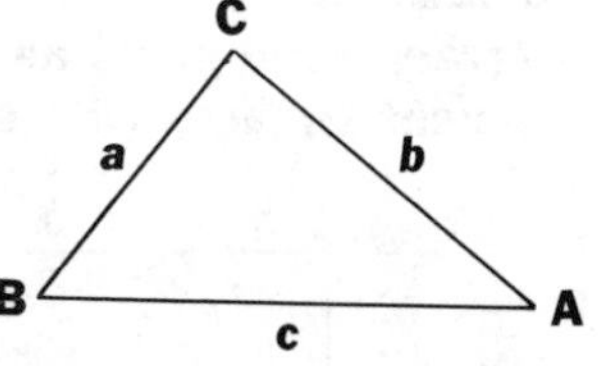

That is, *if* in △ABC, $c^2 = a^2 + b^2$, then △ABC is a right triangle with its *right angle at C*.

Now let's test whether the rope stretchers' 3, 4, 5 triangle formed a right angle. That is, let's see whether $5^2 = 3^2 + 4^2$:

$$\begin{aligned} 5^2 &= 5 \times 5 \\ &= 25 \end{aligned}$$

$$\begin{aligned} 3^2 + 4^2 &= (3 \times 3) + (4 \times 4) \\ &= 9 + 16 \\ &= 25 \end{aligned}$$

(continued)

Since $5^2 = 3^2 + 4^2$, then—according to the converse of the Pythagorean Theorem—the triangle is a right triangle whose right angle is opposite the longest side, the 5-side.

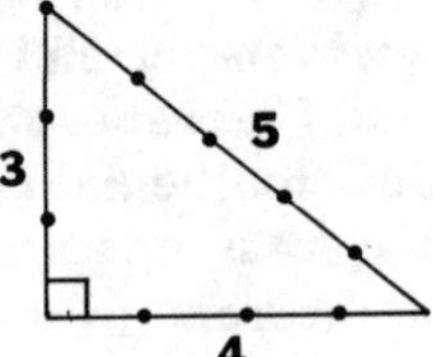

Proof of the Pythagorean Theorem

The theorem's fame led to hundreds of proofs over the centuries, including one by a man who later became President of the United States—President James A. Garfield (1831–1881).

The informal proof presented below is not hard to follow. It depends on:

- the use of several cardboard cutouts
- knowing that the area of a square with side *a* is a^2
- knowing that congruent figures have equal areas.

The following cardboard cutouts (see Fig. 17-52) are needed:

three squares: one with side *a*; another with side *b*; a third with side *c*.

eight congruent triangles, each with legs *a* and *b* and hypotenuse *c*.

PROCEDURE:

1. Arrange the parts as shown in Fig. 17-52, forming two congruent squares with side $a + b$.

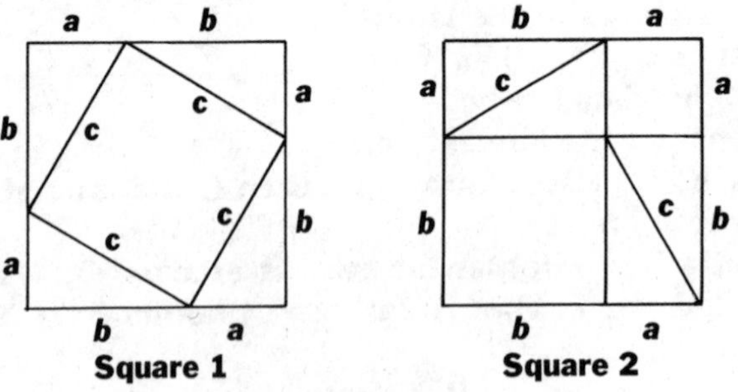

Figure 17-52

2. Remove the four congruent triangles from square 1, and the four congruent triangles from square 2. This leaves Fig. 17-53.

(continued)

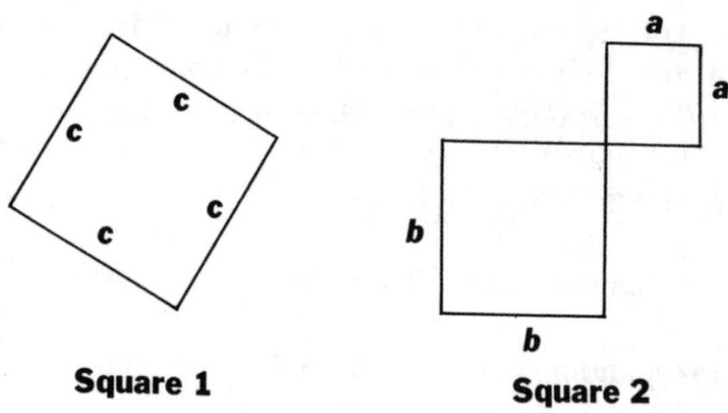

Figure 17-53

3. Since equal areas were removed from equal areas, the resulting areas are equal; that is,

$$c^2 = a^2 + b^2$$

SIMILAR FIGURES

Figures having the same shape and size are congruent. Figures having the same *shape* (but not necessarily the same size) are **similar**. Below are several pairs of similar figures:

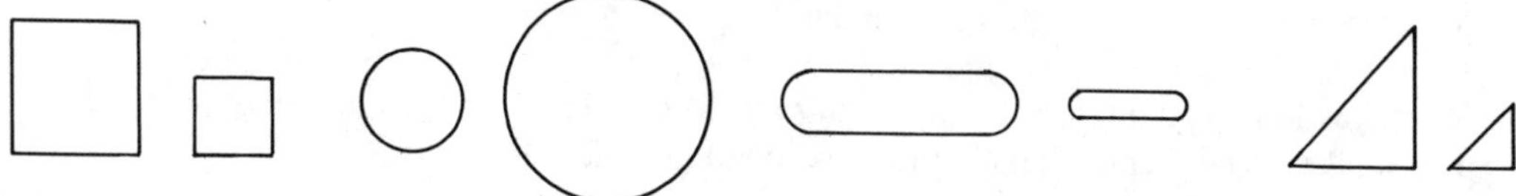

Similarity is the concept behind photographic reductions and enlargements; it guarantees that a scale model of a house or ship looks like the actual structure.

To understand the concept of similarity, we need to know what a ratio means. A **ratio** is the answer obtained when you compare two numbers by dividing one by the other. The ratio of 3 and 2 is 3/2 (or "3 to 2"). If there are 3 girls for every 4 boys in a class, the ratio of girls to boys is 3/4. And if Mike is twice as old as Bill, then the ratio of Mike's age to Bill's is 2/1.

The two pictures below are *similar* figures. The enlarged picture looks exactly like the smaller one, except that it's longer and wider.

Measure the length of each picture and you will find that the larger one is twice as long as the smaller one. That is, the ratio of the two lengths is 2/1.

Measure the width of each picture and you will find that the larger picture is also twice as wide as the smaller one. That is, the ratio of their widths is also 2/1.

If you measure corresponding parts inside each picture, you will again find each part in the larger picture to be twice as large as the corresponding part in the smaller one. The ratio of *all* corresponding parts remains the same: 2/1.

CONDITIONS FOR SIMILARITY

What makes two figures similar? There are two conditions:

1. The corresponding angles of the two figures must be equal.
2. The ratios of their corresponding sides must be equal.

For instance, look at rectangles A and B in Fig. 17-54. Decide whether they are similar by checking whether the two conditions are met:

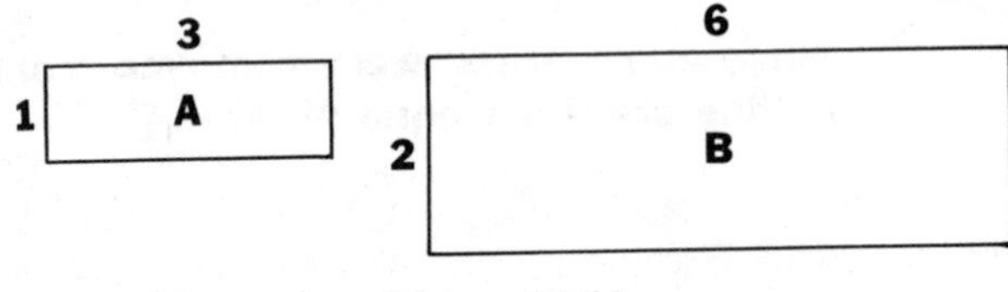

Figure 17-54

1. Since A and B are rectangles, all their angles measure 90° and are, therefore, equal. This satisfies condition 1.
2. The ratio of their widths is 1/2; the ratio of their lengths is 3/6 = 1/2. That is, each side of rectangle A is half as big as the corresponding side of rectangle B. This satisfies condition 2.

Therefore, the two rectangles are similar.

Now look at the two rectangles C and D in Fig. 17-55 and decide whether they are similar. Again, check against the two conditions for similarity:

1. The corresponding angles are equal since all are right angles. This satisfies condition 1.
2. The ratio of the widths (C to D) is 2/1, but the ratio of their lengths is 4/3. The two ratios are different. Therefore, rectangles C and D are *not* similar because condition 2 is not met.

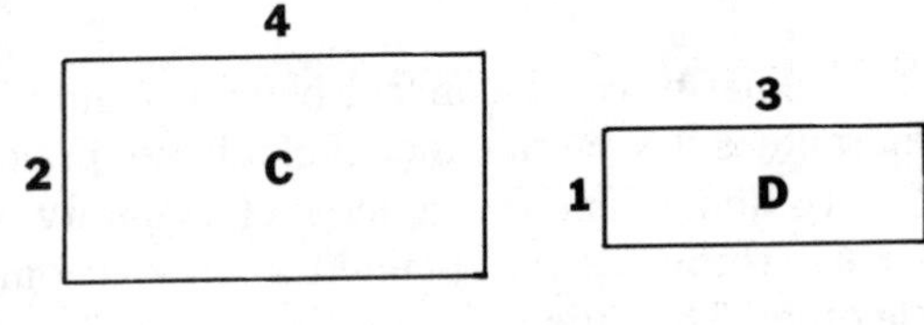

Figure 17-55

EXERCISES

1. Are the following figures *always* similar?

   (a) any two squares (b) any two circles (c) any two rectangles
   (d) any two triangles

2. Are congruent figures *always* similar? Why?

3. Can similar figures be congruent? Explain your answer.

SYMMETRY

Natural and manmade objects abound with examples of symmetry: leaves, butterflies, flowers, vases, wallpaper designs, snow crystals, living things, works of art.

What do the pictures below have in common?

Each figure can be folded in half so the two halves match each other exactly. Such figures are **symmetrical**. The line that divides the figure into two matching halves is a **line of symmetry**.

Fig. 17-56 is symmetrical because it can be folded in half so that the two halves match each other exactly. Line AB is the line of symmetry, dividing the figure into two matching halves.

Figure 17-56

Some figures have two lines of symmetry. Both lines AB and CD divide Fig. 17-57 into two matching halves. Fig. 17-58 also has two lines of symmetry—lines LM and PR.

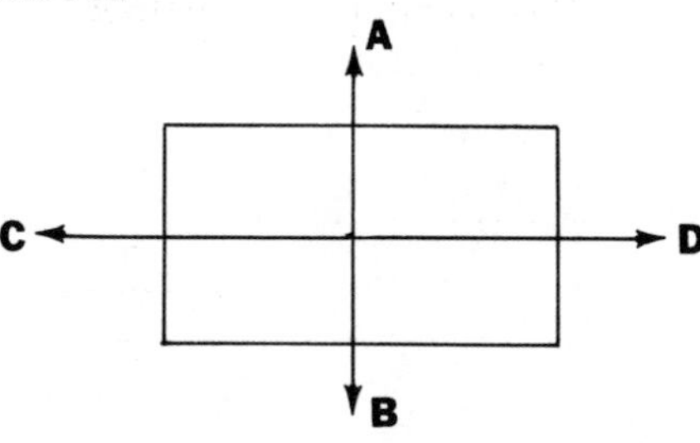

Figure 17-57

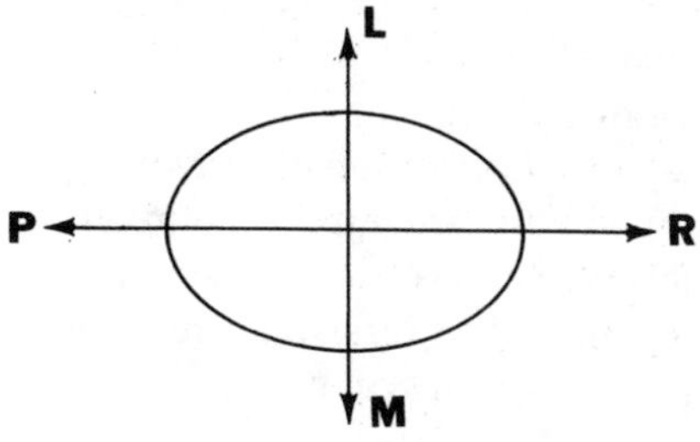

Figure 17-58

A square has four lines of symmetry, as seen in Fig. 17-59. A circle has infinitely many lines of symmetry—every line that passes through its center.

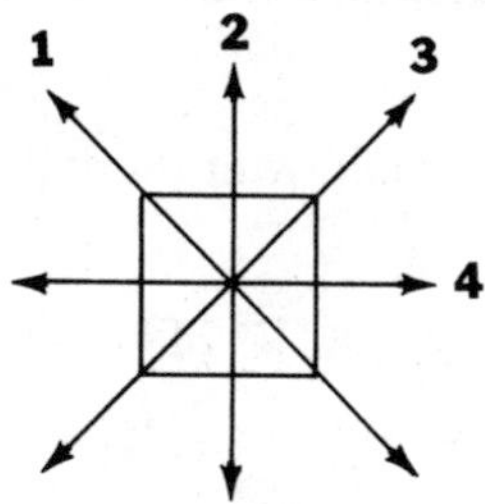

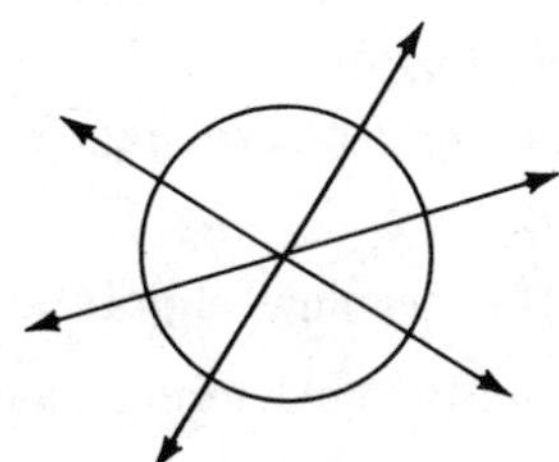

Figure 17-59

EXERCISES

1. How many lines of symmetry, if any, do you see in each of the following figures?

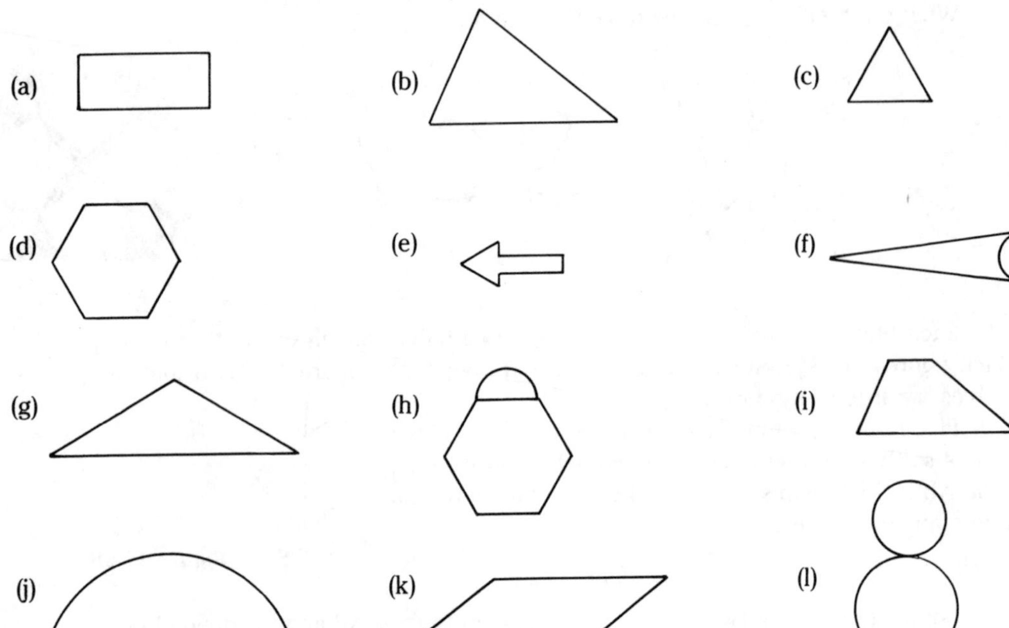

2. Draw a figure with:
   (a) exactly one line of symmetry
   (b) exactly two lines of symmetry
   (c) exactly three lines of symmetry
   (d) more than three lines of symmetry
   (e) no lines of symmetry

HIGHLIGHTS OF GEOMETRY

Geometric Figures

The study of geometric figures should leave your child with the following understandings:

- The meanings of **point**, **line**, **segment**, and **plane** (see page 208).
- Two points determine a line; three points determine a plane.
- **Plane** figures have two dimensions—length and width; **space** figures have three dimensions—length, width, and height.
- The size of an angle is determined by the amount of rotation of one of its sides.

- **Parallel** lines do not intersect.
- **Perpendicular** lines intersect at right angles.
- A **square** is a special kind of rectangle.
- A **rectangle** is a special kind of parallelogram.
- A **parallelogram** is a special kind of quadrilateral.

Your child should also be able to

- Identify acute, obtuse, and right angles
- Use a protractor to measure and draw angles
- Identify triangles, rectangles, squares, and parallelograms, and list several characteristics of each figure
- Identify isosceles, equilateral, and right triangles
- Identify the radius and diameter of a circle
- Identify cylinders, cones, and spheres, and list several characteristics of each figure
- Identify cubes, rectangular prisms, and pyramids, and list several characteristics of each figure
- Draw a picture of a square, rectangle, triangle, and parallelogram
- Explain and use the following vocabulary:

segment	equilateral triangle	circle
parallel	isosceles triangle	radius
perpendicular	right triangle	diameter
protractor	hypotenuse	plane figure
angle	quadrilateral	space figure
right angle	rectangle	cube
acute angle	square	rectangular prism
obtuse angle	parallelogram	pyramid
polygon	pentagon	cylinder
triangle	hexagon	cone
		sphere

Measuring Lengths (Linear Measurement)

The study of linear measurement should leave your child with these understandings:

- The distance around a figure is its **perimeter**.
- The formulas for finding perimeters are:

  Rectangle: $P = 2l + 2w$ (page 223)

  Square: $P = 4s$ (page 223)

  Triangle: $P = a + b + c$ (page 223)
- The distance around a circle is its **circumference**.
- The formula for finding the circumference is:

$$C = 2\pi r \text{ or } C = \pi d \text{ (page 224)}$$

- There is no *exact* value for π. Convenient values used are: $\pi = 3.14$ or $\pi = \frac{22}{7}$.

Your child should also be able to

- Find the perimeters of triangles, rectangles, and squares
- Find the circumference of a circle, knowing its radius or diameter
- Explain and use the following vocabulary:
  perimeter
  circumference
  π (pi)

Measuring Area

The study of *area* should leave your child with these understandings:

- The **area** of a figure means the number of square units contained in the surface of the figure.
- A **square unit** is a square measuring 1 unit on each side.
- Commonly used square units are: square inch, square foot, square yard, square mile, square centimeter, square meter, and square kilometer.
- The formulas for finding areas are:
  Rectangle: $A = lw$ (page 229)
  Square: $A = s^2$ (page 230)
  Triangle: $A = \frac{1}{2}bh$ (page 231)
  Circle: $A = \pi r^2$ (page 233)

Your child should also be able to

- Find the areas of rectangles, squares, and triangles
- Find the area of a circle
- Explain and use the following vocabulary:
  area
  square inch
  square unit

Measuring Volume

The study of *volume* should leave your child with these understandings:

- The **volume** of a figure is the number of cubic units the figure holds.
- When you are painting the outside of a box and you want to know how much paint is needed, you're concerned with *area*. When you want to know how much can be stored in the box, you're concerned with *volume*.
- A **cubic unit** is a cube measuring 1 unit in length, 1 unit in width, and 1 unit in height.
- The formulas for finding volume are:
  Rectangular prism: $V = lwh$ (page 234)

Cube: $V = s^3$ (page 235)

Cylinder: $V = \pi r^2 h$ (page 235)

Also, your child should be able to

- Find the volumes of rectangular prisms, cubes, and cylinders
- Explain and use the following vocabulary:
  volume
  cubic inch
  cubic unit

Congruent and Similar Figures

The study of *congruence* and *similarity* should leave your child with these understandings:

- **Congruent** figures have the same size and shape.
- If two congruent figures are placed one on top of the other, they will match perfectly.
- Two triangles are congruent if the three sides of one triangle are equal to the corresponding sides of the other.
- **Similar** figures have the same shape, but not necessarily the same size.
- An enlargement or reduction of a picture is similar to the original picture.

Your child should also be able to

- Identify congruent and similar figures
- Explain and use the following vocabulary:
  congruent figures
  similar figures

Symmetry

The study of *symmetry* should leave your child with these understandings:

- If two halves of a figure match each other exactly, then the figure is **symmetrical**.
- The line that divides the two matching halves is their **line of symmetry**.
- Leaves, a heart, butterflies, and some vases are examples of symmetry.
- Some figures, like a square, have more than one line of symmetry.

Your child should also be able to

- Tell whether a shape is symmetrical, and locate its line(s) of symmetry
- Explain and use the following vocabulary:
  symmetry
  line of symmetry

QUESTIONS ON THE CHAPTER

Geometric Figures

1. We can draw one line segment through two points, three segments through three points, and six segments through four points, as shown below.

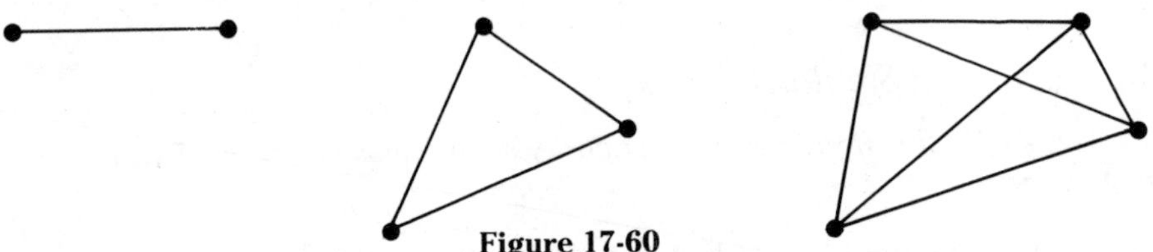

Figure 17-60

How many line segments can you draw through
(a) five points? (b) six points?

2. Through how many degrees does the minute hand of a clock turn in
(a) 15 minutes? (b) 20 minutes? (c) a half-hour? (d) $\frac{3}{4}$ of an hour?

3. Name all the angles in Figure 17-61.

4. Name all the triangles in Figure 17-62.

5. Name all the geometric shapes you see in Figure 17-63.

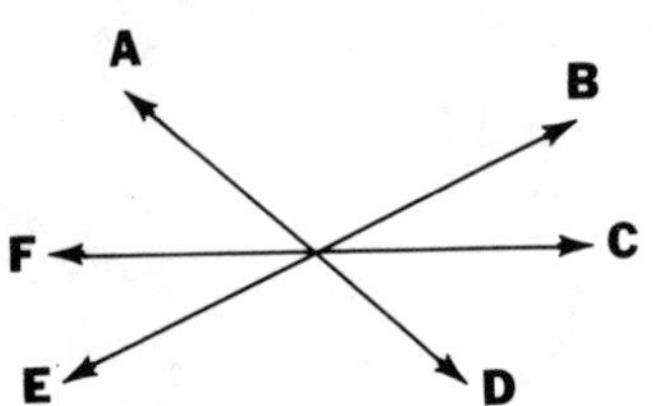

Figure 17-61

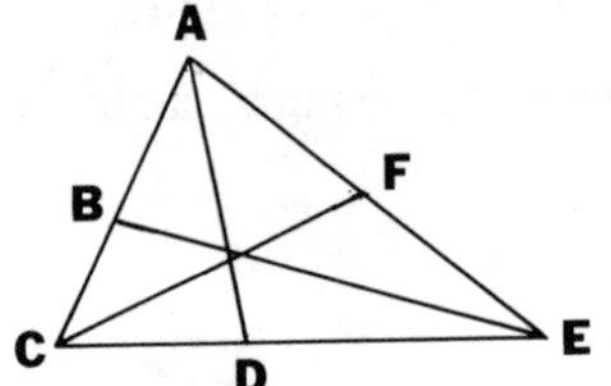

Figure 17-62

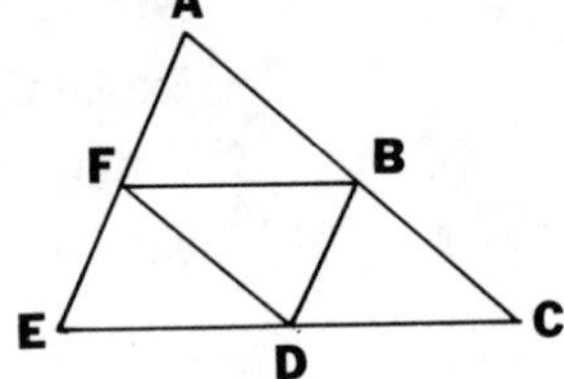

Figure 17-63

6. What kind of triangle is one that has:
(a) sides measuring 5 in., 6 in., and 5 in.?
(b) sides measuring 8 cm, 8 cm, and 8 cm?
(c) angles measuring 60°, 90°, and 30°?

7. Can a triangle have three sides of the same length and a right angle? Why?

8. If one of the two equal angles of an isosceles triangle measures 45°, what is the measure of the other two angles? What kind of triangle is it?

9. Here is a list of geometric figures: square, rectangle, parallelogram, trapezoid, equilateral triangle. Below is a list of properties. For each property, name *all* the figures for which the property is true.
(a) The opposite sides are equal.

(b) The angles are all right angles.
(c) All the angles are equal.
(d) All the sides are equal.
(e) The diagonals are equal.
(f) The opposite sides are parallel.
(g) At least one pair of opposite sides are parallel.

10. The midpoints A, B, C, D of the sides of a rectangle are joined. Can you explain why the figure ABCD is a parallelogram? Can you give a reason why *all* four sides of this parallelogram are equal? [A parallelogram with sides that are all equal is called a **rhombus**.]

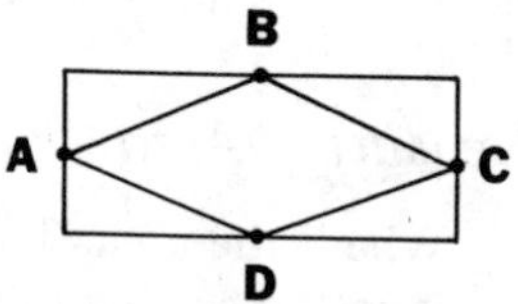

11. Figure 17-64 shows a 2-inch cube. How many cuts will be required to divide the cube into 1-inch cubes?

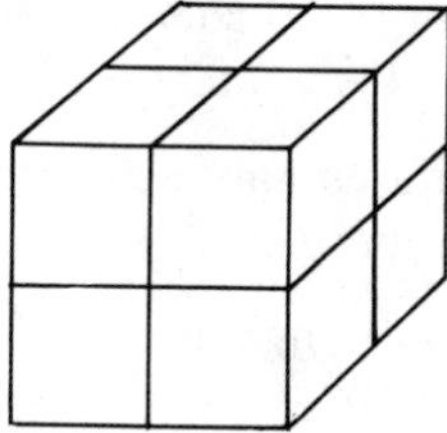

Figure 17-64

12. Below are pictures of paper cutouts that can be folded at the creases. What geometric figure can you form with each shape?

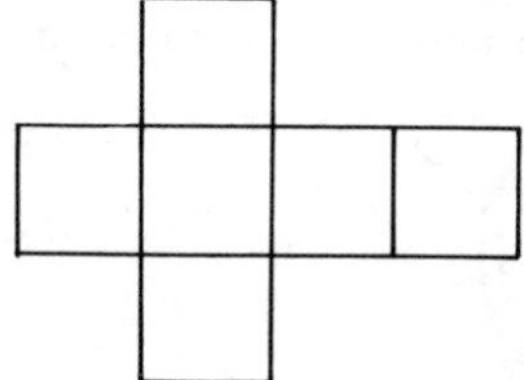
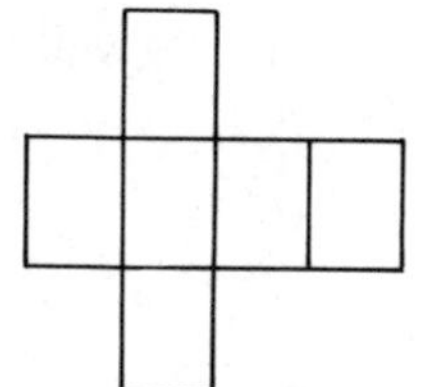
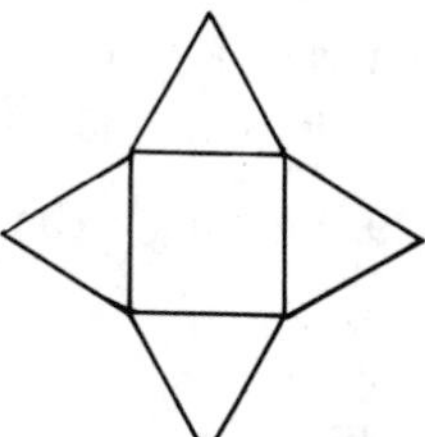

13. Draw an angle and a circle that intersect in (a) one point (b) two points (c) three points (d) four points

NOTE: *Use a calculator freely for some of the calculations in the problems that follow.*

Measuring Length

14. How many feet of molding are needed around a window measuring 34 inches by 52 inches?

15. The perimeter of a *regular* pentagon is 265 feet. How long is each side? [REMINDER: In any *regular* polygon, all sides have the same length.]

16. What is the circumference of a circle whose diameter is:
(a) 6 in. (b) 35 cm (c) $3\frac{1}{2}$ ft (d) 5.25 yd

17. Find the circumference of a circle with a radius of:
    (a) 25 yd (b) 49 ft (c) $5\frac{1}{2}$ cm (d) 10.35 in.

18. What distance do you move in one turn of a ferris wheel if you sit 15 feet from the center?

19. The diameter of the earth is 7918 miles. What is its circumference?

20. If the circumference of a tree is 44 inches, what is its diameter?

Measuring Area

21. What's the area of a driveway measuring 10 feet by 25 feet? How much would it cost to cement this driveway at $1.75 a square foot?

22. How many bricks are needed to construct a walk 50 feet long and 5 feet wide, if 7 bricks are needed for every square foot of walk?

23. The area of a living room floor is 391 square feet; the room is 17 feet wide. How long is the floor?

24. If a rectangle is twice as long as another rectangle, but both have the same width, how do their areas compare?

25. What happens to the area of a triangle if
    (a) its height is doubled?
    (b) its base is doubled?
    (c) both its height and base are doubled?

26. How is the area of a square changed if its side is doubled?

27. A triangle and a parallelogram have equal areas and equal bases. How are their heights related?

28. The length of a rectangle is twice its width. If its area is 98 square meters, find its length and width.

29. Find the area of each figure:

(a)

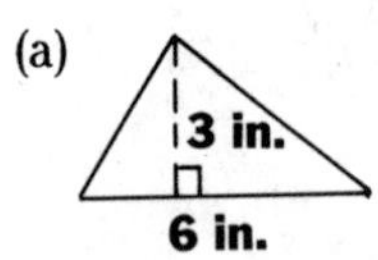

(b)

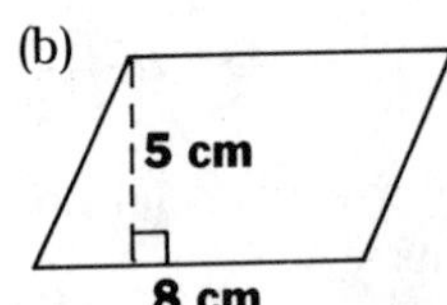

(c)

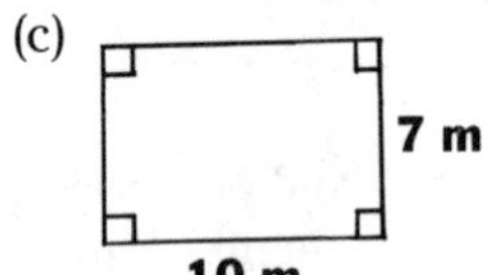

(d)

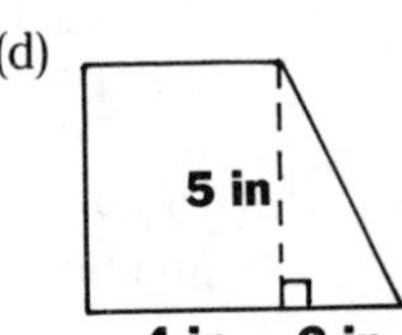

30. How many square inches of tin are in the top of a circular can whose radius is 2 inches?

31. How many times larger is the area of a circular rug 12 feet in diameter than one 6 feet in diameter?

32. What is the area of
    (a) a penny? (b) a quarter? (c) a half-dollar?

33. Find the area of the shaded region:

(a)

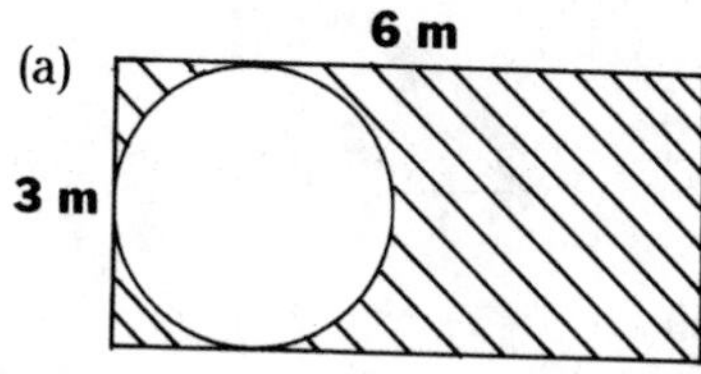

(b)

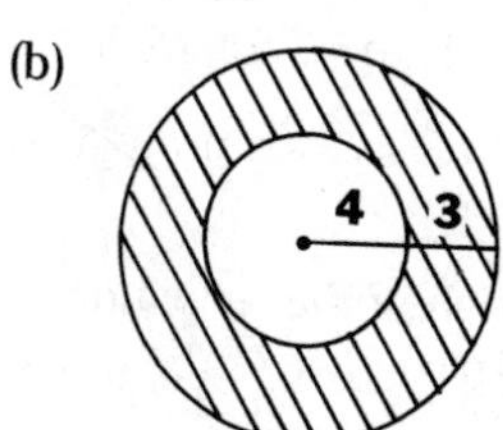

34. If the area of a circle is equal to its circumference, how big is its radius?

Measuring Volume

35. A swimming pool is 40 feet wide, 80 feet long, and has an average depth of 6 feet. How many cubic feet of water does it hold?

36. How many cubic feet of storage space are in a freezer 7 feet by 5 feet by 4 feet?

37. How many cubic feet of space are in a bin 7 feet by 7 feet by 7 feet?

38. What happens to the volume of a box if
    (a) one of its dimensions is doubled?
    (b) two of its dimensions are doubled?
    (c) all three of its dimensions are doubled?

39. A sandbox measures 1.5 meters by 1.5 meters and is .3 meters deep. How many cubic meters of sand will it hold?

40. Compare the volumes of two cans of orange juice, each 8 inches high, one with a 3-inch radius and the other with a 6-inch radius.

41. A baseball has a diameter of 7.4 cm. How many cubic centimeters does it contain?

42. What is the effect on the volume of a sphere when you double its radius? When you triple its radius?

Congruent and Similar Figures

43. Why is △ABC congruent to △CDA in parallelogram ABCD?

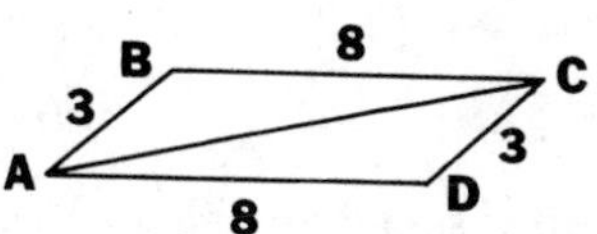

44. Name six matching parts that would be equal in any two congruent triangles.

45. If AB = AC and BD = DC, why is △ABD congruent to △ACD?

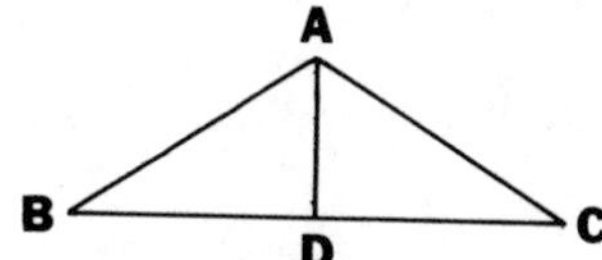

46. How can you tell whether two geometric figures are similar?

47. Are *any* two equilateral triangles similar?

48. If these two triangles are similar, find the two missing lengths.

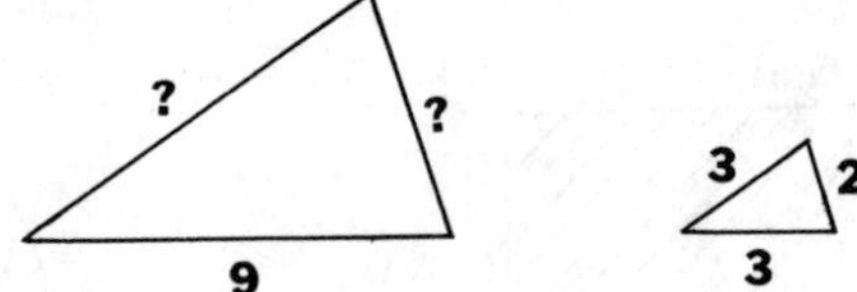

49. Which of the following will always be similar?
    (a) triangles (b) circles (c) squares (d) rectangles (e) spheres

50. A photo is 5 cm wide and 8 cm long. It is enlarged so that it is 24 cm long. How wide is the enlarged photo?

Symmetry—Miscellaneous

51. Which letters of the alphabet, when capitalized, have line symmetry? Locate their line(s) of symmetry.

52. True or false? Give reasons for your answers.
    (a) Congruence is a special kind of similarity.
    (b) A triangle may be similar to a rectangle.
    (c) All equilateral triangles are similar.
    (d) All equilateral polygons are similar.
    (e) If two triangles are congruent to a third, then the first two triangles are congruent to each other.
    (f) If two angles of one triangle are equal to two angles of another triangle, then the third angles are also equal.
    (g) The measure of each angle of an equilateral triangle is 30°.
    (h) If two polygons are similar, then they have the same number of sides.
    (i) Every square is a rectangle.
    (j) Every rectangle is a parallelogram.
    (k) A circle is a polygon.

REVIEW

1. Round 385,179 to the nearest hundred.

2. $\frac{1}{3} + \frac{2}{5} + \frac{5}{6} = ?$ 3. $\frac{7}{5} - \frac{3}{4} = ?$ 4. $\frac{3}{4} \times \frac{4}{5} \times \frac{2}{7} = ?$

5. $\frac{2}{3} \div \frac{3}{8} \times 5 = ?$ 6. $.09 - .002 = ?$ 7. $.3\overline{)43.551}$

8. After spending $35 for slacks and $12 for a shirt, John has $18 left. How much money did he have to begin with?

9. $\frac{1}{6}$ of what number is 14?

10. Which is smaller: .53 or .529?

11. Use the short method to find
    (a) 3.4829×1000
    (b) $18.93 \div 1000$

12. Find $12\frac{1}{2}$% of $80.

13. 60% of what number is 96?

14. 12 is what percent of 72?

15. What part of his income of $32,500 a year does a man spend on housing if his monthly rent is $475?

16. A girl receives a grade of 85% on a test containing 20 questions. How many answers did she get wrong?

17. Two thirds of the human body is water. If a man's body contains 150 pounds of water, how much does he weigh?
18. Which is larger and by how much: 3.12 or .978?
19. If 3% of a city's population is 82,500, what is the total population of the city?
20. *Estimate* the product of 306 × 792. Then calculate the exact answer and see how close your estimate was.

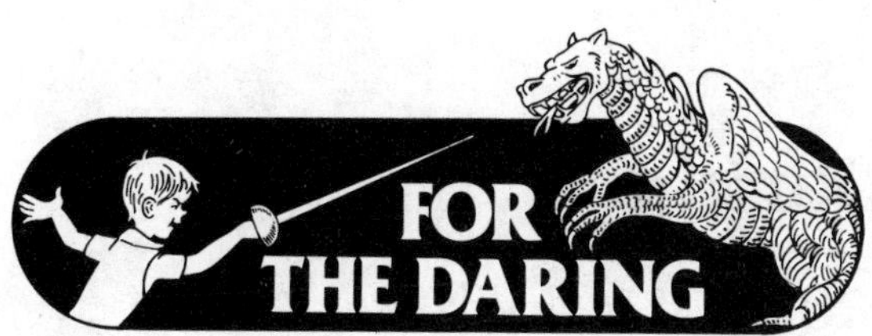

1. A carpenter has a piece of wood in the shape of a right triangle with legs 21 inches and 28 inches.

He cuts from it a square piece as shown in the figure at the right.

Find the length of a side of the square piece.

21 in.

28 in.

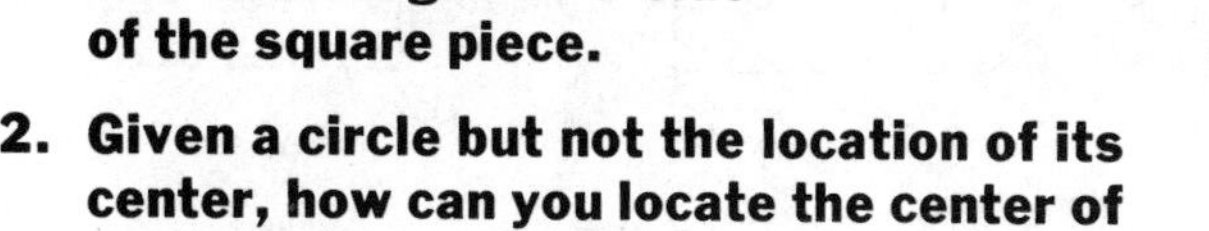

2. Given a circle but not the location of its center, how can you locate the center of the circle?

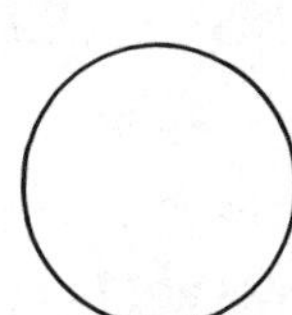

3. Without taking your pencil off the paper, draw four line segments that pass through every point:

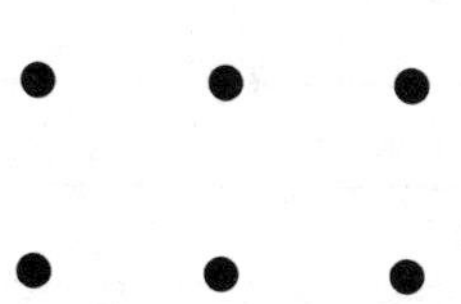

(The answers appear on page 261.)

Answers

For the Daring

Chapter 5 PROBLEM SOLVING

1. Draw a diagram reflecting the facts given in the problem:

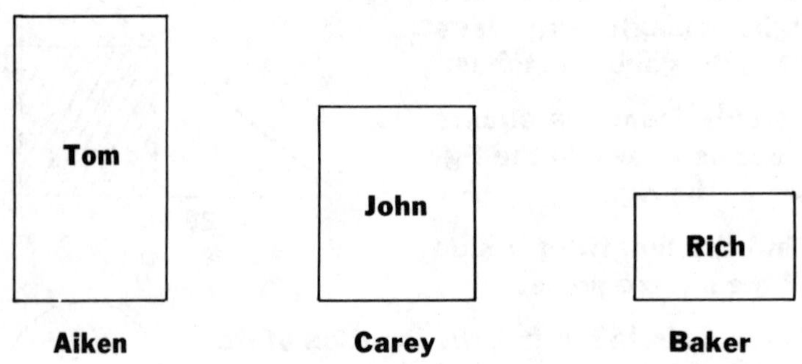

The names are John Carey and Rich Baker.

2. Organize your information. (See Problem 10.)
An X indicates what each person does *not* like.

	Pizza	Veal	Chicken	Fish	Pasta	Steak
Liz	X					X
Arin		X				
Benj			X		X	
Jerry	X	X				
Judy						

Only *fish* has no negative votes.

3. $1^2 + 2^2 + 3^2 + 4^2 + 5^2 + 6^2 + 7^2 + 8^2$
$= 1 + 4 + 9 + 16 + 25 + 36 + 49 + 64$
$= 204$ squares
(See Problem 14.)

Chapter 9 COUNTING

1. Since each of the 3 appetizers can go with each of the 4 entrees, there are 3×4, or 12 possible appetizer-entree combinations.

   Since each of these 12 combinations can go with each of the 2 desserts, there are 12×2, or 24 possible appetizer-entree-dessert combinations.

   You can, therefore, make *24* different meals with the choices offered.

2. Since there are five people, there are 5 ways of filling the *first* position in the line. After the first position is filled, there are four people left from whom to fill the second position. So there are 4 ways of filling the *second* position in the line.

   The third position can be filled by any of the three people left after the first two places have been filled. So there are 3 ways of filling the *third* position.

   There are now two people left from whom to fill the fourth position. So there are 2 ways of filling the *fourth* position.

   After the first four positions have been filled, there is one person left with whom to fill the fifth position. So there is 1 way left of filling the *fifth* position:

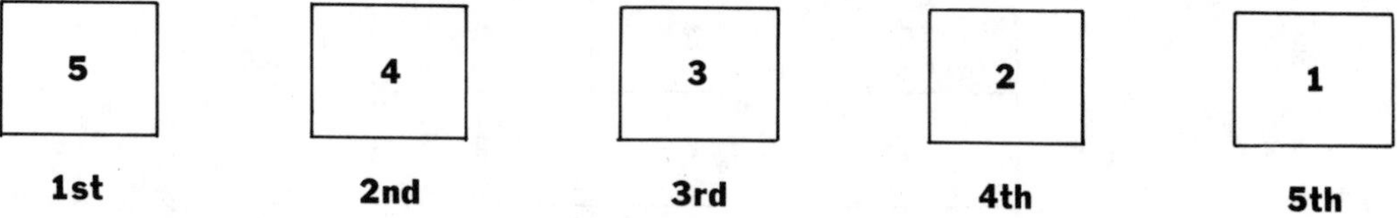

   Since each of the 5 ways to fill the first position can go with each of the 4 ways to fill the second position, there are 5×4, or 20, different ways of filling the first two positions.

   Extending this reasoning to the next three positions, we conclude that there are

$$5 \times 4 \times 3 \times 2 \times 1 = 120$$

   different ways of filling all five positions in the line with the five people.

3. The first digit can be selected in 10 different ways (by selecting any of the ten digits 0 to 9). The second digit can likewise be selected in 10 different ways, as can the third digit:

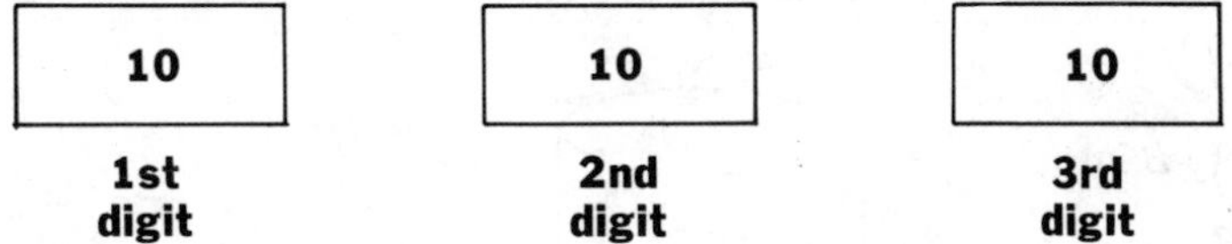

   Since each selection of the first digit can go with each selection of the second digit, and each of these can go with each selection of the third digit, the total number of possible three-digit numbers is

$$10 \times 10 \times 10 = 1000$$

Chapter 10 PLACE VALUE

Four weights are needed: 1 pound, 2 pounds, 4 pounds, and 8 pounds. (Notice that 1, 2, 4, 8 are the first four place values in the binary system.)

Chapter 11 ADDING AND SUBTRACTING WHOLE NUMBERS

1.

8	3	4
1	5	9
6	7	2

2	9	4
7	5	3
6	1	8

4	3	8
9	5	1
2	7	6

2. *One* of them is not a nickel, but the other is.
3. Bottle—$1.25; cork—$.25.
4. Suppose the people involved were Mr. Jones, his son, and his grandson. There are two fathers and also two sons, but there are only three people.
5.

10	3	8
5	7	9
6	11	4

Chapter 12 MULTIPLYING WHOLE NUMBERS

1. If the bottle is filled in an hour, then it must have been half full one minute before, or in 59 minutes.
2. The first person must answer "yes" regardless of whether he is a liar or a truth-teller. Since the second person told the truth about the first one's response, *he* must be the truth-teller and his companion the liar.
3. Call the 8-pound can *A* and the 5-pound can *B*.

   (a) Fill *B*.

   (b) Empty *B* into *A*.

   (c) Fill *B*.

   (d) Pour *B* into *A* until *A* is full. There are 2 pounds left in *B*.

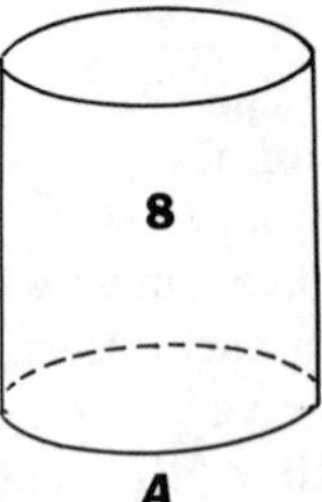

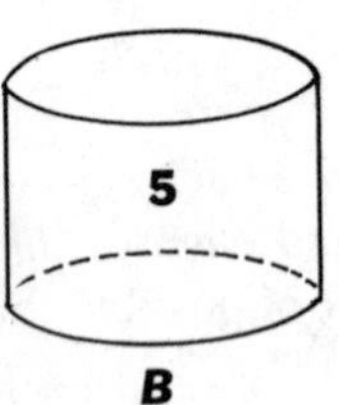

Chapter 13 DIVIDING WHOLE NUMBERS

1. The four piles contain the following numbers of counters: 7, 3, 4, 6
2. 6 dozen dozen = $6 \times 12 \times 12 = 864$
   a half-dozen dozen = $6 \times 12 = 72$

3. $9 + \frac{99}{9}$
   $[9 + \frac{99}{9} = 9 + 11 = 20]$

4. $(3 \times 3 - 3) \div 3$
   $$[(3 \times 3 - 3) \div 3 = (9 - 3) \div 3 = 6 \div 3 = 2]$$

5. A = 5
   B = 4
   C = 6
   D = 3
   F = 8
   M = 1
   P = 7
   R = 2

$$\begin{array}{r} 954 \\ 3\overline{)2862} \\ \underline{27} \\ 16 \\ \underline{15} \\ 12 \\ \underline{12} \\ 0 \end{array}$$

Chapter 14 FRACTIONS

1. Let m = the amount of money the boy had at the beginning.

   Then $\frac{1}{3}m$ = the amount of money he *spent*, leaving $\frac{2}{3}m$.

   If he then lost $\frac{2}{3}$ of the remainder, he lost $\frac{2}{3}$ of $\frac{2}{3}m$, or $\frac{4}{9}m$.

   Having spent $\frac{1}{3}m$, and having lost another $\frac{4}{9}m$, then he no longer has $\frac{1}{3}m + \frac{4}{9}m$, or $\frac{7}{9}m$.

   What he has left is the original amount, m, *less* $\frac{7}{9}m$; that is, $m - \frac{7}{9}m$, or $\frac{2}{9}m$.

   If he had \$12 left, then
   $$\frac{2}{9}m = \$12$$
   $$m = 12 \times \frac{9}{2}$$
   $$m = \$54$$

2. Let A = the number of *acceptable* recorders.
   Then $25 - A$ = the number of *defective* recorders,
   $\$25 \times A$ = the amount paid for A acceptable recorders at \$25 each, and
   $\$10 \times (25 - A)$ = the amount penalized for $(25 - A)$ defective recorders at \$10 each.

   If the worker was paid a net of \$380, then
   $$\begin{aligned} 25A - 10(25 - A) &= 380 \\ 25A - 250 + 10A &= 380 \\ 25A + 10A &= 380 + 250 \\ 35A &= 630 \\ A &= \frac{630}{35} \\ A &= 18 \text{ acceptable recorders} \end{aligned}$$

3. There are infinitely many fractions between $\frac{1}{4}$ and $\frac{1}{2}$, as there are between *any* two fractions. The reason is that between $\frac{1}{4}$ and $\frac{1}{2}$ there is at least one other fraction—the *average* of the two fractions:
   $$\frac{\frac{1}{4} + \frac{1}{2}}{2} = \frac{\frac{3}{4}}{2} = \frac{3}{4} \div 2 = \frac{3}{4} \times \frac{1}{2} = \frac{3}{8}$$

Then there is at least one fraction between $\frac{1}{4}$ and $\frac{3}{8}$—the average of *these* two fractions:

$$\frac{\frac{1}{4}+\frac{3}{8}}{2}=\frac{5}{16}$$

Since we can continue the process of taking averages indefinitely, the result is infinitely many fractions between $\frac{1}{4}$ and $\frac{1}{2}$. We can see this on the number line:

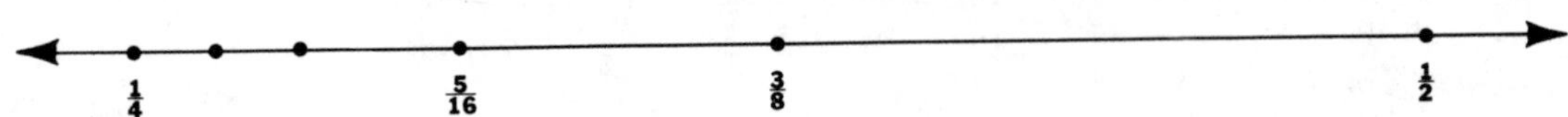

4. Jones—4; Smith—1; Carter—8; Baker—9.
5. The least common multiple of 4, 2, and 5 is 20. This means that all three dogs barked at the same time every 20 minutes. So at 12 midnight—120 minutes after they began with a simultaneous bark—all three dogs barked at once.

Chapter 15 DECIMALS

1. 35 miles. The joggers meet in an hour, giving the horsefly one hour to travel.
2. Theoretically, it will never reach the wall since it is always half of *some* distance away.
3. *NOTE:* The strategy used to find the LCD in Chapter 14 (page 153) can be used to solve this problem.

   - Counting by *threes* and getting a remainder of *two* would hold for any of the following numbers of pages:

     5, 8, 11, 14, 17, 20, 23, 26, 29, 32,
     35, 38, 41, 44, (47), 50, 53, 57, . . .

   - Counting by *fives* and getting a remainder of *two* would hold for any of the following numbers of pages:

     7, 12, 17, 22, 27, 32, 37, 42, (47), 52,
     57, 62, . . .

   - Counting by *sevens* and getting a remainder of *five* would hold for any of the following numbers of pages:

     12, 19, 26, 33, 40, (47), 54, 61, 68, . . .

   Since *47* satisfies all three conditions, there are 47 lines on the page.

Chapter 16 PERCENT

1. (a) yes (b) No, a 15% discount is larger.
2. (a) $6+\frac{6}{6}$ (b) $44 \div 4$ (c) $\frac{5}{5} \times 5$
3. Suppose the three couples are:
   Mr. and Mrs. Carter; Mr. and Mrs. Floyd; Mr. and Mrs. Jordan

Crossings	**Left on other side of river**
(1) Mrs. Floyd and Mrs. Jordan cross, and Mrs. Floyd returns with the boat.	(1) Mrs. Jordan

(2) Mrs. Carter and Mrs. Floyd cross, and Mrs. Carter returns. — (2) Mrs. Jordan, Mrs. Floyd

(3) Mr. Floyd and Mr. Jordan cross, and Mr. and Mrs. Floyd return. — (3) Mr. and Mrs. Jordan

(4) Mr. Carter and Mr. Floyd cross, and Mrs. Jordan returns. — (4) Mr. Carter, Mr. Floyd, Mr. Jordan

(5) Mrs. Floyd and Mrs. Jordan cross, and Mr. Carter returns to pick up his wife. — (5) Mr. and Mrs. Carter, Mr. and Mrs. Floyd, Mr. and Mrs. Jordan

Chapter 17 GEOMETRY

1. 12 inches. By the Pythagorean Theorem you can show that:
   (a) the measurements in △ ABC are as shown in the diagram.
   (b) DEFC is a square within the right △ ACB (with ∠C as the right angle).

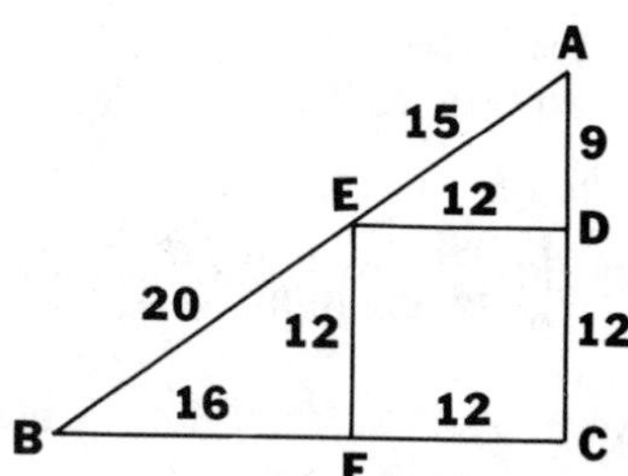

2. Draw any triangle in the circle. Draw the perpendicular bisectors of any two sides of the triangle. The point at which the bisectors intersect is the center of the circle.

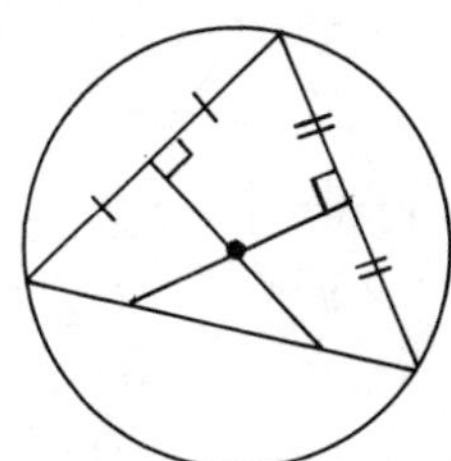

3.

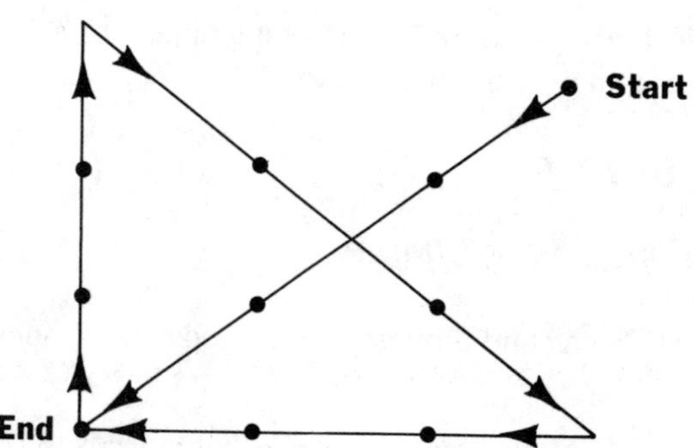

Questions on the Chapters

Chapter 6 CALCULATORS

Pages 32–33

5. (a) any number from 54 to 63.
   (b) by subtracting any number from 300 to 363

6. $123456 \times 9 + 7 = 1{,}111{,}111$
   $123456789 \times 9 + 10 = 1{,}111{,}111{,}111$
   Any number in the series will have as many 1s in the answer as the number added.

8. (a) $42 \times 23 = 966$ (b) $27\overline{)432}$ = 16

9. (a) Instead of pressing 5, you pressed 9.
   (b) Instead of pressing 2, you pressed 1.

10. (a) 72×4; 6×48; 12×24
    (c) $365 \times 24 \times 60 \times 82 = 43{,}099{,}200$ in a year;
    $43{,}099{,}200 \times 35 = 1{,}508{,}472{,}000$ in 35 years

Chapter 9 COUNTING

Page 61: Questions on the Chapter

1. You match, one-to-one, the objects of a set with the set of counting numbers:
   ```
   *  *  *  *
   ↕  ↕  ↕  ↕
   1  2  3  4
   ```

2. To tell how many objects there are in a set, and to tell the position of an object in a set

3. That there are *no* objects in the set

4. Both have the same number of objects.

5. There are as many oozies as boozies—because there exists a one-to-one matching between both sets.

12. For every object in set A there is a matching object in set B, and for every object in set B there is a matching object in set A.

Chapter 10 PLACE VALUE

Page 74: Questions on the Chapter

3. In 32, there are 3 tens and 2 ones; in 87, 8 tens and 7 ones; in 60, 6 tens and 0 ones.

6. In 129—1 hundred, 2 tens, 9 ones; in 682—6 hundreds, 8 tens, 2 ones; in 340—3 hundreds, 4 tens, 0 ones; in 501—5 hundreds, 0 tens, 1 one.

7.

10	1
1	8

10	1
9	7

10	1
4	0

100	10	1
1	7	5

100	10	1
5	7	2

100	10	1
9	0	8

8. (a) 1 (b) 10 (c) 100 (d) 10

9. (a) true (b) false (c) true

13. (a) 13, 14, 15, 16——18, 19, 20, 21
    (b) 61, 62——64, 65, 66, 67——69, 70, 71, 72
    (c) 372, 373, 374, 375——377, 378, 379, 380

14. making the value of each digit in a number depend on the position it occupies in the number

15. infinitely many

16. (a) To tell us there are *no* objects in a set
    (b) To serve as a placeholder, to avoid confusion

17. one of the basic symbols we use to represent numbers

18. ten: 0, 1, 2, 3, 4, 5, 6, 7, 8, and 9

19. a written (or physical) symbol to represent a number

20. The *decimal system* is the system we use in our everyday arithmetic to represent numbers. It's called by that name because it uses *10* basic symbols.

21. 100 10 1

Chapter 11 ADDING AND SUBTRACTING WHOLE NUMBERS

Page 95

1. (a) $6 - 4 + 3 = 5$ (b) $1 + 7 + 9 - 5 = 12$

3. (a)

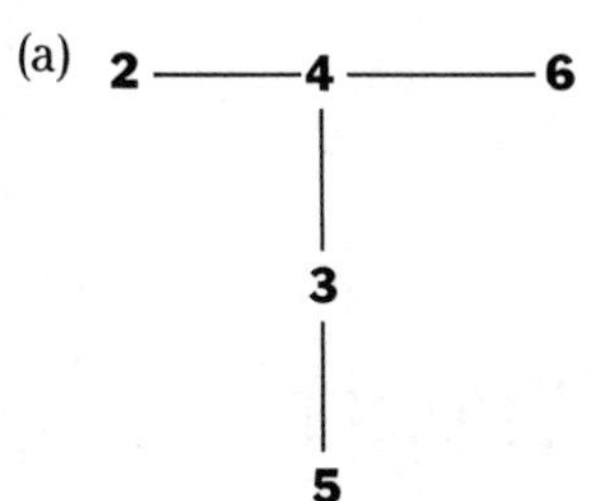

(b)

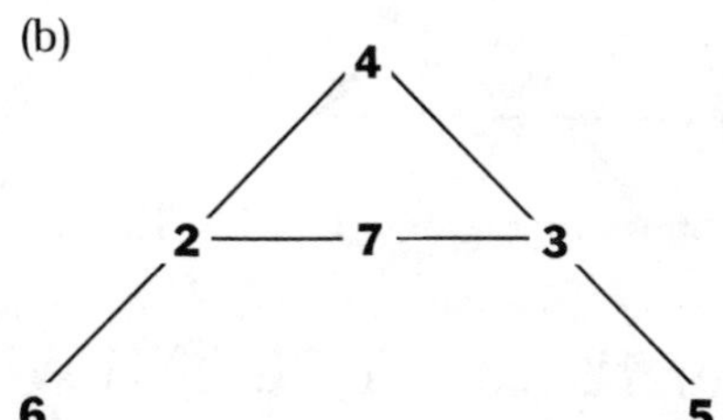

4. (a) $\begin{array}{r} 27 \\ +\ 56 \\ \hline 83 \end{array}$ (b) $\begin{array}{r} 56 \\ +\ 39 \\ \hline 95 \end{array}$

   (c) $\begin{array}{r} 48 \\ -\ 32 \\ \hline 16 \end{array}$ (d) $\begin{array}{r} 65 \\ -\ 27 \\ \hline 38 \end{array}$

Page 96: Questions on the Chapter

1. 4
2. 3
3. 2 + 2
4. 9

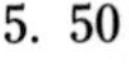

5. 50
6. 46
7. 28
8. 35
9. 33
10. 10
11. 30
12. 30
13. 33
14. 35
15. 28
16. 35
17. 37
18. 42
19. 33
20. 28
21. 28
22. 28
23. 146
24. (a) 13
    (b) 7
    (c) 8 + ?
    0 8 13
    8 + 5 = 13
    (d) ? −4
    0 7 11
    11 − 4 = 7
    (e) 4
25. (a) 62 (b) 167 (c) 584
    (d) 480
26. (a) 25 (b) 141 (c) 35
    (d) 227 (e) 193 (f) 386
27. (a) 13, 15, 17 (b) 22, 29, 37
28. 24 years younger
29. 14 years
30. 25
31. (a) true (b) true (c) false
32. (a) 2 (b) 5 (c) 41
33. 17
34. □ = 8, △ = 5
35. 1913
36. 271
37. 105
38. 66
39. 19
40. (c) 46

Chapter 12 MULTIPLYING WHOLE NUMBERS

Page 111

3. (i) (1, 24), (2, 12), (3, 8), (4, 6)
   (j) $4 \times 3 = 12$
   (k) $5 + 5 + 5 + 5 = 20$
4. (d) $9 + 9 + 9 + 9 + 9$; $5 + 5 + 5 + 5 + 5 + 5 + 5 + 5 + 5$;
   4×9
   (e) 48
6. 4
7. 2 and 12
8. 6

Page 120: Questions on the Chapter

1. 96
2. 92
3. 210
4. 730
5. 804
6. 630
7. 2015
8. 1220
9. 756
10. 1134
11. 5400
12. 12,960
13. 2430
14. 7293
15. 1530
16. 4301
17. 11,016
18. 1495
19. (a) $2 \times 3 = \mathit{1} \times 6$
    (b) $4 \times 6 = 8 \times \mathit{3}$
20. 5 and 9
21. (a) 2 [2 × 6, 3 × 4]
    (b) 3 [2 × 12, 3 × 8, 4 × 6]
    (c) 0
22. 962
23. $350
24. 522
25. 5200
26. 315
27. (a) true (b) true (c) true
28. (a) 20 (b) 90 (c) 40

Page 122: Review

1. 165
2. 188
3. 52

Chapter 13 DIVIDING WHOLE NUMBERS

Page 138: Questions on the Chapter

1. (a) $27 \div 9 = 3$ (b) $27 \div 3 = 9$
2. $9 \times 8 = 72$
4.

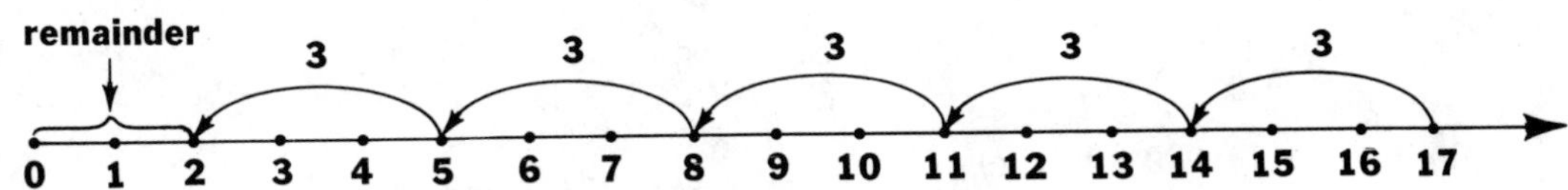

$17 \div 3 = 5$ r2

(a) the number of spaces between the point 0 and the point 17
(b) the number of spaces you jump at a time in moving from 10 to 0
(c) 5 r2 means that to cover the distance of 17 spaces, you made 5 jumps in groups of 3 and were left with a remainder of 2 spaces.

5. 9 columns
6. (a) 9 (b) 1 (c) 0 (d) no whole-number answer

7. (a) 23 (b) 29 (c) 213 (d) 181 (e) 89
8. (a) 26 r4 (b) 853 r2
9. (a) 4 (b) 8 (c) 27 (d) 47 r24
10. $2083 r4 ($2083.33)
11. (a) 52 (b) 21 r7 (21.41)
12. $41 r7 ($41.88)
13. (a) 18 (b) 414 (c) 23 (d) 207 (e) 4140
14. □ = 18; △ = 4
15. 82
16. 61° r1 (61.14°)
17. 19

Page 139: Review

2.

a 5	6	■	b 1	8	■
0	■	c 3	5	■	d 6
■	e 2	9	■	f 3	4
g 4	7	■	h 1	8	■
5	■	i 2	0	■	j 9
■	k 3	6	■	l 7	2

Chapter 14 FRACTIONS

Page 151: Exercises

1. (a) $\frac{2}{4}$ (b) $\frac{3}{5}$ (c) $\frac{7}{10}$ (d) $\frac{2}{3}$ (e) $\frac{3}{8}$
2. (a) 4 (b) 3 (c) 5 (d) 10 (e) 15
3. (a) $\frac{1}{3}$ (b) $\frac{1}{2}$ (c) $\frac{2}{5}$ (d) $\frac{1}{3}$ (e) $\frac{2}{7}$

   (f) $\frac{2}{5}$ (g) $\frac{2}{3}$

Page 152: Exercises

1. (a) $\frac{4}{5}$ (b) $\frac{6}{7}$ (c) $\frac{7}{9}$ (d) $\frac{3}{3}$ (e) $\frac{5}{4}$
2. $\frac{3}{6}$ or $\frac{1}{2}$
3. $\frac{5}{7}$

Page 154: Exercises

1. (a) $\frac{7}{10}$ (b) $\frac{11}{12}$ (c) $\frac{7}{10}$ (d) $\frac{17}{24}$
2. $\frac{11}{15}$

Page 155: Exercises

1. (a) $3\frac{3}{4}$ (b) $3\frac{5}{6}$ (c) $5\frac{3}{8}$

2. (a) $\frac{9}{4}$ (b) $\frac{8}{5}$ (c) $\frac{14}{3}$ (d) $\frac{22}{7}$ (e) $\frac{76}{9}$ (f) $\frac{53}{10}$
3. (a) $2\frac{2}{3}$ (b) $5\frac{1}{2}$ (c) $3\frac{1}{4}$ (d) 2 (e) $1\frac{1}{8}$
4. $4\frac{3}{4}$
5. $10\frac{1}{4}$

Page 157: Exercises

1. (a) $\frac{1}{3}$ (b) $\frac{3}{5}$ (c) $\frac{2}{9}$ (d) $\frac{11}{16}$ (e) $\frac{1}{8}$
2. (a) $2\frac{1}{2}$ (b) $1\frac{1}{12}$ (c) $\frac{5}{6}$ (d) $5\frac{2}{5}$
3. $1\frac{7}{8}$ inches
4. $4\frac{9}{10}$ miles

Page 161: Exercises

1. (a) 4 (b) 8 (c) 9 (d) 14 (e) 36
2. (a) $\frac{1}{8}$ (b) $\frac{3}{32}$ (c) $\frac{1}{2}$ (d) $\frac{1}{2}$ (e) $\frac{1}{12}$
3. (a) $2\frac{1}{5}$ (b) $1\frac{4}{9}$ (c) 186 (d) $\frac{7}{10}$ (e) $3\frac{3}{32}$
4. $6\frac{1}{2}$ yards
5. 110 miles

Page 162: Exercises

1. (a) 2 (b) $2\frac{4}{5}$ (c) $\frac{3}{20}$ (d) 18 (e) 1 (f) 13 (g) $\frac{5}{7}$ (h) $1\frac{17}{28}$
2. 12 strips
3. 52 miles per hour

Page 167: Questions on the Chapter

1. (a) $\frac{3}{5}$ (b) $\frac{1}{6}$ (c) $\frac{2}{3}$
2. (a) $\frac{5}{40}$ (b) $\frac{6}{8}$ (c) $\frac{8}{20}$
3. (a) $\frac{1}{3}$ (b) $\frac{1}{2}$ (c) $\frac{2}{7}$ (d) $\frac{3}{8}$
4. (a) $\frac{9}{4}$ (b) $\frac{8}{5}$ (c) $\frac{25}{8}$ (d) $\frac{22}{9}$
5. (a) $2\frac{2}{3}$ (b) $5\frac{1}{2}$ (c) $3\frac{1}{4}$ (d) $1\frac{2}{5}$ (e) $2\frac{3}{8}$
6. (a) $1\frac{1}{4}$ (b) $\frac{7}{8}$ (c) $1\frac{17}{24}$ (d) $8\frac{2}{3}$ (e) $5\frac{4}{5}$ (f) $1\frac{19}{24}$
7. (a) $\frac{2}{5}$ (b) $\frac{1}{3}$ (c) $\frac{27}{40}$ (d) $\frac{1}{20}$ (e) $\frac{3}{16}$ (f) $2\frac{2}{5}$ (g) $2\frac{2}{3}$ (h) $2\frac{1}{2}$ (i) $3\frac{1}{4}$ (j) $2\frac{1}{5}$ (k) $3\frac{5}{12}$ (l) $\frac{17}{24}$
8. (a) $\frac{1}{15}$ (b) $\frac{15}{32}$ (c) 3 (d) $1\frac{1}{2}$ (e) $3\frac{1}{2}$ (f) 2 (g) $5\frac{21}{40}$
9. (a) $\frac{4}{9}$ (b) $2\frac{2}{15}$ (c) 2 (d) $\frac{7}{16}$ (e) 20 (f) $6\frac{2}{3}$ (g) $\frac{5}{6}$ (h) 3 (i) $\frac{1}{8}$ (j) 1
10. (a) $\frac{1}{2}$ (b) $\frac{4}{5}$ (c) 2 (d) $\frac{1}{2}$ (e) $\frac{7}{5}$ (f) $\frac{5}{3}$ (g) $\frac{8}{7}$
11. (a) $\frac{4}{4}$ (b) $\frac{5}{3}$
12. $\frac{3}{8}$, $\frac{7}{12}$, $\frac{3}{5}$, 2
13. $\frac{5}{12}$
14. $5\frac{5}{8}$ feet
15. $8\frac{13}{16}$ inches
16. $12\frac{3}{4}$ pounds
17. $\frac{5}{8}$ yards
18. $18\frac{3}{4}$ feet
19. 15 cups of flour and $7\frac{1}{2}$ cups of sugar
20. $5\frac{17}{19}$, or nearly 6 pieces of paper
21. \$$1\frac{5}{7}$ (or \$1.71)
22. $48\frac{8}{9}$ (or 48.89 miles/hour)
23. (a) $\frac{7}{16}$ (b) $\frac{11}{15}$
24. $\frac{1}{6}$
25. $\frac{7}{20}$
26. $\frac{15}{8}$
27. (a) Yes. If in $\frac{2}{3}$ you *subtract* 1 from the numerator and denominator, you get $\frac{2-1}{3-1} = \frac{1}{2}$, which is not equal to $\frac{2}{3}$.

    (b) Yes. If in $\frac{2}{3}$ you *add* 1 to the numerator and the denominator, you get $\frac{2+1}{3+1} = \frac{3}{4}$, which is not equal to $\frac{2}{3}$.

    The only operations that don't change the value of a fraction are multiplying and dividing the numerator and denominator by the same number.

28. The greater fraction is the one with the smaller denominator.

Page 169: Review

1. (a) 10 (b) 1 (c) 100
2. 35
3. \$993$\frac{3}{4}$ (or \$993.75)
4. 8
5. (a) 7 (b) 7 (c) 6

Chapter 15 DECIMALS

Page 175: Exercises

1. (a) three tenths (b) eight hundredths (c) five thousandths (d) six and thirty-eight hundredths (e) three hundred forty-eight thousandths (f) two and seven tenths (g) four and one ten-thousandth
2. (a) .4 (b) 2.3 (c) .05 (d) .015 (e) 1.25

Page 178: Exercises

1. (a) .9 (b) 1.4 (c) 2.38 (d) 1.5
2. (a) .5 (b) .19 (c) .008 (d) 1.65

Page 179: Exercises

1. (a) 1.8 (b) .015 (c) .0002 (d) 3.0 (e) \$.0354
2. (a) 2.75; 27.5; 275 (b) \$12.80; \$128; \$1280 (c) \$1; \$10; \$100 (d) .03; .3; 3 (e) 39; 390; 3900

Page 182: Exercises

1. (a) 20 (b) .005 (c) .625 (d) 200 (e) 30
2. (a) 2.5; .25; .025 (b) .42; .042; .0042 (c) 6.25; .625; .0625 (d) .003; .0003; .00003 (e) .0008; .00008; .000008

Page 183: Exercises

1. (a) .5 (b) 3.5 (c) 62.1
2. (a) .65 (b) .21 (c) 6.01 (d) .10

Page 188: Questions on the Chapter

1. (a) .7 (b) 1.5 (c) 1.03 (d) 16.5 (e) .636 (f) 5.08
2. (a) .7 (b) 1.6 (c) 2.25 (d) \$3.28 (e) 1.445 (f) 16.625
3. (a) .21 (b) .48 (c) .288 (d) 15 (e) 30.1 (f) .0268
4. (a) 2.5 (b) 3 (c) 8.5 (d) 2450 (e) 500 (f) 1230 (g) 30 (h) 156.7 (i) 56.67
5. 4.98 inches
6. \$6.81
7. 3.4°
8. \$336.70

9. $21.06
10. $1791.67
11. 2.6 miles per hour
12. (a) .625 (b) .325 (c) .9375 (d) 2.375 (e) 5.333
13. (a) $\frac{24}{100} = \frac{6}{25}$ (b) $\frac{65}{1000} = \frac{13}{200}$ (c) $\frac{1}{1000}$
14. (a) .024
15. (a) 2.8 (b) both have same value (c) .2
16. (c) .25
17. (a) 15.1; 15.07 (b) .1; .13 (c) .0; .01

Page 189: Review

1.

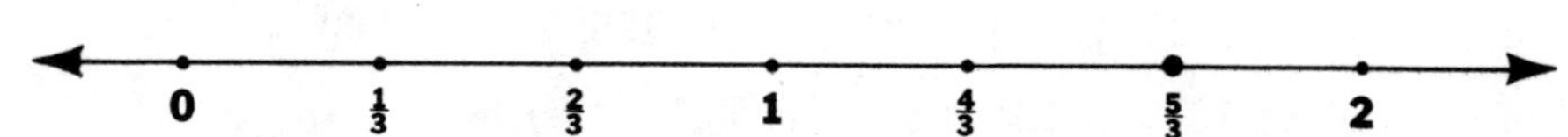

2. $\frac{2}{3}$
3. $\frac{41}{24} = 1\frac{17}{24}$
4. $\frac{4}{3} = 1\frac{1}{3}$
5. $1\frac{1}{4}$
6. the answer you get when the two numbers are multiplied
7. quotient
8. 15
9. $\frac{9}{12} = \frac{3}{4}$
10. 108

Chapter 16 PERCENT

Page 196: Exercises

1. (a) 38% (b) 9% (c) 85%
2. (a) .29 (b) .98 (c) .09 (d) 1.30 (e) .0325
3. (a) 45% (b) 9% (c) 37.5% (d) 250%
4. (a) $\frac{25}{100} = \frac{1}{4}$ (b) $\frac{80}{100} = \frac{4}{5}$ (c) $\frac{4}{100} = \frac{1}{25}$ (d) $\frac{62.5}{100} = \frac{625}{1000} = \frac{5}{8}$
   (e) $\frac{125}{100} = \frac{5}{4}$
5. (a) 30% (b) 40% (c) $37\frac{1}{2}$% (d) $33\frac{1}{3}$% (e) 2%

Page 198: Exercises

1. 8
2. 3
3. 7
4. $5.10
5. 20

Page 198: Exercises

1. $33\frac{1}{3}$%
2. 75%
3. $28\frac{4}{7}$%
4. $18\frac{3}{4}$%
5. 125%

Page 199: Exercises

1. 200
2. 60
3. 800
4. 21
5. 48

Page 200: Questions on the Chapter

1. $\frac{6}{100}$; .06; 6%
2. (a) .10; $\frac{1}{10}$ (b) .30; $\frac{3}{10}$ (c) .25; $\frac{1}{4}$ (d) .125; $\frac{1}{8}$ (e) $.33\frac{1}{3}$; $\frac{1}{3}$
3. (a) 20% (b) $16\frac{2}{3}$% (c) 60% (d) $66\frac{2}{3}$% (e) $87\frac{1}{2}$%
4. (a) 12.5% (b) $62\frac{1}{2}$% (c) $66\frac{2}{3}$% (d) 100%
5. (a) Out of every 100 parts making up the shirt material, 55 are cotton and 45 are polyester.
   (b) Out of every 100 people at the ball game, 65 were men and 35 were women.
   (c) Out of every 100 hours, the girl spends 25 in school and 30 sleeping. She spends 45 hours, or 45%, on other activities.
6. 27.5%
7. $\frac{1}{5}$
8. $\frac{1}{8}$
9. 40%
10. (a) *All* students attended class.
    (b) No, because there can't be more students present than the number of students there are in the class.
    (c) Yes, because the cost of a sweater can be increased by any amount. Increasing it by 150% means increasing it by $1\frac{1}{2}$ times its original cost. If the original cost was $10, then increasing it by 150% means increasing it by $15, making the new cost $25.
11. (a) 9 (b) $6 (c) 100.8
12. (a) 25% (b) 80% (c) 100% (d) 125%
13. (a) 200 (b) 52 (c) 15 (d) 7000
14. 15 correct answers
15. $12.50
16. 3
17. 60% men; 40% women
18. won 60%, lost 40%
19. $16

20. 30 members

22. Yes. (a) 15% of 42 = .15 × 42 = 6.3
(b) 42% of 15 = .42 × 15 = 6.3

Page 201: Review

1. (a) 6 (b) 8
2. 1.93
3. $\frac{1}{12}$
4. $\frac{3}{7}$
5. $18.39
6. (a) 1.4 (b) 400 (c) 6
7. $6.19
8. (a) 15 (b) 18 (c) $\frac{1}{3}$

Chapter 17 GEOMETRY

Page 209: Exercises

1. (a) segment DG (b) ray AR (c) line MT (d) line m
3. in one point
4. yes
5. no
6. (a) a line segment (b) a point (c) two intersecting lines
(d) a plane (e) two parallel planes (f) two perpendicular planes
7. (a) and (b) are both true since a line contains infinitely many points.
(c) True, since it extends indefinitely in both directions.
(d) False; a line *segment* has two endpoints.
(e) True. In either case, the endpoints are the points P and Q.
(f) True. This can be seen in this diagram:

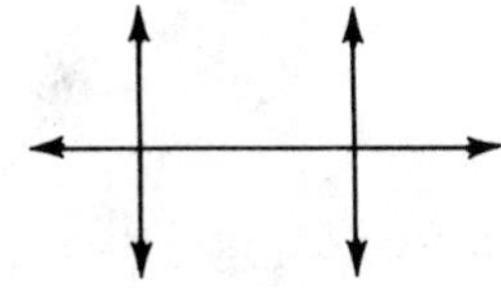

Page 212: Exercises

1. (a) right angle SRT (b) acute angle A
(c) right angle TPQ (d) obtuse angle M
2. (a) Angles R, S, T, Q are all right angles.
(b) Angles A, B, C are all acute.
(c) Angles D, E, F, G, H, J are all obtuse.
(d) Angles F and L are right angles; K is acute; G is obtuse.
(e) Angles X and Z are acute; Y is obtuse.
3.

(a)

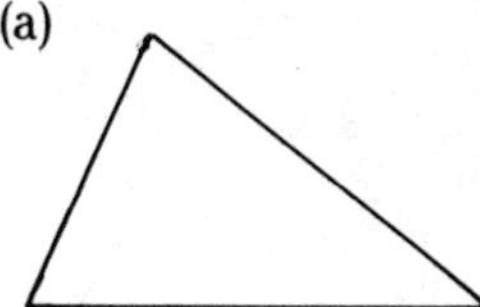

(b)

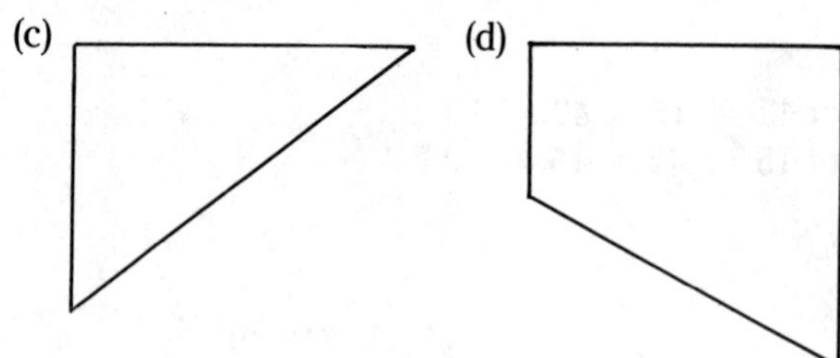

5. (a) 90° (b) 60° (c) 180°

Page 215: Exercises

2. yes
3. 10; 15

Page 217: Exercises

1. They contain three sides and three angles.
2. three: △ SRT, △ SRP, △ TRP
3. six: ADEH, ABGH, BCFG, CDEF, ACFH, BDEG
4. yes
5. no
6. yes
8. yes
9. yes
10. 1
11. 30°
12. 90°; a right triangle

Page 220: Exercises

1. (a) OE, OD (b) CE, DP (c) DE (d) CD (e) AB (f) CE (g) O
2.

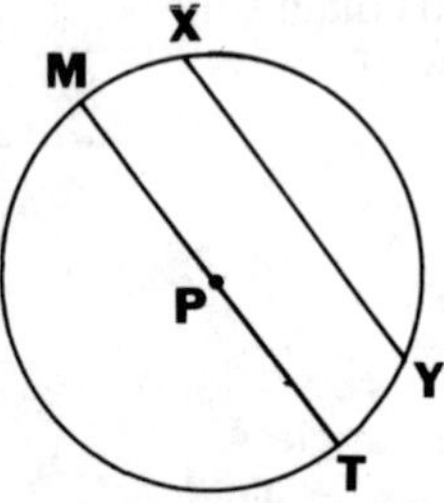

3.

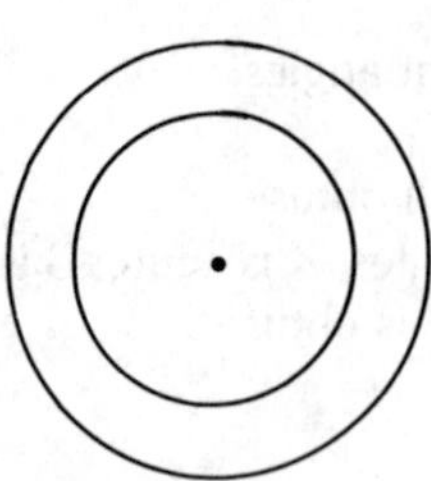

4. (a) 10 in. (b) 42 cm (c) 9 yd
5. (a) 3 cm (b) 9 ft (c) 1.9 in.

Page 221: Exercises

2. (a) sphere (b) point (c) cylinder (d) cylinder (e) cylinder (f) sphere (g) cone

Page 222: Exercises

2. (a) rectangular prism (b) rectangular prism (c) cube (d) pyramid
3. All faces of a cube are squares; the faces of a rectangular prism are rectangles.
4. (a) cube (b) cone (c) sphere (d) cylinder (e) pyramid
   (f) rectangular prism (g) rectangular prism (h) pyramid (i) cube

Page 224: Exercises

1. (a) 16 cm (b) 58 ft (c) 11 yd (d) $17\frac{1}{2}$ in.
2. 166 ft
3. (a) 28 m (b) 48 ft (c) 23 in.
4. (a) 46 in. (b) 295 ft (c) $21\frac{11}{12}$ in. (d) 16 ft (e) $33\frac{1}{2}$ in.
5. (a) 42 cm (b) 9.75 in. (c) $16\frac{1}{8}$ ft
6. 3 yd
7. 19 cm

Page 226: Exercises

1. (a) 25.12 cm (b) 72.22 yd (c) 132 in. (d) 18.21 ft
2. (a) 12.56 m (b) 56.52 in. (c) 88 ft
3. 62.80 in.
4. d = 13.38 cm; r = 6.69 cm

Page 232: Exercises

1. An *inch* is a line *segment* measuring an inch. A *square inch* is a *square* measuring 1 inch on each side. The same distinction holds for a *centimeter* and a *square centimeter*.
2. (a) length (b) area (c) area (d) length (e) area
3. (a) 48 sq cm (b) 99 sq mi (c) 50 sq in. (d) .06 sq km
4. (a) 16 sq in. (b) 36 sq cm (c) 12.25 sq in. (d) $5\frac{1}{16}$ sq ft
5. (a) 35 sq in. (b) 70 sq yd (c) 80 sq cm (d) 39 sq ft (e) $39\frac{7}{8}$ sq ft
   (f) 53.01 sq in.
6. (a) 48 sq in. (b) 20 sq yd (c) 7.5 sq cm (d) 1.5 sq ft (e) 14 sq in.
   (f) $14\frac{7}{16}$ sq in.
7. 16 sq m

Page 234: Exercises

1. (a) 28.26 sq in. (b) 154 sq m (c) 706.5 sq ft (d) 3846.5 sq yd
   (e) 19.63 sq cm
2. (a) 50.24 sq m (b) 78.5 sq in. (c) 63.59 sq ft
3. 314 sq in.

Page 235: Exercises

1. An *inch* is a line *segment* measuring an inch; a *square inch* is a *square* measuring 1 inch on each side; a *cubic inch* is a *cube* measuring 1 inch on each edge—that is, 1 inch in length, width, and height.
2. (a) 40 cu in. (b) 280 cu cm (c) 360 cu ft
   (d) 70 cu yd (e) 102 cu m
3. (a) 125 cu ft (b) 2744 cu in. (c) 15$\frac{5}{8}$ cu ft (d) .027 cu cm
4. 144 cu ft
5. 21.95 cu cm

Page 236: Exercises

1. (a) 141.3 cu ft (b) 6154.4 cu in. (c) 4019.2 cu cm (d) 461.58 cu m
2. (a) 4.19 cu m (b) 33.49 cu cm (c) 113.04 cu in. (d) 4186.67 cu yd
3. 12717 cu in., or 7.36 cu ft
4. 904.32 cu in.

Page 244: Exercises

1. (a) yes (b) yes (c) no (d) no
2. yes, because they have the same shape
3. yes, if their corresponding parts are equal

Page 246: Exercises

1. (a) 2 (b) 0 (c) 3 (d) 6 (e) 1 (f) 1 (g) 1
   (h) 1 (i) 0 (j) 1 (k) 0 (l) 1

Page 250: Questions on the Chapter

1. (a) 10 (b) 15 *NOTE:* Can you detect a pattern from the results for 2 to 6 points? [As the number of points is increased by 1, the number of segments increases by 2, 3, 4, 5 . . . .]

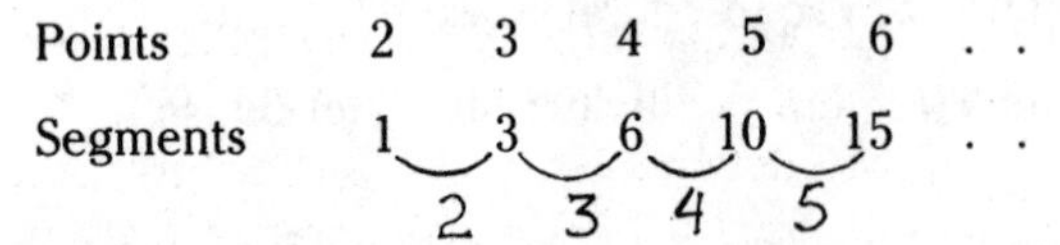

2. (a) 90° (b) 120° (c) 180° (d) 270°
3.

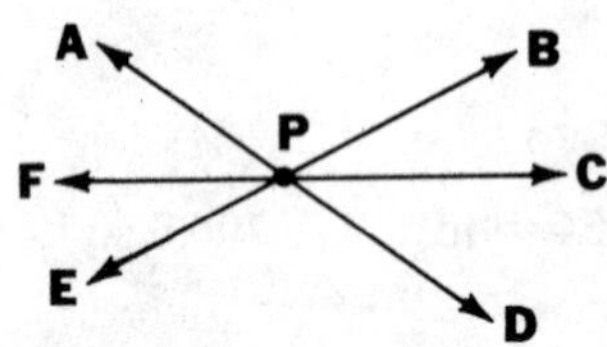

Angles: APB, APC, APE, APF, BPC, BPD, BPF, CPD, CPE, DPE, DPF, EPF

4.

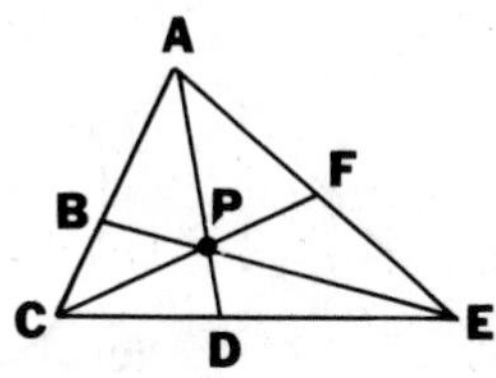

Triangles: ABE, ACD, ACE, ADE, APB, APC, APE, APF, BCE, BEA, BPA, BPC, CDA, CDP, CEF, DEP, EFP

5. Triangles: ACE, ABF, BCD, BDF, DEF
Parallelograms: FBDE, FBCD, ABDF
Trapezoids: FBCE, ABDE, DFAC

6. (a) isosceles (b) equilateral (c) right

7. No, because if all three sides are equal then it must be an equilateral triangle with each angle measuring 60°.

8. 45° and 90°. It's a right isosceles triangle.

9. (a) square, rectangle, parallelogram
(b) square, rectangle
(c) square, rectangle, equilateral triangle
(d) square, equilateral triangle
(e) square, rectangle
(f) square, rectangle, parallelogram
(g) square, rectangle, parallelogram, trapezoid

10. *NOTE:* The explanation below—though long—is not difficult to understand. With patience, fortitude, and the determination *not* to be intimidated, you'll make it!

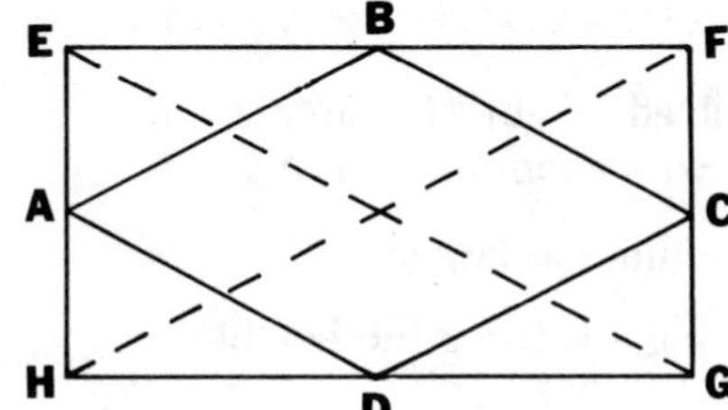

Draw diagonals EG and FH.

In △ EFG: B is the midpoint of EF, and C is the midpoint of FG.

In △ EHG: A is the midpoint of EH, and D is the midpoint of GH.
By Theorem 4 on page 217, BC and AD are parallel to diagonal EG and half its length. This means that BC and AD are parallel and equal to each other. Therefore, by Theorem 3 on page 219, ABCD is a parallelogram.

Similarly, AB and DC are parallel to diagonal HF and half its length. Since the two diagonals are equal, AB, BC, CD, and AD are equal, and the figure is a *rhombus.*

11. Three cuts:

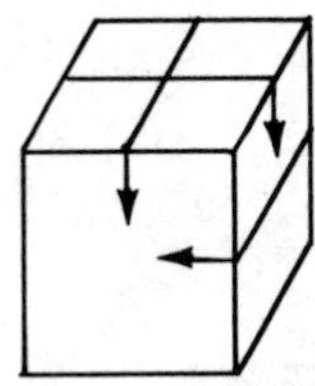

12. (a) cube (b) rectangular prism (c) pyramid

13.

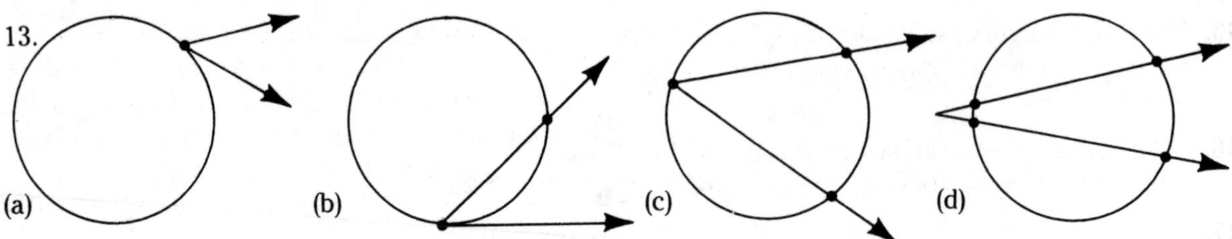

14. 14.33 ft

15. 53 ft

16. (a) 18.84 in. (b) 109.90 cm (c) 11 ft
(d) 16.49 yd

17. (a) 157 yd (b) 307.72 ft (c) 34.54 cm
(d) 65 in.

18. 94.20 ft

19. 24,862.52, or about 25,000 miles

20. 14.01 in

21. 250 sq ft; $437.50

22. 35.71, or about 36 bricks

23. 23 ft

24. The area of the first rectangle is twice the area of the second.

25. (a) The area is doubled. (b) The area is doubled. (c) The area is *four* times as big.

26. The area is made *four* times as large.

27. The height of the triangle is twice the height of the parallelogram.

28. The width is 7 meters and the length is 14 meters.

29. (a) 9 sq in. (b) 40 sq cm (c) 70 sq m
(d) 25 sq in.

30. 12.56 sq in.

31. four times larger

32. *NOTE:* The *diameters* of the three coins are as follows:
penny: 19.05 mm; quarter: 24.26 mm; half-dollar: 30.56 mm (a) 285.18 sq mm
(b) 462.01 sq mm (c) 733.12 sq mm

33. (a) Subtract the area of the circle from the area of the rectangle:

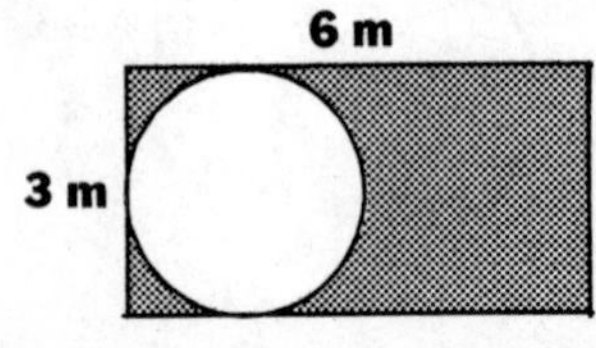

Since the diameter of the circle is 3 m, its area is

$$A = \pi \times (1.5)^2$$
$$A = 7.07 \text{ sq m}$$

The area of the rectangle is

$$A = 3 \times 6$$
$$A = 18 \text{ sq m}$$

The area of the shaded region is: $18 - 7.07 = 10.93$ sq m

(b) Subtract the area of the small circle ($r = 4$) from the area of the larger circle ($r = 7$):
Area of larger circle:

$$A = \pi \times 7^2$$
$$A = 153.86 \text{ sq units}$$

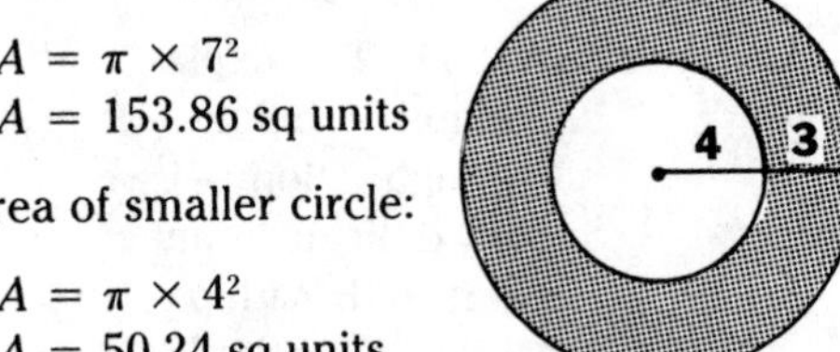

Area of smaller circle:

$$A = \pi \times 4^2$$
$$A = 50.24 \text{ sq units}$$

The area of the shaded region is
$153.86 - 50.24 = 103.62$ sq units

34. 2 units [$A = \pi r^2 = 4\pi$; $C = 2\pi r = 4\pi$]

35. 19,200 cu ft

36. 140 cu ft

37. 343 cu ft

38. (a) The volume is doubled.
(b) The volume is multiplied by 4.
(c) The volume is multiplied by 8.

39. .68 cu m

40. The 6-inch can contains *four* times as much juice as the 3-inch can.

41. 212.07 cu cm

42. Doubling the radius makes the volume *8* times as large. Tripling the radius makes the volume *27* times as large.

43. The three sides of △ ABC are equal to the matching sides of △ CDA. (Note that AC is a common side to both triangles.)

44. the three sides and the three angles

45. The three matching sides of the two triangles are equal. (Note that AD is a common side to both triangles.)
46. if their corresponding angles are equal, and the ratios of their corresponding sides are equal
47. yes
48. 9 and 6
49. circles, squares, spheres
50. 15 cm
51. The answer will vary with the way the letters are formed:
    A, B, C, D, E, H, I, M, O, T, U, V, W, X, Y
52. (a) True. Congruent figures have not only the same shape, but also the same size.
    (b) False. The two shapes cannot be the same.
    (c) True. All their angles measure 60°, and the ratios of their sides will always be the same.
    (d) False. An equilateral triangle cannot have the same shape as a square.
    (e) True. Both triangles have the same size and shape as the third triangle.
    (f) True, since the sum of the angles of each triangle is 180°.
    (g) False. The measure of each angle of an equilateral triangle is 60°.
    (h) True. They can't have the same shape if they have different numbers of sides.
    (i) True, since all its angles are right angles and its opposite sides are parallel.
    (j) True, since its opposite sides are parallel.
    (k) False, since a polygon is composed of line segments.

Page 254: Review

1. 385,200
2. $1\frac{17}{30}$
3. $\frac{13}{20}$
4. $\frac{6}{35}$
5. $8\frac{8}{9}$
6. .088
7. 145.17
8. $65
9. 84
10. .529
11. (a) 3482.9 (b) .01893
12. $10
13. 160
14. 16.7%
15. $\frac{57}{325}$, or about $\frac{1}{6}$
16. 3
17. 225 pounds
18. 3.12 is larger by 2.142.
19. 2,750,000
20. *Estimate:* About 300 × 800, or 240,000

    Exact: 306 × 792 = 242,352

BIRD ARRAY

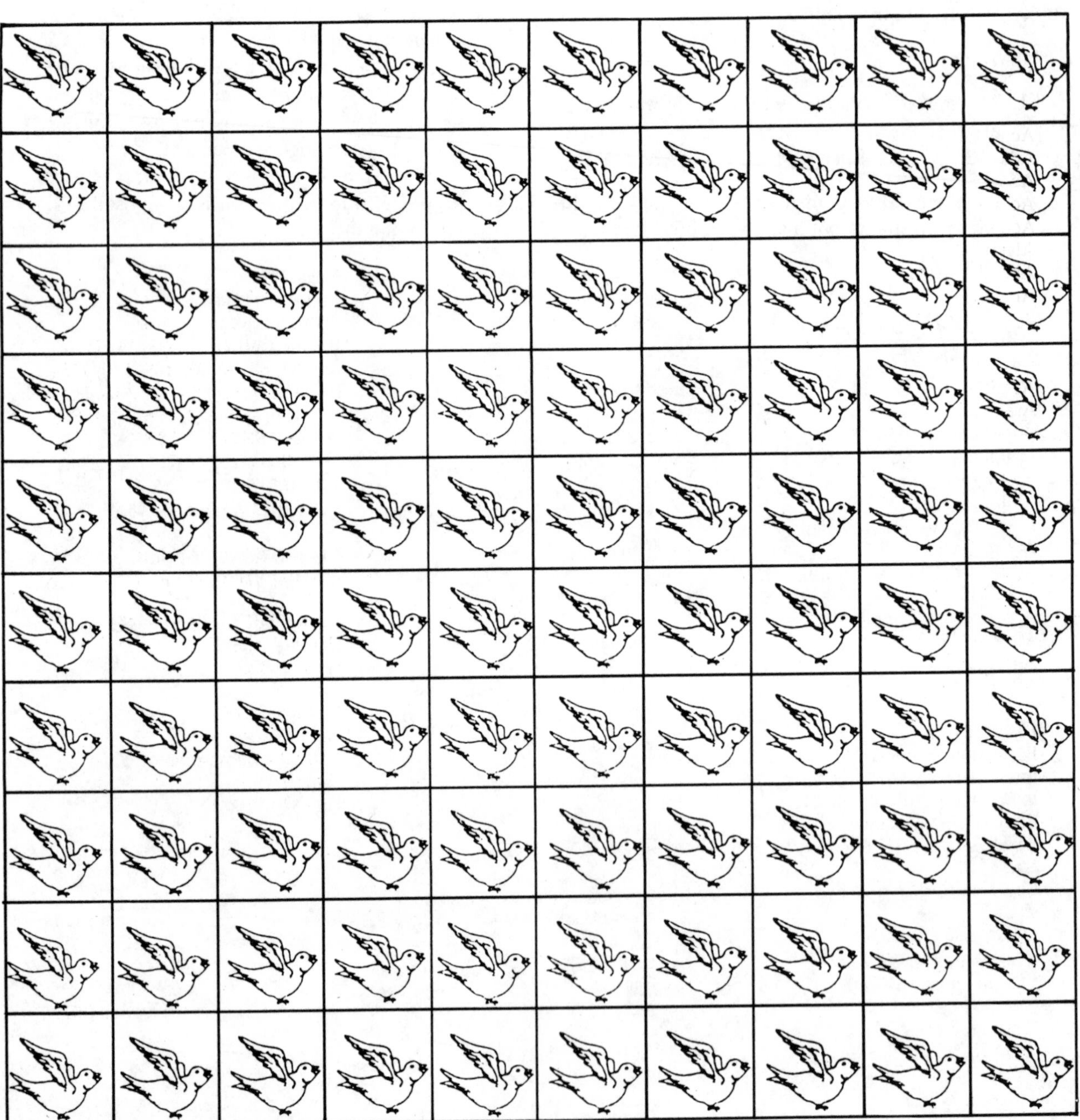

Index